QUANTITATIVE ANALYSIS FOR MARKETING MANAGEMENT

Quantitative Analysis for Marketing Management

WILLIAM R. KING

Associate Professor
Graduate School of Business
University of Pittsburgh

McGRAW-HILL BOOK COMPANY
NEW YORK • ST. LOUIS • SAN FRANCISCO
TORONTO • LONDON • SYDNEY

QUANTITATIVE ANALYSIS FOR MARKETING MANAGEMENT

Library of Congress Catalog Card Number 67-21595

34605

1 2 3 4 5 6 7 8 9 0   MAMM   7 4 3 2 1 0 6 9 8 7

TO FAY, JIM-JIM, SUZIE, AND CEZE

. . . the reasons for my being

. . . the antitheses of my reason

Preface

An author should begin a book by stating what the book is and what its objectives are. This is not a typical book in marketing management. Perhaps, therefore, it is best first to describe what this book *is not* and those objectives to which it does *not* subscribe.

Between these covers very little will be found concerning such topics as marketing organization and the postdecision processes of human motivation and control. Also, scant attention is paid to marketing principles, descriptions of the marketing system, and discussions of the latest fads in marketing. Although an introductory chapter is devoted to material which serves to place the remainder of the work in perspective, students wishing a comprehensive treatment of such topics are referred to the many excellent volumes which abundantly fill the marketing sections of libraries.

This book does not purport to teach a student to be a marketing manager. To become an effective modern marketing manager, one must combine some intellect and interest with a thorough knowledge of basic management principles and an understanding of the values and limitations of the processes of analysis which can be brought to bear on marketing decision problems. A book can do little to develop any of these faculties except the last. It is to this objective that this book subscribes.

Perhaps the greatest value which the typical student will derive from this text is an understanding of the analytic point of view. This apparently innocuous objective is actually the highest goal to which a text in quantitative analysis may aspire, for the most significant difference between "traditional" marketing analysis and "modern" analysis is precisely in the viewpoint of the individuals who perform the analysis. The tools and techniques of quantitative analysis are important; but the scientific method (point of view) is paramount. The marketing manager who knows no mathematical techniques but possesses a viewpoint which will permit him to obtain a new perspective on old marketing problems is more likely than the mathematician or technician to achieve the delicate blend of analysis, judgment, and intuition which is becoming necessary for successful marketing management.

This is a book about the *decisions* with which the marketing manager is constantly faced. Moreover its focus is on the *analysis* of those decision problems with whatever tools are available—qualitative or quantitative, for one cannot really separate the two. Most often, because purely qualitative approaches are taught elsewhere, the emphasis is on the quantitative; hence the title, *Quantitative Analysis for Marketing Management*. The approach taken to the quantitative analysis of marketing decision problems is a scientific one which is built upon a conceptual framework of *decision theory*. A great deal of attention is paid to the practical problems of applying that framework in the real world, for it is only through an impact on the real world that marketing analysis bears fruit.

The presentation is focused for the modern analytic marketing manager and marketing analyst. Much material which has previously been discussed only in technical papers is presented here in a manner which will be palatable to the less technically oriented. In many chapters the technical reader will find that work which has recently appeared in the literature has been purposefully omitted. The reasons for these omissions are intrinsic to the objectives of the book. These objectives are oriented toward a demonstration of the utility of the scientific point of view in marketing management and not toward providing the reader with the latest research models and approaches. Basic and time-tested models which serve to illustrate the analytical structure and methodology have therefore been relied upon heavily. In taking this approach, the author hopes to bring an understanding of the underlying ideas of quantitative analysis into the realm of the marketing manager. In another sense the emphasis on the practical problems involved in performing analysis and in applying analytic results to the real world represents an attempt to bring the real world to the analyst.

Indeed, the philosophy which underlies the entire book is that *the manager and analyst must work together closely in all phases of the analysis of marketing decisions*. The time is past when a manager can make effective decisions on the sole basis of judgment and experience. Most modern managers recognize that this is so. But we are also rapidly approaching the end of another phase of managerial problem solving. That phase, which is best exemplified by the work of operations researchers in the 1950s and 1960s, involved the detection of a problem by a manager who called in an analyst to solve it. The analyst proceeded to construct and solve a model which provided the manager with the "answer" to his dilemma. The manager then acted on the answer in a fashion which was largely dictated by his degree of faith in the individual analyst. If he had confidence in the analyst, he adopted the proposed solution. If he did not have faith, he politely thanked him for his trouble, paid his fee, and acted on the basis of his own intuition.

Today our rapidly changing environment requires that the manager-analyst relationship be closer and more effective than this. In fact, one person may be both manager and analyst. The important thing is that the manager must be an essential part of all phases of problem analysis from the original formulation to the final solution and implementation. One of the basic objectives of this book is to demonstrate why this is so and how such a relationship can be effectively carried on.

The book emphasizes the role of models in marketing analysis. Because they are so often misunderstood, it is important that they be presented in a context which is familiar to the student. To avoid the errors of the supersalesman who oversells the customer only to leave him disappointed with the product, great care is taken to point out the limitations of models and the assumptions, implications, and practical difficulties in application which are features of each of the specific models presented. The logical and mathematical models which are used deemphasize the mechanical and memory-based aspects of mathematics. Whenever simple functional forms are descriptive, if not realistic, simplicity, rather than realism, has been chosen, with appropriate mention of the real-world counterparts which might be encountered. The examples used in the early chapters are overly simple, yet chosen to depict real marketing decision problems.

The level of previous mathematical training required for the book is not high; algebra is a requisite for all of the chapters, and elementary calculus will always suffice. Most of the essential material draws only on algebra for a complete understanding of the basic methodology and its applicability. Those who are more technically oriented will wish to pursue those aspects requiring calculus. A number of technical notes are included to present specialized material which is of primary interest to the analyst. The level of mathematics required by each of these notes is specified in a footnote. Since the technical notes have greatest appeal to the analytically oriented and are in no case essential to the understanding of the basic material in the chapter, this feature should provide an effective apparatus to permit students and teachers to utilize those portions best suiting their needs. The necessary ideas of mathematics which the typical student may not have at his fingertips are presented in the context of the decision problems in which they have proved most useful. These treatments are accomplished at an intuitive, rather than formal, level so that the student without a formal background is accommodated. In using the text with technically oriented students, the author has found that this feature is appealing to them also, since it generally provides an alternative viewpoint to that they have seen elsewhere.

The material presented is drawn from the original work of many who have labored in various fields. Where the work of others has been used as background, appropriate credit is given. Since their work has generally been significantly altered and reorganized to conform to the pattern of the text, any blame for deficiencies or omissions is mine. Some of my original work has not previously appeared elsewhere in published form. That portion which has been developed through my industrial consulting experience has been disguised in form, but not substance, for the sake of industrial security.

This book has been used, in one or another of several draft forms, as the basis for a number of classes in marketing management and quantitative marketing analysis at both the undergraduate and introductory graduate levels while I served as Assistant Professor of Operations Research at Case Institute of Technology. The contributions of the varied groups of students whose responses to the earlier versions served to improve a vastly imperfect attempt at presenting the basic ideas of quantitative marketing analysis are gratefully acknowledged. Perhaps one of the significant aspects of this contribution is that it arose spontaneously from student groups as different as inexperienced, but stimulating, college juniors and experienced high-level managers who had returned to formal schooling. As a result, the changes which have been made served both to eliminate logical errors and unplanned implications and to introduce more effectively the practical problems of a manager into the models used.

Any work appearing under the authorship of an individual slights some who have contributed greatly to its completion. In this case there is a long list of those to whom a gesture of appreciation, however heartfelt, is meager reward for the help and encouragement provided. Jack Little of MIT first aroused my interest in marketing. He and Russ Ackoff, of the University of Pennsylvania, have had such an influence on my thinking that any of my writings is necessarily permeated with their ideas. Russ Ackoff, Glen Camp, Vern Mickleson, Stuart Cooke, Burt Dean, and Dan Teichroew, all, at the time, of Case Institute of Technology, provided me with opportunity and encouragement in the development of the material herein. Jack Coleman and Bill Converse of the Air Force Institute of Technology provided both opportunity and encouragement during the writing process. Gordon Paul of LSU and Bill Cox of Western Reserve critically reviewed the manuscript and suggested changes which contributed substantially to the present worth of the book. Judy Johnson and Betty Uther typed the manuscript with their usual competence with some aid, during periods of extreme panic, from my wife Fay. And, those who are in the background of all of my undertakings—Fay, Jim-Jim, Suzie, and Ceze—provided joy in moments of displeasure, warmth in moments of uncertainty, and a level of continuous activity whose net effect on the completion of the book is beyond the scope of the present measurement capabilities of science.

William R. King

Contents

4 *Marketing Models*   *83*

5 Product Decisions 111

6 Market Analysis and Test Marketing 166

10 *Promotional Decisions* 430

11 *Distribution Decisions* 512

1

Marketing and Marketing Analysis

All of the decisions and activities involved in the planning and execution of the operations necessary to provide consumers with goods which satisfy their needs comprise the content of the field called *marketing*. The purchase of this text at the college bookstore, closing the mortgage loan on a young couple's first home, the decision by a manufacturer to begin producing electronic ovens for the home, the call of an "Avon lady" on a suburban housewife, and the repair of a leaky faucet by the local plumber, each constitutes an element of marketing.

The Importance of the Marketing Function

Traditionally the activities of a typical business enterprise have been functionally categorized into production, marketing, and finance. In this context the importance of the marketing function to the business organization is readily apparent. Marketing is the primary income-producing portion of most business enterprises. Until the product leaves the realm of finance and production and enters the marketing process, the producer has incurred only costs while obtaining no returns.

The marketing process is also important in the satisfaction of the needs and desires of the consumer. Were it not for advertising, for example, consumers would be dependent on word-of-mouth transfer of information concerning the benefits of new products and services. Whatever we may think of the quality of individual advertising messages, the overall benefit from advertising gained by the consuming public cannot be denied.

Similarly, other aspects of the marketing process serve to put a product into the hands of the potential buyer at a *place* which is convenient (e.g., the shopping center rather than the production plant), at a *time* which is agreeable to him (e.g., when he wishes to consume it rather than when it is produced), and in a *form* which is convenient (e.g., six-packs of beer rather than individual bottles). In addition, the marketing process assures the consumer of convenient legal *ownership* of the product he desires (e.g., he may purchase directly from the druggist rather than negotiate with the producer).

The Dynamics of Marketing Development

Despite the importance of the marketing function both to the producer, to whom it provides the dollars needed to continue operations, and to the consumer, to whom it provides products and services which satisfy needs and desires, the development of the U.S. marketing system has been criticized as less than a model of efficiency. One reason for this, of course, is that this nation has never had a planned and controlled economy. A lack of such overall planning almost invariably implies that a system will contain redundancies and inefficiencies. In a free society, however, we view these inefficiencies as part of the price we must pay for economic freedom.

Perhaps a more significant reason for fat and inefficiency in the development of our marketing processes, however, is that U.S. marketers have enjoyed long periods of enormous latent demand for almost every variety of goods and services. At some stages of our history the most meager planning and foresight were sufficient to ensure a degree of success for a product faced with a virtually insatiable consumer demand. Witness, for example, the post World War II period, when demand for autos and refrigerators—indeed, almost anything but buggy whips and swords—literally overwhelmed our production capabilities.

It is to the everlasting credit of American marketing executives that even during such periods of infinite demand, product changes and improvements were made and new marketing techniques were instituted, often at the loss of short-run profit. If, for example, the production and sale of 1942 autos in the guise of postwar models had been continued long after 1946, profits of auto companies during this period would undoubtedly have been greater. However, even though design and retooling in the face of unfilled demand resulted in some loss of sales, the actions taken by industrial enterprises, such as the automobile manufacturers, have contributed significantly to the continued growth and prosperity of the American economy. In the context of the very different marketing environment of today, it is clear that the innovative ideas and actions of that postwar period were, to a large extent, the direct antecedents of the attitudes and methods of modern marketing management.

Realizing this, one might be led from questions regarding the absolute inefficiencies in the U.S. marketing system to amazement at the relatively high effectiveness and efficiency which have been achieved by dynamic planning and foresight. Indeed, since the standard of living which the world's various political and economic systems can provide is a vital weapon in the war for the minds of men which has been in progress over the past several decades, it is not too illogical to conclude that the dynamic character of the U.S. marketing system has provided, and is providing, the impetus necessary for the survival of our economic system.

The Modern Marketing Environment

The present state of marketing activity in the United States is indeed different from that which existed for long periods of American history. No longer will the worker-consumer purchase anything with four walls and a roof to house his family or be satisfied with the seasonal availability of fruits for his table. The modern consumer is affluent and sophisticated. His family's basic needs are well provided for and he has ample free time. This provides him both with a need for recreational and other leisure-time products and with an opportunity to shop for the items which compete for his discretionary dollars.

From the point of view of the marketing manager, competition for the consumer's dollar is more keen today than ever before. Not only must huge steel companies vie with each other for the orders of the auto manufacturers, but they must simultaneously compete with the producers of aluminum, fiber glass, and materials which did not even exist a decade ago. On the retail level the era of the shopping center and discount house has brought drastic changes in merchandising. Department stores, which were traditionally located in downtown areas, have opened suburban branches, and the emphasis on low prices and minimal service has relegated specialty stores to a less important role.

The era of a production-oriented economy operating to satisfy an existing demand is undoubtedly gone forever. Today's marketing-oriented system operates to create demands which it then fulfills. Offered everything from hula hoops and skateboards to patriotically styled coffins in which he may depart this earth, the modern consumer is called upon to perform his function and *consume*, and the marketing manager, who once largely oversaw the distribution of a product, must determine to some degree both what will be demanded and how to fulfill the demand effectively.

The Marketing System

As a first step in understanding and analyzing a complex system with which we are all acquainted in our role as consumers, we may examine the company (manufacturer) in isolation with the consuming public.

The typical manufacturer distributes his product to consumers through a network of wholesalers and retailers, some of whom may deal exclusively with one company but most of whom handle several competitive products as well. Purchase of the product by the ultimate consumer is promoted through advertising agencies that plan and place advertisements through television, radio, newspapers, point-of-sale displays, and other specialized media. Cooperative effort may be carried on by the manufacturer and retailer through local newspaper advertisements of a single brand name available at a local merchant's establishment or by a manufacturer and a wholesaler who distribute a brand name in a given area.

Direct personal selling by the manufacturer's salesmen, who may call on both wholesalers and retailers to encourage them to carry an assortment of sizes, styles, and grades in inventory and to lobby for advantageous display space, represents a dual-purpose point of contact. On the one hand, this contact provides sales pressure emanating from the manufacturer. On the other hand, it provides the manufacturer with timely information concerning sales (since his records usually provide him only with data on shipments to wholesalers rather than sales to consumers), returns, complaints, and other matters which may affect his future decisions concerning products, prices, and the host of other variables which are subject to his control.

Figure 1-1 is a simple illustration of the company in isolation with the consuming public. The lines connecting the various elements are lines of contact. The contact may be physical, as in the case of the shipment of the product, or abstract, as in the case of the exposure of a consumer to a television commercial.

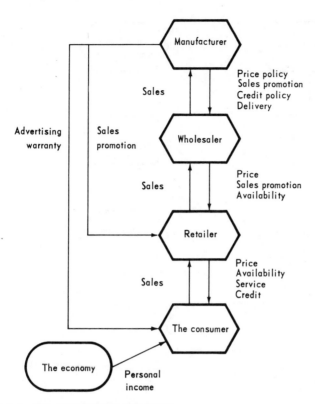

FIGURE 1-1   THE COMPANY IN ISOLATION

The entries on each line represent some of the modes of contact. The element labeled "The economy" represents such factors as the hourly pay of workers, the labor situation, the weather, and other outside influences on the consumer's purchases (or the decisions made by the organizations in the marketing system).

Concurrently with all of this activity, competitors, in league with competitive advertising agencies, are carrying on similar promotional efforts directed largely at the same targets—the body of wholesalers, retailers, and ultimately the consuming public. It is the interaction of the efforts of all of those elements which finally determines the degree of success which will be achieved by each.

A more comprehensive diagram of the marketing system includes such competitive elements, as shown in Figure 1-2. Each contact line in this diagram has largely the same modes of contact as the comparable line on the previous diagram.

Such a complex system is inevitably fraught with a myriad of decision problems which must be faced at each level: "how much to buy," "what price to charge," "when to advertise," "in which media," etc. In each decision situation the decision maker must

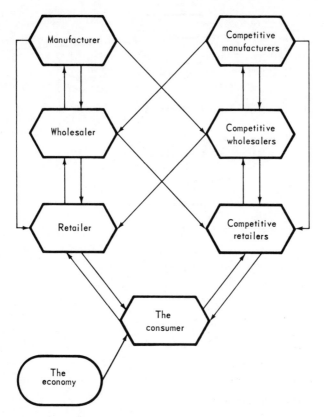

FIGURE 1-2   THE MARKETING SYSTEM

consider not only the potential direct effects of the actions he may take, but the effects of actions of each of the other elements of the system as well. The "best" answers to such questions are not at all obvious. Each question is itself a complex entity which requires careful analysis and thought on the part of the marketing manager.

The Need for Scientific Analysis

The phenomenal success of our system in pushing to the edge of space while simultaneously maintaining a huge military establishment and supplying the wants of an affluent society requires the marketing manager to "run faster and faster just to stand still." He must face an unending sequence of decision situations, any one of which may be of such great importance as to spell doom or success for a product.

Traditionally the marketing decision maker has resolved decisions involving product selection, pricing, distribution channels, and the like by calling on the same experience and intuition, or horse sense, for the operation of the marketplace which led to success in the past. The modern marketing executive is fortunate, indeed, if he has precedents to use as a background for current problems and a keen intuition which enables him to avoid the pitfalls with which the system abounds, but in the cold, uncompromising environment which he must operate, the manager often finds that he needs cold, uncompromising analysis of his decision problems. For, indeed, his problems are so complex as to be beyond the scope of human intuition and subjective analysis.

It would be foolish, however, to argue that marketing decision problems became complex last night, or at any recent point in time, or, indeed, over the past two decades. The increase in complexity has been a gradual one, resulting directly from our ever-expanding society. Our present-day perception of the degree of this increase in complexity is amplified by the relatively slow concurrent development, within the marketing environment, of new techniques for handling ever more complex problems.

One significant factor which has come into play almost overnight has served to focus the attention of marketing executives on their need for problem analysis. That factor—the moronic[1] electronic computer with its associated fast data processing equipment—has proliferated to the extent that the data now available to the marketing decision maker are almost inexhaustible. Virtually instantly the modern marketing executive can have at his fingertips data on sales, market shares, costs, and descriptions of the potential consuming public which only a few years ago would have been too costly to consider garnering from the reams of diverse reports in which they were originally compiled.

Faced with this complexity and availability of data, and knowing that the techniques necessary to make use of the data are becoming available to his competitors, the executive has little choice but to undertake an analysis of decision problems outside the context of his judgment and intuition. This kind of analysis is generally associated with the application of the *scientific method* to decision problems.

Of course, this does not mean that the decision maker cedes his responsibility for making decisions to some mystical scientific

---

[1] The adjective "moronic" is applied to the computer by the author in a defensive reaction to the widespread association of the word "brain" with the computer. Whatever the capacity of the computer as we know it may be, it is in no sense a brain, but merely a fast and efficient processor which must be ordered to perform each and every action in a degree of detail which would normally be associated with the intellect of a submoron.

process or that his judgment and intuition do not play a major role in decision making. The scientific analysis of marketing problems is a logical and objective process which is to some degree independent of the subjectivities and intangibles which must be incorporated into the decision process. As such, the scientific approach is an effective *complement* to the judgment and intuition of the decision maker.

The Value of Scientific Analysis

To justify the use of scientific analysis as an aid to marketing decision making which is preferable to witchcraft or coin tossing, one must argue on some basis other than the inherent "goodness" of science. Science, like motherhood and all else in this complex world, has both positive and negative aspects and does not therefore justify itself simply by existing.

One can, of course, point to the successful applications of the scientific method to military and industrial decision problems and argue the likelihood of further success as greater knowledge is gained. Indeed, the gains achieved in the past as a direct result of scientific analysis are impressive, and perhaps this is sufficient evidence to warrant further attention.

To proceed a bit further along a pragmatic line, one might recognize that the same objective input information is available to the scientific analyst as to the witch, gambler, or intuitive decision analyst. The scientific approach has the additional virtue of guaranteed logic and consistency. As we shall see later, the analytic process is quite similar to the mental steps a human is likely to follow in his subjective analysis of a decision problem. The primary difference, then, is that the totally subjective process has no guarantee of logic and consistency. In addition, since the scientific process merely serves as a complement to the subjective processes of the decision maker, nothing is lost and something may be gained by making use of it.

Another value of scientific marketing analysis is its reproducibility. Scientific analysis is a logical process which is well suited to being carried out with pencil and paper (and sometimes slide rule, desk calculator, or electronic computer). In any event, the assumptions, logical steps, and conclusions are always clearly spelled out and recorded. Thus, the analysis may always be resurrected, after the decision to which it contributed is made and the results are observed, and the analytic procedure itself can be evaluated. If this testing proves the worth of the analysis, the same procedure can be applied again with greater assurance that the results will be desirable. A purely subjective approach has no such permanent value. Once a subjectively approached decision problem is solved, little knowledge has been gained about the *method* which was used to arrive at the decision. In fact, since there is no natural "record keeping" associated with nonscientific approaches, it is usually impossible to reconstruct even a portion of the process for, indeed, most of it is mental and any attempt to review it is fraught with rationalization. Each decision is therefore a separate entity, and the analysis of each new problem must begin at the beginning, even if the problem is of a repetitive nature. In scientific analysis, formal "learning" occurs and may be drawn upon again and again as new decision problems arise.

The reproducibility of scientific analysis also yields benefits which transcend the particular people who perform the analysis. Anyone with a proper background can be taught the elements of a formal analytic approach, but the procedures of subjective analysis generally cannot be communicated easily. Any "learning" which results from the nonscientific approach accrues *only* to the individuals who directly participate in the process. If they are unavailable in the future, the organization must begin the trial-and-error process anew—a costly procedure which largely negates the primary advantage of human beings over other forms of life, their ability to build on foundations of knowledge laid by their predecessors.

The importance of the nonscientific aspects of the marketing decision maker's tasks, however, is not necessarily lessened through the use of scientific analysis. One of the primary features of the scientific approach is its degree of abstraction—the omission of certain aspects of the real-world problem which the decision maker faces. Such omissions mean that only a part of the real-world problem is treated scientifically. The decision maker must then integrate the results of scientific analysis with the significant intangibles which are not part of the formal analysis in order to arrive at a "best" decision. In doing so, he must call upon the same levels of judgment, intuition, and experience which are used by the "traditional" manager. The difference is that the scientifically oriented manager draws upon both subjective and objective analyses—each in the areas in which it is most useful. This effective blending of the objective and the subjective is probably the single quality which best defines a "modern" marketing manager.

The view we shall take of modern marketing management and the role which scientific analysis plays in it is therefore a simple one. The process of scientific analysis is viewed as a logical and consistent method of reducing a large part of a complex decision problem to simple outputs which the manager can use, in conjunction with other factors, in arriving at "best" decisions. It permits him to focus the analytic resources at his disposal on the aspects of the problem where they are most effective. He is therefore able to utilize efficiently both scientific and nonscientific analysis to best advantage. Such an integration can hardly be worse than, and is potentially far superior to, a purely subjective approach to decision making.

The Role of Scientific Marketing Analysis

The current need for scientific analysis of marketing decisions emanates largely from three sources: the increasing complexity of the system, the relatively slow development of new methods of analysis within the context of marketing, and the advent of the electronic computer and other data processing equipment.

The necessities of survival in World War II simultaneously forced scientists into the analysis of real-world decision problems and military managers into a recognition of the potential value of scientific analysis. After the war it was recognized that the methods of analysis which could be applied to a problem of locating a radar station could equally well be used in a problem of plant location. Hence, the transition of scientific analysis to industrial decision problems occurred rapidly. Applications, largely confined to the production segment of businesses, began to be made, and a breed of individual who was proficient in the scientific analysis of decision problems—the staff operations research specialist—began to have detectable impact on the outcomes of business decision problems.

Such individuals typically call themselves operations researchers, management scientists, systems analysts, or something similar. They share in common, however, an interest in problems and their solutions—not necessarily financial problems or production problems per se. This interest has naturally turned to the complex problems of the marketing environment. Concurrently, the marketing manager has begun to recognize his need for help in analyzing marketing problems. The result has been a significant increase in the number of marketing problems which have been subjected to analysis by methods which some might say were borrowed from other business functions, but are in reality the general scientific approach applied to decision situations—whatever may be the particular context in which they exist.

To apply scientific methods to the analysis of decision problems, one must be "quantitatively oriented," for one of the primary characteristics of scientific inquiry is the process of quantitative measurement. The marketing manager has for a long time, of course, had quantitatively oriented staff specialists to call on for aid. These "marketing researchers," however, have been concerned largely with the collection and analysis of marketing data rather than the analysis of *problems*. In analyzing data, the marketing researcher has been quantitative in the sense that he works with numbers, but the term "quantitative analysis" has come to mean something far more than a reliance on numbers. Today the term relates to a body of scientific methodology which draws on numerical data but is oriented toward problem solving rather than information analysis. In large part the role of the remainder of this book will be to demonstrate this methodology and the ways in which it is applied.

It should be noted that the emphasis placed on the decision-oriented analyst does not exclude the traditional analyst from an important role in the marketing decision process. One way of viewing his activities is seen in Figure 1-3.

In this simple feedback loop the decision maker calls on the decision analyst for predictions or recommendations concerning the potential results of his alternative actions. In turn, the decision analyst must use data which have been compiled and evaluated by the traditional marketing researcher from the raw data supplied in the marketplace. The decision maker then takes some action and some results are observed. At this point the process begins again: the results are evaluated and compared with the predictions, better predictions and recommendations are developed, and the decision maker is given new inputs on which to base subsequent actions.

As we have previously noted, the role of the decision maker is still of paramount importance. The predictions or recommendations which serve as an input to the decision maker must be complemented by his judgment and intuition. In no sense is the role of scientific analysis a substitute for experience or subjective analysis. One may view the output of the decision-oriented analyst as

FIGURE 1-3   MARKETING ANALYSIS IN THE DECISION MAKING PROCESS

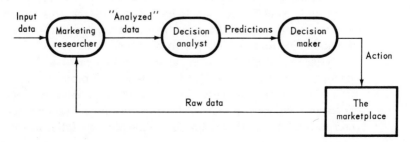

one of several inputs to the subjective decision process of the decision maker. This input, being the result of a logical and consistent analysis, greatly simplifies a complex problem since it incorporates the net result of many factors into a single input.

In determining the price to charge for a product, for example, data on costs and potential demand at various prices are analyzed and incorporated into a mathematical prediction of the profits which should be attained at various prices. The price predicted to be the one which maximizes profit is presented to the decision maker along with an explanation of the predictive technique, its assumptions, and its omissions. The decision maker must weigh the prediction in the light of his experience with similar products and circumstances and estimate the effect of factors which have been omitted. The result of his action—the setting of a price—is reflected by sales data generated in the marketplace. These data are analyzed and used to revise predictions of the best price. On the basis of the revised predictions, the decision maker may reevaluate the situation and choose a different price. The process may then continue, the price being reevaluated as frequently as desired or as frequently as it is practicable to make price changes.

As one might expect, the terminology applied to the various analytic functions in any particular organization may be quite different from that used here. In fact, it is not at all necessary that different individuals perform each of the roles outlined in Figure 1-3. The marketing manager may be called on to do all of these tasks, or he may have aid available. In every organization, however, the decision maker finds it is necessary that certain tasks be performed—the collection and analysis of market data, the prediction of the possible outcomes of various actions, and the evaluation and comparison of these actions. If the manager is fortunate enough to have researchers and analysts to aid him in such predecision tasks, he may devote more time to the postdecision elements of his job—the organization and motivation of the human beings who will carry out his decisions.

In many instances, little specialized staff talent is available to the manager on a continuing basis; in this case the manager may be called on to render choices which are most easily made with the aid of scientific analysis. If he is to compete in this environment effectively, he must have a basic understanding of scientific methodology and of the capabilities and limitations of scientific analysis, for although the manager does not need to be a trained scientist, he must understand the scientific process well enough to make use of analytic aid when it is potentially fruitful, to know in advance the general kind of results analysis may produce, and to place boundaries on the applicability of the answers provided by analysis. In this way the modern manager can effectively blend his own strong points with the strong points of scientific analysis to arrive at overall "best" decisions.

In any case, it should be emphasized that this approach *does not* envision the role of the manager as a passive one. Too often quantitative analysts have acted as though problems could be given to the analyst by the manager, solved by the analyst, and then given back, as complete solutions, to the manager. Such a procedure is relatively inefficient and ineffective, particularly in the marketing environment.

We shall demonstrate in later chapters how the manager is an intrinsic part of the scientific approach to problem solving which is proposed here. In several phases of using the methodology to be demonstrated, the manager is essential. Indeed, as we have pointed out, the modern marketing manager may himself use the scientific approach for solving marketing problems without calling on experts for anything other than input information on costs, sales revenues, profits, etc.

The remaining chapters of this book are devoted to demonstrating the conceptual basis which may be applied to the analysis of marketing decisions, its logical consistency and efficiency, and particular applications to specific marketing decision problems. The objective is twofold: first, to demonstrate the value of scientific marketing analysis to the marketing manager; secondly, to provide the manager or novice analyst with a solid foundation for applying this kind of thinking to real-world marketing problems. In attempting to accomplish these objectives, we shall often imply that the decision maker and analyst are two different individuals. In many cases, as we have pointed out, this is neither the existing situation nor necessarily the best way to conduct the affairs of marketing management. Clarity of presentation is the main motivation for this implication. When analyst and manager are two individuals, the full richness of the process is easily demonstrated. When one individual is both decision maker and analyst, the process itself may be simpler, but it is more difficult to demonstrate.

REFERENCES

Alderson, W., and S. J. Shapiro (eds.): *Marketing and the Computer,* Prentice-Hall, Inc., Englewood Cliffs, N.J., 1963.

Beckman, T. N., and W. R. Davidson: *Marketing* (7th ed.), The Ronald Press Company, New York, 1962.

Churchman, C. W., R. L. Ackoff, and E. L. Arnoff: *Introduction to Operations Research,* John Wiley & Sons, Inc., New York, 1957.

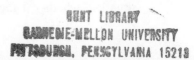

CHAPTER *2*

Marketing Decisions

The typical marketing manager is faced with an unending series of choices: which product to market? how much to spend on advertising? what price shall be charged? shall distribution be direct to retailers or through wholesalers? etc. Some of these choices are easily made, as in the case of the producer of a new soft drink who decides to charge a dime for his product because that price is "traditional" and an easy-to-pay price for such an item is practical. Other choices are not so simple; for example, the same marketing manager must choose whether to advertise most heavily during the summer, when he knows sales of his product will be at a peak, or in the winter, in order to stimulate sales during the slow season.

In any event, the choices the marketing manager must make are those which cause the fortunes of the business enterprise to rise and ebb. They are important choices and therefore they should be carefully scrutinized. In the previous chapter we have argued that scientific analysis should be at least a part of the scrutiny. The first step in applying scientific analysis to marketing decisions lies in understanding that creature with which we deal—the *decision problem*.

Decision Problems

The alternative the marketing decision maker chooses in a decision situation, whether it involves product selection, pricing, or advertising, interacts with the general state of things as described by the block labeled "The economy" in the diagrams of the previous chapter. This interaction, together with the actions which may be taken by competitors, determines the *outcome* of the decision situation— the state of affairs attained as a result of the chosen alternative, outside influences, and competitive actions. In the pricing situation, for example, the chosen price interacts with the economic status of consumers and the prices of competitors to result in an outcome which might be described in terms of the number of units sold, the share of market attained, revenue, or any of several other possible *outcome descriptors*.

Outcomes of marketing decision situations represent the states by which the success or failure of the chosen actions is gauged. In truth, the aggregate of the outcomes of the many decision situations which face a business enterprise during a single year determines the success or failure of operations for the year and, in the long run, the success or failure of the enterprise itself.

The range of outcomes which may result from any particular decision situation is often quite wide. For instance, in the choice of a new parlor game to be marketed, the possible alternatives and outcomes might range from a complicated game of strategy, which will interest only a few intellectuals and result in low revenues, to another Monopoly, which will generate profits for decades. Because of this wide range there is great pressure on the marketing decision maker to choose the *best*, or at least a *good*, alternative— whether a game to market, a price to charge, or a medium for advertising. To make such a choice intelligently, the executive must be able to sift the important factors which bear on the problem from the maze of complex trivia with which all significant decision situations abound.

Perhaps, however, the first faculty to be developed by anyone who is to make choices in any environment is the *ability to recognize problems*. For although it seems obvious that a problem must be recognized before it can be solved, it is sometimes easy to be led blindly down the path of convenience and tradition without realizing that alternatives exist and that a choice is available. The advertiser who continues to utilize television spot announcements because "this is the way that it is done in the widget business," even though widget sales are poor, is an all too typical example of a decision maker who has failed to recognize the existence of a problem.[1]

[1] This is not to say, of course, that there is necessarily a better way to advertise; but other ways might be considered even though it has not previously "been done that way in the widget business."

Any decision problem, whether in marketing or in the daily life of an individual, involves several important elements. First, some person or some group must be faced with the problem—the *decision maker*. The term *decision maker* as used here should not be interpreted with the usual connotation of a forceful dynamic activist as opposed to one who procrastinates. The scientific meaning of the term does not imply evaluation of the personal qualities of the human(s) who may fill the role. In the scientific sense *a decision maker is an entity, either an individual or a group, who is dissatisfied with some existing state or with the prospect of a future state and who possesses the desire and authority to initiate actions designed to alter this state.* For example, the marketing vice-president who is dissatisfied with a downward sales trend is potentially a decision maker in the scientific sense. He must, of course, actually possess both the desire and the authority to alter promotional expenditures or take other actions designed to increase sales if he is to function as a decision maker.

The decision maker's desires to achieve some state of affairs—his *objectives*—are the reason for the existence of a problem. Often these are expressed as a wish for either the attainment of some new state, such as higher profits, or the retention of some existing one, such as "our image as the industry's leader." Usually, however, the objectives of the decision maker are expressed as some combination of achievable goals and retentive constraints on the pursuit of those goals, e.g., to maximize profits while simultaneously maintaining an image level.

To pursue his objectives meaningfully, the decision maker must have available *alternative actions* which can influence the state of affairs he wishes to achieve. These available actions, together with a *state of doubt* as to which alternative is best, constitute the heart of any decision problem.

The failure of potential decision makers—those dissatisfied individuals with authority to act—to recognize the existence of alternatives is perhaps the most prevalent sin of omission among executives who base their thinking solely on subjective experience, judgment, and intuition. Theirs is the failure to recognize and consider alternatives which were never before used and are therefore likely to be beyond the scope of experience and sound subjective evaluation. In effect, the potential decision maker who is dissatisfied but recognizes no alternatives to "the way in which it is done" abdicates his responsibility, since the existence of alternatives implies the potential existence of a problem. *The failure to recognize the alternatives, and hence the problem, has the same result as would a conscious choice of the status quo from a range of possible alternatives.* The manager should, then, constantly be on guard for decision problems appearing from nowhere with the development of new techniques, policies, or actions of competitors. It should always be borne in mind that one never really shirks a decision, since the state which may be retained by the apparent avoidance of alternatives is itself an outcome of the unrecognized problem. As such, it is always a prime candidate for subsequent comparison with outcomes which might have resulted from unrecognized alternatives.

One situation in which new alternatives become available in advertising is worthy of note. Certain products traditionally shunned national magazine advertising because such appeals could not be directed on a local level; for instance, they could not be advertised as "made in Texas by Texans" because the message would probably not be effective among the many non-Texans it would reach. The advent of regional editions of some national magazines has made it possible to direct different messages to different audiences, as in the case of radio, television, and local newspapers. The advertising manager who has failed to consider this new alternative in more than an offhand fashion, while the company is less than satisfied with product sales, presents an example of a potential decision maker who has failed to recognize a problem.

Identifying Decision Problem Elements

All of these elements—the decision maker, objectives, alternatives, and a state of doubt—must exist in some context. The *environment*, or set of external factors which affect the potential outcomes of a decision problem, often presents great practical difficulties for the analyst who is attempting to identify the relevant aspects of a decision problem.

The organizational environment, for instance, is formally structured as a none too simple network of lines of authority, flows of information, and areas of responsibility. The operating structure—the ways in which people actually go about directing the affairs of a business and communicating within it—is often vastly different from, and usually more complex than, the formal structure.

The person to whom the organization chart ascribes responsibility and the person who actually influences the day-to-day activities of the enterprise may not be identical. Often the distinction is so subtle as to defy detection, sometimes even by the persons involved. The case of the newly promoted executive who, unfamiliar with his new task, continues to retain the comfortable prerogatives of his old position is a case in point as is the comic-strip boss who plays golf while ceding his duties to subordinates. At the other extreme is the clerk's alteration of actions ordered by his superiors on the basis of knowledge generated in the field which was "obviously unknown" to the executive who selected the course of action.

If the decision maker is a group rather than an individual, it is necessary to identify not only the individuals who comprise the group but also the *process* by which they arrive at decisions. Often it is difficult to subject this process to any degree of formalization. The analyst who queries a group concerning the process will often, after some initial hesitation, receive the democratic response "majority rules." To most groups this seems to be both a way around the embarrassing position of appearing so disorganized as to have no decision-making procedure and a reasonably accurate description of the informal argumentative process which they actually use. The presentation of the paradox of preferences which is inherent in the rule of the majority often produces a restatement of their procedure which is more accurate, but harder to formulate.

The *voting paradox* involves the consensus choice, say of three individuals for three alternatives *A*, *B*, and *C*. To demonstrate this, consider Table 2-1 as giving the preference order of the three

TABLE 2-1   VOTING PARADOX

	Preference order→		
Individual	*Most preferred*	*Second most preferred*	*Least preferred*
1	*A*	*B*	*C*
2	*C*	*A*	*B*
3	*B*	*C*	*A*

for the alternatives. It is clear that *A* will be preferred to *B* by the group since both individuals 1 and 2 prefer it to *B* and the "majority rules." Similarly, *B* will be chosen in preference to *C* by the group by a vote which goes against the preferences of the middle individual. As a result, since *A* is preferred to *B* and *B* is preferred to *C* by the group, one might infer logically that alternative *A* should be preferred to *C*. After all, if one likes chocolate ice cream better than vanilla, and vanilla better than strawberry, the logical conclusion is that chocolate is preferred to strawberry. A glance at the table of individual preferences indicates that this apparently logical deduction is not valid for this group. In fact, two of the three individuals prefer *C* to *A*, and hence the group, by virtue of majority rule, prefers *C* to *A*. This potential violation of logic in group preference relations under the rule of the majority is one reason that most informally organized groups actually make decisions in more subtle ways.

Objectives are often another gray area which requires a good deal of soul-searching and tactful examination. Stated objectives of increasing sales, profits, or market share are often complemented by unstated desires relating to the stability of management, market leadership, or image. As an example of a failure to consider objectives relating to image, consider the analyst who spent weeks formulating a decision problem in which the specified alternatives involved either increasing promotional effort, discontinuing sales, or making no change in each of a number of market areas, only to find that the company's president would not consider withdrawing from any market area, regardless of the consequences. Although it had never been openly stated, the president's view of his product as the industry's leader and his estimate of the loss of face inherent in pulling out of an area were of such overriding importance as to negate mere consideration of sales and profits.[2]

The identification of relevant alternatives and outside influences is often no less difficult than the identification of the real decision maker and his objectives. The former may often involve developmental aspects—the invention of new feasible alternatives; the latter may require thorough study of the internal elements and power structure of the firm, competitors and their objectives, the government, and various uncontrollables such as the political climate or the weather.

This delineation of some of the difficulties which may be involved in identifying the important elements of a problem should not be construed to imply that every dissatisfied individual has a problem which should be analyzed or even that a dissatisfied person necessarily has a problem. The picnicker's dissatisfaction with rainy weather appears to describe a problem situation, but since for most practical purposes he can do nothing to affect the weather, he has no alternatives and hence no problem.

In marketing, the effectiveness of the organization is measured by the ability to recognize real problems—those states of dissatisfaction in which there are alternatives which can influence the attainment of desired outcomes. Indeed, it is true of all of life to say that one progresses by a series of choices which may be made either implicitly or explicitly. Those explicit choices which represent the solutions to decision problems will invariably lead to better

[2] It is true that it may be argued that the short-term image objective is really a long-term profit objective and that the president's objectives are perfectly consistent and rational. The point made here, however, is that the analyst must ferret out such objectives before he can adequately formulate the decision problem.

long-run results than will implicit choices made by ignoring the available alternatives.

Solutions to Decision Problems

To solve a decision problem, it is necessary that the decision maker choose the *best* of the available alternatives. In very simple decision problems this is equivalent to saying that he should choose an alternative which leads him to a state which is at least as preferable as any other state. In more complex problems the idea of a best alternative is somewhat more subtle. We may define a *problem solution* as the delineation of the best of a set of alternative actions.

This apparently simple and straightforward statement has astonishing practical ramifications. What is meant by *best*, for instance? How is the best alternative to be found? Is the alternative which I choose as best necessarily the same as my superior would choose? Such questions are answered by the normative conceptual framework which is developed in Chapter 3 and utilized in particular marketing contexts in subsequent chapters.[3] In the remainder of this chapter we seek to define more clearly the role of best alternatives in the marketing process.

The *form* of the problem solution is of some practical significance to the marketing process. In any particular decision problem which. requires that a concrete action be taken, the form of the action is usually apparent. If we are to set a price, the form of the solution is a quantity expressed in dollars and cents. If the problem is a choice between advertising media, the form is the qualitative delineation of one or more media. However, the solution to a decision problem which is attained through scientific analysis may well be in even more useful form. The solution may be a *decision rule*—a simple rule which prescribes how the known elements of the problem situation should be combined in order to arrive at the specific "best" action. For example, a rule specifying how one could combine data on the relative costs of various advertising media in a way to reveal which medium is best to use would be a decision rule which prescribed the determination of a best action. One may consider either the decision rule or the best action as the problem solution in such a situation, for they are really the same thing, the best action being simply a specific realization of the application of the rule. The advantage of having the problem solution in the form of a decision rule is clear, since the rule may be directly applied to each of a series of similar problem situations.

[3] The framework is normative in that it prescribes how best choices *ought* to be made rather than describing how choices *are* made.

If a manager is faced with deciding the quantity of raw materials to be purchased on the first of each month, for example, he may be able to apply the same decision rule over and over each time the problem arises, recognizing that the known elements of the problem such as costs, demand levels, etc., change in numerical value from month to month. This is the feature of scientific analysis—the practicability of obtaining problem solutions in the form of decision rules—that has led to the computerization and mechanization of decision making which are so often discussed in nonscientific literature.

The concept of searching for and selecting the best alternative, however defined, is itself subject to controversy at the practical level. Professor Herbert Simon of Carnegie Tech has proposed in his *principle of bounded rationality* that humans seldom attempt to find the best alternative in a decision situation. Rather, they select a number of *good enough* outcomes and an alternative which is likely to achieve one of these outcomes. In searching for a new product, for instance, the typical marketing executive makes no attempt to enumerate all possible products in order to select the best, but rather he decides what he wishes to achieve with a new product and selects one which is likely to satisfy those desires.

This *descriptive* (how people *do* act) concept has *normative* implications (how people *should* act), for one might argue the irrationality of complete rationality; i.e., the completely rational man should evaluate all alternatives and choose the best, and yet if the time and money involved in doing so are prohibitive, it is irrational to do so. In any event, the conceptual framework to be developed in the next chapter is so versatile as to provide a basis even for deciding when not to seek the best alternative. Hence, while we should not always unhesitatingly apply it directly to any marketing problem at hand, we shall often find it useful as a basis for *deciding how to decide.*

Managerial Decision Problems

Some problems are relatively simple and need not be subjected to extensive analysis. The decision maker who is standing in the middle of a street with a car bearing down on him at 60 miles per hour is in a problem situation, for he desires to avoid injury and may select from at least two alternatives: to run to either the right or the left curb. Any prolonged contemplation or analysis on his part, however, will lead to an outcome which is clearly not best for him. The need in this situation is for a *quick and accurate decision*—the kind that is the forte of some traditional managers and is the very cornerstone of combat management in the military, where battles are often won or lost simply by initiative rather than by the choice of the best alternative.

In more complex situations, when time and resources for analysis are available, the choice of the best, or at least a good alternative, is of paramount importance. In particular the degree of complexity involved in *managerial decision problems* is most often so great as to preclude anything but the most cursory subjective analysis. In such problems the outcome is rarely influenced to any great extent by inordinate speed or initiative but rather by the quality of the alternative chosen.

To identify this level of complexity, one needs only to consider how industrial enterprises have decentralized since the industrial revolution. The division of the management function into categories called personnel selection, sales, quality control, and the like, and the concurrent geographic dispersal of operations to reach new markets and sources of raw materials have created organizational entities with *conflicting objectives*. To demonstrate this, consider a common managerial problem involving the determination of which products are to be produced and in what quantity. The production department of the enterprise would undoubtedly prefer that few products be produced in rather large quantities in order to minimize the number of costly machine setups necessary to convert from production of one product to production of another. Such a policy would lead to large inventories of a few products. Sales personnel, on the other hand, desire to have many different products in inventory so that they may promise early delivery on any product. The people in finance recognize that large inventories tie up money which could be invested elsewhere, and so they want low total inventories. The personnel manager desires constant production levels so that he will not continually be hiring new workers for short periods of peak production and laying them off in slack periods. One could go on to identify objectives of almost every functional unit of an organization relative to this decision problem. As demonstrated, these objectives each conflict to some greater or lesser degree—low inventory levels versus high inventories, many products versus few products, etc.

This *conflict of objectives* and a concern for the *overall effectiveness of the organization* are the primary aspects of a managerial decision problem, for the manager desires to optimize (to select a best alternative) with respect to the whole organization rather than to make one functional unit appear superior at the expense of others and, perhaps, at the expense of the overall performance of the organization. Of course, a managerial decision problem may exist at levels other than that of corporate management, since the decentralization of industrial enterprises to dozens of subordinate levels implies the existence of conflicting objectives and desires for coordinated effectiveness at every level.

The Systems Approach

The advent of managerial decision problems involving the overall effectiveness of an organization and conflicting objectives of functional units has given rise to a new concept in studying organizations and their problems—*the systems approach*. A *system* is "an organized or complex whole; an assemblage or combination of things or parts forming a complex or unitary whole." We are all familiar with such typical systems as the solar system, the Mississippi River system, and the nervous system.

In business the natural tendency of the manager to focus attention on a particular specialized function, rather than on the complex system composed of the business organization and its environment, has often led to an avoidance of true managerial decision problems. Even though problems which involve complex business systems are usually difficult to solve and, in fact, often difficult to formulate in understandable form, they are nonetheless the truly important problems of business. The business enterprise will not progress if actions are constantly taken to enhance the performance of one department at the expense of another. Such actions are called *sub-optimum*. It is the desire of the scientific approach to develop *optimum* solutions to problems, those which are best for the organization as a whole rather than for a single portion of the organization. The systems approach to management is therefore incompatible with the idea of operating a functional unit "just as you would operate the family grocery store," for there may be occasions when a small decrease in the apparent performance of one functional unit results in a large increase in performance of the total organization.

In the production-marketing context which was previously discussed, for example, the production of a wide range of products with high machine teardown and setup costs between products may result in increased total sales since no customers who desire quick delivery will be turned away because of low inventories of the product they want. The increased sales may more than offset the increased production setup costs, resulting in a higher profit for the enterprise. In this case the apparent high costs incurred by the production department are not at all indicative of poor performance, but they are indicative of the systems approach to the solution of decision problems.

To develop optimum solutions to managerial problems, it is necessary to view the organization in as large a context as possible. In solving a marketing problem, for example, one needs to consider the effect on production, finance, quality control, and other relevant organizational entities, while simultaneously weighing the relative influence of competitors, the government, suppliers, and other elements of the system. To do this in its entirety, of course, might require infinite analytic resources. In practice, one accounts for limited analytic capability by abstracting much of the real-world complexity of the problem and considering systems which may involve less than the total organization. At all times, however, *the effect of making these approximations should be evaluated.* In effect it is sometimes necessary to view several subsystems rather than a single large system. In doing this, the interrelations of the subsystems are often initially neglected. After some tentative solutions are obtained relevant to the subsystems, the effect of these relationships may be estimated and the solutions adjusted accordingly.

Interdisciplinary Teams

The idea of studying as large a system as is feasible has led to a quiet revolution in problem solving. The systems approach, or way of viewing problems, has resulted in an understanding that no single individual is ever fully qualified to formulate and solve a decision problem. All of the disciplines which utilize the scientific method have developed techniques and ways of formulating problems which can prove useful in entirely different environments; e.g., the ideas of fluid mechanics or electrical flow may prove useful in analyzing the flow of products through a distribution system. As a result of the recognition that no specialist has a monopoly on problem-solving ability, the analysis of business decision problems has more and more become the province of teams composed of individuals with varied backgrounds. These individuals bring the point of view of their experience to bear on the problem, often with results which are significantly superior to those a single individual might be expected to produce. In addition, the nature of managerial decision problems implies systems which have psychological, sociological, and physical aspects. Hence, what better way to study complex systems made up of interrelated parts than with teams made up of individuals who can bring the knowledge of the related disciplines to bear?

We have already pointed out that the blocks in Figure 1-3 which appear to represent the places of three different people in the decision making process actually represent functions which may be performed by one or more individuals. Here, we should point out that the practical evidence which is available concerning the results of scientific analysis of business decision problems suggests that a group of people may well be the best entity for performing these functions. It is not necessarily the case that each individual in a group which analyzes a problem will clearly fall into one of the categories which were there termed "marketing researcher," "decision analyst," and "decision maker." Any number of individuals may perform parts of each of these tasks, the ultimate result being a choice of a best alternative by the group.

Of course, this presents an additional difficulty because, as we have already pointed out, the ways in which groups do, or should, arrive at decisions are not clearly understood. The seeming inconsistency in the use of interdisciplinary teams is mitigated by the practical success of such groups in problem solving. Perhaps a group which is composed of individuals with different backgrounds, but is closely bound by an understanding of, and belief in, basic scientific methodology, is an entity quite different from the typical group which sets about the task of making a decision.

The Decision Making Process

To gain further insight into marketing decision problems and their solution, one may view the *process* of problem solving, i.e., the steps which must be followed, implicitly or explicitly, in arriving at a problem solution.

Figure 2-1 outlines a sequence of activities which begin with the input of information. In terms of a new product pricing decision, this may be data on production costs, anticipated distribution costs, and the marketing results of similar products. The decision maker (or analyst) *analyzes* these data and uses them as a basis

FIGURE 2-1   ELEMENTS OF THE DECISION MAKING PROCESS

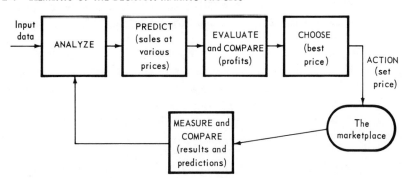

for *predicting* some future state—the sales which the new product can be expected to achieve at various price levels, for instance. These outcomes are then *evaluated* and *compared,* perhaps in terms of profit, and the best alternative is *chosen.* The price is then set and the product distributed. Out in the hard cold world some actual outcome—sales and profit—is achieved which is likely to be different from that which was predicted. These results are *measured* and *compared* with the predictions through the lower *feedback loop* so that revised predictions may be made and a reconsideration of the pricing action undertaken.

All of these facets of the decision process—the analysis of input information, prediction of outcomes, evaluation of alternatives, comparison of alternatives, selection of best alternative, measurement of results, and comparison of actual and predicted results to gain new input information—must be carefully considered by the decision maker and analyst. The particular roles played by the analyst and decision maker in this process may be better placed in focus at this point by comparison of Figure 2-1, which emphasizes the decision making process, with Figure 1-3, which emphasizes the roles of the individuals who may conduct the various activities. Each of the aspects of the process will be taken up in later chapters.

Product Management

In discussing the systems approach, we would be remiss if we did not mention the impact of this viewpoint on the *execution* phase of marketing management.

We may view the manager's role as a dual one involving both strategic decision making and execution. In the first phase of his task the manager decides what is to be done. Then he must ensure that effective implementation actions are taken. The latter element, the *execution phase,* is equal in importance to the decision making phase, for good alternatives which are not fully implemented may yield worse results than inferior choices appropriately executed.

In the execution phase of management the systems approach has had as great an impact as in the decision making phase. The advent of the *product manager* reflects the attempt which has been made to structure the organization to carry out decisions in a fashion consistent with the modern systems background used to make the decisions.

The creation of product manager positions resulted from the recognition that top management could not be expected to know all of the details about each of the organization's products. Similarly functional managers properly show more concern for their function than they do for products. Thus the need for a manager who can cut across traditional functional lines to bring together the resources required to achieve *product goals* is clear.

The product manager is able to organize the diverse objectives and motivations of the various functional units of an organization so that the total effort is directed toward the accomplishment of overall product goals. Thus, controllable aspects of the enterprise can be viewed by one person (or group) whose concern is with the product and its contribution to the organization, rather than with one of the specific methods for achieving the goals. In this way effective implementation and control of the product's sales program may be enhanced.

Although the execution phase of management is not the principal concern of this book, we should note the essential unity of the systems approach to decision making and the product management view of the execution process. Both emanate from the systems viewpoint. Because of its concern with product goals and the interrelationship of functions, product management may be thought of as a way of implementing the solutions to decision problems in the same context as that in which the decision problems were formulated and solved—the system.[4]

Summary

In this chapter the basic elements of any decision problem—the dissatisfied decision maker, his objectives, the alternative actions available to him, his state of doubt as to the best alternative, and the surrounding environment—have been treated in the context of marketing management. The meaningful decision problems of the marketing manager typically involved a desire for the optimization of the overall effectiveness of the organization and a conflict of objectives within the organization.

The systems approach to the solution of such managerial decision problems is the concept which is put forth as the methodological goal of problem analysis. This approach involves the study of as large a system as is possible so that the interactions of the various elements can be recognized and used to advantage. The practical limitations of the systems approach have been illustrated along

---

[4] The King reference extensively discusses the impact of the systems approach on both the decision making and execution functions of management. The Luck and Nowak reference discusses the implications of product management to marketing.

with one of its important practical applications to marketing management—"product management."

In the next chapter we shall discuss the analysis and solution of decision problems in marketing. The approach to be taken involves the development of a conceptual framework which is applicable to a wide range of decision situations.

EXERCISES

1. What would be good *outcome descriptions* in the following everyday situations?
   a. You apply for a job.
   b. You go to the supermarket to get a jar of molasses.
   c. You try to borrow as much money as possible from your friends.
   d. You try to borrow $5 to buy a football ticket.
2. Who is the governmental decision maker you would identify with the problem of dealing with an allied government's request that troops be sent to their country to put down an uprising? Who is the decision maker on the question of determining the salaries which our military forces will be paid?
3. A decision maker who is faced with a difficult problem might decide to go and play a round of golf. In what circumstances would this constitute a chosen alternative in the problem and in what circumstances would it not?
4. A successful manufacturer of buggy whips began to notice falling sales about the time that the automobile came into common use. Being dissatisfied, he investigated and found that the decreasing use of horse-drawn vehicles was the reason. Did he have a problem? As best you can, formulate his problem by enumerating what you think would be his objectives, alternatives, etc.
5. A United States citizen is very much dissatisfied with the attitude of the Communist leaders of China. Does he have a problem?
6. Can you determine how the "voting paradox" might apply to the deliberations of the United States Congress?
7. Is it necessarily true that the alternative which is actually chosen by a decision maker is the solution to his decision problem?
8. How would you apply the "principle of bounded rationality" to your search for a job?
9. How great a role do you think the ability to make quick and accurate decisions plays in the success of a marketing manager?
10. As a regional sales manager, you are faced with dealing with a salesman whose performance has dropped off drastically. He is paid on a strict commission basis, and company policy will not allow you to discharge him. Several conversations with him have not proved fruitful. The possibility of transferring him to another region is brought up. You know that if he is away from his home town and family, his performance is likely to drop further. What would

be the systems view of this problem? Is it a managerial decision problem? Why?

11. Give a brief summary of the essential similarities and differences of the systems approach in decision making and in the idea of product management.

REFERENCES

Ackoff, R. L.: *Scientific Method—Optimizing Applied Research Decisions,* John Wiley & Sons, Inc., New York, 1962.

Churchman, C. W., R. L. Ackoff, and E. L. Arnoff: *Introduction to Operations Research,* John Wiley & Sons, Inc., New York, 1957.

Fulmer, R. M.: "Product Management: Panacea or Pandora's Box," *California Management Review,* Summer, 1965, pp. 63–74.

Johnson, R. A., F. E. Kast, and J. E. Rosenzweig: "Systems Theory and Management," *Management Science,* vol. 10, no. 2 (January, 1964), pp. 367–384.

King, W. R.: "The Systems Concept in Management," *Journal of Industrial Engineering* (May, 1957), pp. 270–273.

Luck, D. J., and T. Nowak: "Product Management—Vision Unfulfilled," *Harvard Business Review* (May–June, 1965), pp. 143–154.

March, J. G., and H. A. Simon: *Organizations,* John Wiley & Sons, Inc., New York, 1958.

Optner, S. L.: *Systems Analysis for Business and Industrial Problem Solving,* Prentice-Hall, Inc., Englewood Cliffs, N.J., 1965.

3

A Conceptual Framework for Marketing Analysis

The evaluation and comparison of alternative marketing actions involve a study of the ways in which the elements of a decision situation interact to produce a resulting state of affairs. In this chapter this interaction is studied and conceptualized, resulting in a framework around which decision problems may be analyzed and solved.

The Outcome Array

The *outcome* of a decision situation is determined by the interaction of the aspects which are *controllable* by the decision maker and factors which are not subject to his control—the *uncontrollables*. In a new product situation the aspects which are subject to control might be price, product characteristics, and distribution techniques, for example. The uncontrollables might involve consumer demand, the weather, and competitive reactions. Both kinds of factors will combine to determine the outcome of a decision situation, which might be described in this case in terms of sales, profits, and market share.

The alternative actions available to the decision maker are composed of the controllable aspects of the decision problem. For example, if the controllable aspects of a problem are the unit selling price and the amount of advertising expenditure, the alternative actions, often referred to as *strategies,* would then be various possible combinations of these controllable variables. A list of some of the available strategies in this case might take the form:

Strategy 1: price 50 cents; $3 million advertising
Strategy 2: price 59 cents; $3½ million advertising
Strategy 3: price 60 cents; $3½ million advertising

Two kinds of uncontrollables which bear on the outcome of a decision situation need to be distinguished: those which are the result of the vagaries of "nature," such as the weather and labor strikes, and those which are the result of conscious actions of rational competitors in the marketing environment. The former are referred to as *states of nature* (although they are not always tied to the common interpretation of the word "nature," as is the weather), and the latter are termed *competitive actions.*

In the most general context the selected strategy, the existing state of nature, and the actions taken by competitors interact to produce an outcome of the decision situation.

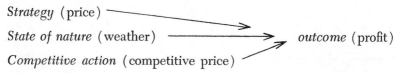

Strategy (price)
State of nature (weather) ⟶ *outcome* (profit)
Competitive action (competitive price)

In the brewing industry, for example, beer sales are highly dependent on the weather. In a pricing decision problem in this environment, the price charged for a bottle of beer, the competitor's prices, and the temperature and humidity represent respectively the strategy, competitive action, and state of nature interacting to produce some outcome which may be measured in terms of the brewer's profit.

These interactions may be conceptualized in a three-dimensional *outcome array* in which each element represents the outcome associated with a particular combination of strategy, state of nature, and competitive action. The entire array displays all possible outcomes which might result from the various combinations of the three interacting elements. Figure 3-1 shows such an array in which the various strategies are labeled $S_1, S_2, \ldots, S_n$, the states of nature $N_1, N_2, \ldots, N_m$, and the competitive actions $C_1, C_2, \ldots, C_p$. The entry in each element of the array is some description of the specific outcome, represented as O_{ijk}. For some purposes this may simply be a verbal description, such as "sales good, returns about 10 percent, and profit high." For most analyses, however, it is preferable that outcomes be stated in terms of more precise outcome descriptors, e.g., "sales $100,000, market share 30 percent, returns 10 percent, and profit $8,000."

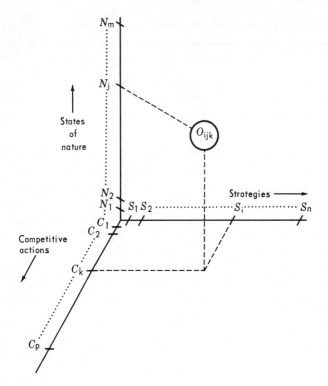

FIGURE 3-1   A THREE-DIMENSIONAL OUTCOME ARRAY

Since much of scientific analysis presumes the existence of numerical measures, the outcome descriptions should, at best, be quantitative. From a practical standpoint, however, it must be recognized that no universally valid numerical measures exist for many important outcome descriptors (image, for example) and that numerical measurements are not always available to the decision maker even though measures may exist.

In any case, *the outcome descriptions will normally be stated in terms of measures related to the attainment (or degree of attainment) of the decision maker's objectives.* If, for example, the set of objectives relates to profit, competitive position, and the company's image, the possible outcomes might be described in terms of dollars of profit, market share, and a qualitative description of the image level achieved.

Simplifying the Outcome Array

Most frequently, for purposes of analysis, the three-dimensional outcome array is simplified and presented in two dimensions. This may be accomplished in any of several ways.

STRATEGY—STATE OF NATURE OUTCOME ARRAYS

If competitive actions are not significant enough to be considered as a major element of the problem situation, they may be entirely ignored. Such a situation might occur in the case of a distributor's determination of the quantity in which he should stock a particular item in a market area in which there is little price competition. The distributor might decide to consider only his various strategies (purchase quantities) and the relevant states of nature (which might be the various levels of consumer demand for the product) and to ignore competitive actions on the grounds that competitors are not likely to make any changes which will affect the outcome of his decision problem. This situation might be described by a two-dimensional outcome array, shown as Table 3-1, in which the

TABLE 3-1   STRATEGY—STATE OF NATURE OUTCOME ARRAY

Strategy	State of nature			
	N_1 (demand for 1 unit)	N_2 (demand for 2 units)	N_3 (demand for 3 units)	N_4 (demand for 4 units)
S_1: purchase 1 unit	1 unit sold	1 unit sold 1 sale lost	1 unit sold 2 sales lost	1 unit sold 3 sales lost
S_2: purchase 2 units	1 unit sold 1 unit unsold	2 units sold	2 units sold 1 sale lost	2 units sold 2 sales lost
S_3: purchase 3 units	1 unit sold 2 units unsold	2 units sold 1 unit unsold	3 units sold	3 units sold 1 sale lost

decision maker may choose to purchase either one, two, or three units during a given period—his strategies. Demand may be placed on him to supply from one to four units—the states of nature—resulting in outcomes which may be described in terms of the number of units sold, the number of units unsold, and the number of lost sales due to an inadequate purchase quantity. For example, if the decision maker chooses strategy S_1 (to purchase one unit) and if state of nature N_3 (demand for three units) occurs, the outcome will be that he sells the one unit which he has stocked and loses two potential sales (since demand existed for a total of three units and he had only one in stock). Similar interpretations may be given to each of the other outcomes in this outcome array.

STRATEGY—COMPETITIVE ACTION OUTCOME ARRAYS

If the vagaries of nature are deemed insignificant to the overall outcome of the problem situation, they may be ignored and the

strategies arrayed in two dimensions with the possible actions
of competitors. In bidding for a contract, for example, only the
price bid by competitors and the decision maker's bid significantly
influence the outcome—the contract award. In the case of only
a single competitor, a strategy–competitive action array such as
Table 3-2 is most useful.

TABLE 3-2   STRATEGY–COMPETITIVE ACTION OUTCOME ARRAY

	Competitive action		
Strategy	C_1 (*competition bids* $225,000)	C_2 (*competition bids* $275,000)	C_3 (*competition does not bid*)
S_1: we bid $200,000	We get contract	We get contract	We get contract
S_2: we bid $250,000	Competition gets contract	We get contract	We get contract
S_3: we bid $300,000	Competition gets contract	Competition gets contract	We get contract
S_4: we do not bid	Competition gets contract	Competition gets contract	Neither gets contract

In this situation we consider either not bidding or making three
feasible bids of $200,000, $250,000, or $300,000. If our competitor
takes each of the actions described by the column headings, the
outcome is the contract award as stated in the table.

STRATEGY–GENERALIZED STATE OF NATURE OUTCOME ARRAYS

Alternatively, if the number of states of nature and competitive
actions to be considered is not too great, these two factors may
be combined in all possible ways to produce a single set of *generalized states of nature* which includes both aspects. If, for example,
there are only two states of nature, N_1 and N_2, and two possible
competitive actions, C_1 and C_2, they may be combined into the
four generalized states:

N_1C_1
N_1C_2
N_2C_1
N_2C_2

These generalized states of nature should be interpreted as encompassing the simultaneous occurrence of each of the components;
e.g., N_1C_2 means that the set of environmental factors N_1 occurs
in conjunction with the action C_2 being taken by one's competitor.

An outcome table using such generalized states in a problem involving good or bad weather (N_1 or N_2) and a 5-cent or 10-cent competitive price (C_1 or C_2) is shown as Table 3-3. Note that all possible combinations of the states of nature and competitive prices which might occur are entered as column headings in this table.

TABLE 3-3 STRATEGY–GENERALIZED STATE OF NATURE OUTCOME ARRAY

Strategy	N_1C_1 (good weather; 5¢ competitive price)	N_1C_2 (good weather; 10¢ competitive price)	N_2C_1 (bad weather; 5¢ competitive price)	N_2C_2 (bad weather; 10¢ competitive price)
S_1	O_{111}	O_{112}	O_{121}	O_{122}
S_2	O_{211}	O_{212}	O_{221}	O_{222}
.	.	.	.	.
.	.	.	.	.
S_n	O_{n11}	O_{n12}	O_{n21}	O_{n22}

The exact nature of the controllable elements which comprise the strategies is not specified in this table. Neither is the nature of the outcomes (labeled O_{ijk} to indicate the interaction of strategy S_i, state of nature N_j, and competitive action C_k) indicated. It is clear that this method of presentation is identical to the geometric approach of Figure 3-1. The emphasis here is on the similarity of states of nature and competitive actions, the similarity being that both encompass factors which are *uncontrollable* by the decision maker. This similarity permits the construction of generalized states of nature from all possible combinations of values of the uncontrollable variables.

It should be remembered that, regardless of which of these forms of presentation is chosen and of the degree of influence which each element is expected to have, it is always the interaction of the *three* elements (strategy, state of nature, and competitive action) which jointly determines the outcome of a general marketing decision situation. In each particular situation careful attention should be given to the degree of influence of each factor. More attention will be paid to this point in Chapter 4.

In the remainder of this chapter we shall deal only with arrays of the strategy–state of nature and the strategy–generalized state of nature variety. In speaking of both, we shall simply use the terminology of the former. In subsequent chapters we shall continue this shorthand since the elements which make up the arrays in question will always be clear from the context. Arrays of the strategy–

competitive action variety will be dealt with subsequently in the context of the marketing situations in which they have proved most useful.

Evaluating and Comparing Outcomes

The solution to a decision problem, or the choice of a best strategy, depends upon the outcomes to which each strategy may lead. To compare alternatives meaningfully, one must first be able to *evaluate* and *compare* the outcomes to which the strategies may lead.

The practical problems arising in the seemingly simple process of comparison of outcomes are immense. Suppose, for instance, that a decision maker has objectives with respect to the market share and profits achieved in a particular decision problem. Two possible outcomes are

O_1: market share 10 percent; profit $50,000
O_2: market share 15 percent; profit $40,000

There is no clear basis for choice between these two simple outcomes. In O_1 a larger profit is achieved with a lower market share than in O_2. It is not at all obvious which outcome a decision maker might prefer. In fact, it is apparent that two different individuals or organizations might have vastly different preferences for these outcomes. The marketing manager associated with a newly introduced product might be greatly concerned with penetration of the market and prefer O_2 to O_1, even though less immediate profit is achieved. The manager whose product is "mature" might prefer the other.

Economists have long concerned themselves with a concept called *utility—the capacity of an event, object, or state of affairs to satisfy human wants.* The idea of a basic measure of the degree of satisfaction derived from a state of affairs leads naturally to consideration of using such a measure to evaluate and compare outcomes. If, for example, the outcome O_1 is preferred to O_2, O_1 will possess greater utility than O_2. If one had a measure of the utility of each outcome, he could simply select that with the greatest utility as that which is most preferred.

The Utility Concept

The idea of utility as a basic measure which may be used to evaluate and compare outcomes is defined as a completely *personalistic* concept—in terms of specific individuals or organizations. There is no general or abstract kind of utility—only your utility, my utility, the utility of the General Motors organization, etc. Moreover, the concept of utility is a completely *subjective* one. No one may disagree with my utilities or your utilities, for they are exactly that—yours or mine. They are the formal expression of the desires and needs which exist within the individual or organization, and whether or not they choose to recognize the concept of utility formally, they must use the idea in nearly everything that they do.

If one were able to measure the utility which the decision maker derives from each possible outcome of a decision situation, he would have a single measurement which reflects the totality of the decision maker's objectives and his subjective desires and hopes as to their attainment. In effect, one's utility for each outcome of a marketing decision situation may be thought of as a function $U(\ )$ of that outcome; i.e.,

$$U(O_{ijk})$$

gives the decision maker's or organization's *numerical* utility for the outcome resulting from strategy S_i, state of nature N_j, and competitive action C_k. The function $U(\ )$ is termed the decision maker's *utility function.*

An alternative way of viewing the utility function emphasizes the interaction of a strategy, state of nature, and competitive action in producing an outcome. Symbolically, we may express the utility function as

$$U(S_i, N_j, C_k)$$

or

$$U(S_i, N_{jk})$$

whichever is appropriate. When the utility function is thought of in the latter way—as dependent on *both* the strategy and (generalized) state of nature—it becomes apparent that the "pure" outcome descriptors which were used in the previous outcome arrays are not sufficient to permit the *evaluation* of outcomes. Since each outcome is produced by a strategy and state of nature, its evaluation is dependent on aspects of both. The competitive bidding situation of Table 3-2 offers an illustration. There the "pure" outcome descriptors involved only the contract award, with nothing being said about the strategy which led to the award. Yet clearly, assuming that the contract involves a fixed task to be performed, the decision maker would prefer to be paid $300,000 rather than $200,000. This is equivalent to saying that he places greater utility on being awarded the contract with a bid of $300,000 than on being awarded the contract with a bid of $200,000.

It is possible, though not likely, that the decision maker has objectives only with respect to the award of the contract and not with respect to the return from the contract award. If he is solely concerned with obtaining the award, regardless of the return, his utility function might be that shown as Figure 3-2. This figure demonstrates that the decision maker places ten units of utility on winning the contract, one unit on neither his nor his competitor's being awarded the contract, and zero units on losing the contract to his competitor. This might be a reasonably accurate description of the situation in which large aerospace contractors bid on an initial exploratory contract which could lead to a huge development and production contract at a later date. There, the amount of the contract is insignificant; the important thing is to *win*, so that the possibility of getting the subsequent lucrative contracts will still be open. If this were the case, a utility function such as that in Figure 3-2, in which the utility depends only on the "pure" outcome and not on the strategy which achieves the outcome, might be applicable.

In most such bidding situations the utility function of Figure 3-3 will be more descriptive. There, the utility for winning the contract with a bid of $300,000 is higher than the utility of winning with a $250,000 bid, which is in turn higher than that for winning with a $200,000 bid. The utility of losing the contract to one's competitor (via any bid) is still lowest.

FIGURE 3-2   UTILITY FUNCTION FOR COMPETITIVE BIDDING PROBLEM—SITUATION I

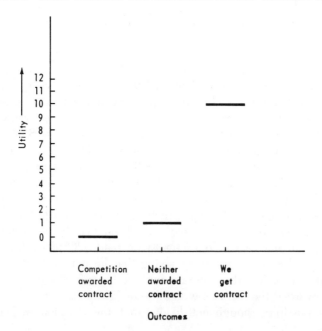

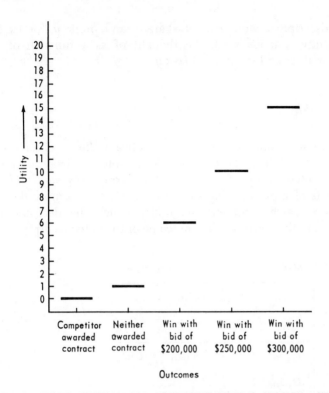

FIGURE 3-3   UTILITY FUNCTION FOR COMPETITIVE BIDDING PROBLEM—SITUATION II

The important point demonstrated here is that outcomes are described in terms related to the decision maker's objectives. Hence, *if there is any measure which is simultaneously related to a strategy or state of nature and the objectives, that measure should be included as an outcome descriptor*. The return from the contract award is such a measure in this example. Often there will be a different cost associated with taking each strategy—if the strategies are simply different advertising expenditure levels, for example. These costs are usually associated with objectives stated in terms of total costs or profits. Hence, they must be incorporated in some way into each outcome description. (We shall elaborate on this in the next chapter.)

One further point should be noted concerning the utility function. Since outcomes are described in terms of variables which are related to the decision maker's objectives, the utility function may be thought of as being defined over the domain of these variables rather than over the outcomes themselves. In other words, we may think of a utility function for market share, if that is the only relevant outcome descriptor, or a multidimensional utility function defined over a number of descriptor variables. If the relevant descriptors were market share x and profit y, for example, the utility function could be thought of as a function of these two variables and symbolized $U(x,y)$.

The Payoff Table

If it were feasible to measure numerically the decision maker's utilities for each of the outcomes in an outcome array, they might be presented in a corresponding utility array, or *payoff table*. The elements of a payoff table such as Table 3-4 are simply the single numbers which represent the utility which the decision maker attaches to the corresponding outcome of the outcome array.

TABLE 3-4   PAYOFF TABLE

Strategy	State of nature			
	N_1	N_2 $\cdots$	N_j $\cdots$	N_m
S_1	$U(O_{11})$	$U(O_{12})$ $\cdots$	$U(O_{1j})$ $\cdots$	$U(O_{1m})$
S_2	$U(O_{21})$	$U(O_{22})$		
$\cdot$	$\cdot$		$\cdot$	
$\cdot$			$\cdot$	$\cdot$
S_i	$U(O_{i1})$ $\cdot$ $\cdot$ $\cdot$ $\cdot$ $\cdot$ $\cdot$ $U(O_{ij})$			$\cdot$
$\cdot$	$\cdot$			$\cdot$
$\cdot$			$\cdot$	
S_n	$U(O_{n1})$ $\cdot$ $\cdot$ $\cdot$ $\cdot$ $\cdot$ $\cdot$ $\cdot$ $\cdot$ $\cdot$ $U(O_{nm})$			

The payoff table is a transformation of the outcome array, the transformation being via the utility function $U(\ )$; i.e., each element of a payoff table is the utility for the outcome which appears in the corresponding position in the relevant outcome array. If strategy S_1 and state of nature N_2 combine to produce an outcome

O_{12}: \$8,000 profit; 25 percent market share

which would appear in the first row and second column of an outcome array, the utility $U(O_{12})$, or $U(\$8,000, 25\%)$, whatever numerical value it might have, would be evaluated from the two-dimensional utility function involving profit and market share. This utility would then appear in the first row and second column of the payoff table. Each of the other outcomes would be evaluated by using the utility function, and the utilities would be entered in the appropriate spaces in the payoff table.

At this point the reader may have been concerned with several points which have been glossed over in the previous discussion. Whose utilities are we talking about when the decision maker is a group of people? How are an individual's personal utilities and organizational utilities related, if at all? How are these utilities measured? Before embarking upon these not at all trivial questions, we shall proceed as though numerical values for the utilities of all outcomes are known, and develop the concepts of normative decision theory. To do this, we shall first need to note some of the basic ideas of probability theory.

Basic Probability Concepts

The basic ideas relating to probability are familiar to most of us. Any event whose outcome is at least partially determined by chance, such as the flip of a coin, may be described in probabilistic terms.

The probability of an outcome of an event is most easily thought of as the long-term percentage of times the various possible outcomes would occur if the event were repeated again and again. In terms of the coin flip, after a long series of flips one might divide the total number of occurrences of the outcome heads by the total number of flips and call the resulting decimal the *probability of heads*. As such, one would likely determine the probability of a head to be near $\frac{1}{2}$. Similarly the probability of a 6 on a single throw of a die would be about $\frac{1}{6}$.

The concept of probability may also be applied to sequences of events. *The knowledge that the probability of a head is $\frac{1}{2}$ will in no way help one to predict what the outcome of a particular flip of a coin will be,* for it will be either heads or tails and one is either right or wrong. But this knowledge does permit one to predict that in a long sequence of flips, the relative frequency of heads will be close to 50 percent.

To apply probabilities to the analysis of marketing decisions, one must recognize that the basic idea is applicable to the uncertain outcomes of events which can influence these decisions—the states of nature. To conclude that the probability is $\frac{1}{3}$ that "June rainfall in Cleveland will exceed 2 inches" implies that an investigation of weather bureau records for many past Junes has indicated a relative frequency of $\frac{1}{3}$ for such a state. If this is so, and if there is no reason to believe that weather patterns now and in the future will differ from those in the past—i.e., if the pattern exhibits *stabil-*

ity over time—one might conclude that the future percentage of occurrence of rainy (over 2 inches) Junes will also be about $\frac{1}{3}$.

Probability, then, is simply a way of dealing with our uncertainties about the future. In attaching probabilities to states of nature or outcomes, we evaluate the likelihood of their occurring and we thereby synthesize our information about future likelihoods into a single number. It should be noted that we may attach probabilities to *either* states of nature or outcomes, whichever may seem to be more useful. For example, the outcome array of Table 3-5 illustrates a situation in which several strategy–state of nature combinations lead to some common outcomes, simply denoted as O_1, O_2, and O_3.

TABLE 3-5   OUTCOME TABLE

Strategy	State of nature		
	N_1	N_2	N_3
S_1	O_1	O_2	O_1
S_2	O_1	O_1	O_2
S_3	O_1	O_2	O_3

If we attach likelihoods (probabilities) of $\frac{1}{2}$, $\frac{3}{8}$, and $\frac{1}{8}$ respectively to the states of nature, we are simultaneously determining that the likelihoods of the various outcomes for each of the strategies are those given by Table 3-6. Note that the table is a kind of presentation which we have not illustrated before. It is a sometimes useful way of illustrating a decision problem in a fashion which emphasizes outcomes rather than states of nature. The $\frac{5}{8}$ entry for (S_1O_1) indicates that S_1 yields O_1 if either N_1 or N_3 occurs. Since either N_1 or N_3 will occur with probability $\frac{1}{2}$ plus $\frac{1}{8}$, the likelihood of occurrence of O_1 is $\frac{5}{8}$. Similarly O_2 occurs for S_1 only under the state N_2; hence the probability of O_2 under S_1 is the same as for N_2—$\frac{3}{8}$. Since O_3 cannot occur if we adopt S_1, its likelihood is zero.

TABLE 3-6   PROBABILITY TABLE

Strategy	Outcome		
	O_1	O_2	O_3
S_1	$\frac{5}{8}$	$\frac{3}{8}$	0
S_2	$\frac{7}{8}$	$\frac{1}{8}$	0
S_3	$\frac{1}{2}$	$\frac{3}{8}$	$\frac{1}{8}$

Several properties of the probabilities for states of nature or outcomes which we shall make use of are implied by the foregoing statements.[1]

1. Each state of nature or outcome must have an associated probability greater than or equal to 0 and less than or equal to 1.
2. The states of nature or outcomes must be defined so that they are *disjoint;* i.e., the occurrence of one precludes the simultaneous occurrence of any other. (For each strategy and state of nature combination, there must be one and only one clearly defined outcome.)
3. The states of nature or outcomes to which probabilities are attached must be an exhaustive listing. (No states of nature or outcomes which might possibly occur can be omitted from our outcome table.)

Taken together, these properties imply that the probabilities attached to all of the states of nature should sum to unity. This is equivalent to saying that it is certain that one of the possible states of nature will occur. Correspondingly, in terms of outcomes, one of the disjoint and exhaustive outcomes will occur for each strategy, and hence the probabilities associated with the possible outcomes *for each strategy* should also sum to 1.

We may make use of two special ideas concerning states to aid us in using probability concepts in marketing decisions. The *negation* of a state A is a state which implies the nonoccurrence of the state A. We symbolize the negation of A by $\bar{A}$. For example, if the state A is "we sell over a million widgets this year," the state $\bar{A}$ is "we do not sell over a million widgets this year." The *intersection* of two states A and B is the state which incorporates the occurrence of both A and B. If A is "we sell over a million widgets this year" and B is "our market share is over 30 percent," the intersection of A and B, symbolized as AB, is "we sell over a million widgets this year *and* our market share is over 30 percent." We have already used this idea in constructing the generalized states of nature in Table 3-3. In that situation one possible state of nature—good weather—and one possible competitive action—5 cents—were combined to form the intersection state N_1C_1—good weather and 5-cent competitive price. Similarly, each of the other generalized states of nature in that table represents an intersection of two states.

The ideas of negations and intersections permit us to define the

---

[1] The reader who is unfamiliar with probability concepts and is bothered by these statements should refer to one of the excellent texts listed in the references.

probabilistic independence of two states A and B. These states are probabilistically independent if and only if the probability associated with each intersection composed of the states and their negations is the product of the probabilities associated with the individual components of the intersection, i.e.,

Prob AB = Prob $(A) \cdot$ Prob (B)
Prob $A\bar{B}$ = Prob $(A) \cdot$ Prob $(\bar{B})$
Prob $\bar{A}B$ = Prob $(\bar{A}) \cdot$ Prob (B)
Prob $\bar{A}\bar{B}$ = Prob $(\bar{A}) \cdot$ Prob $(\bar{B})$

A similar definition holds for the probabilistic independence of a system of more than two states.[2]

To illustrate this, consider the states to be defined as follows:

A: rainy weather
B: sales at least $1 million

Consequently, the states $\bar{A}$ and $\bar{B}$ are, respectively, nonrainy weather and sales under $1 million. If the probabilities are

Prob (A) = 0.4
Prob $(\bar{A})$ = 0.6
Prob (B) = 0.7
Prob $(\bar{B})$ = 0.3

then, for the states to be independent, the probability of the intersection "rainy weather and sales at least $1 million" must be 0.28 (0.4 times 0.7), and the probabilities of the other intersections must be $P(A\bar{B})$ = 0.12, $P(\bar{A}B)$ = 0.42, and $P(\bar{A}\bar{B})$ = 0.18. Conversely, if the probabilities of the intersections are those given, the states are probabilistically independent.

RANDOM VARIABLES

The basic concepts of probability lead naturally to the idea of a *random variable*.

Whenever outcomes subject to chance may be described in terms of a numerically valued variable, the variable may be thought of as taking on its various values with given likelihoods (probabilities).

---

[2] See either of the probability texts in the references. The definition of probabilistic independence given here is a general one. In the case of only two states A and B, it is sufficient to show that

Prob (AB) = Prob $(A) \cdot$ Prob (B)

to know that A and B are independent. Conversely, if one knows A and B to be independent, he can infer that the probability associated with each of the four intersection states is multiplicative as shown above.

For example, under strategy S_3 in Table 3-6, O_1 occurs with probability $\frac{1}{2}$, O_2 with probability $\frac{3}{8}$, and O_3 with probability $\frac{1}{8}$. If we associate a numerically valued variable, say $5, $10, and $15 of profit, with each of these outcomes, we may focus our attention on the values of the variable and their associated probabilities. In other words, we have defined a random variable, profit, which takes on the values $5, $10, and $15 according to the likelihoods of Table 3-7.

TABLE 3-7   A SIMPLE RANDOM VARIABLE

Values of random variable (Profit)	$5	$10	$15
Probability of various values	$\frac{1}{2}$	$\frac{3}{8}$	$\frac{1}{8}$

In doing this, we have simply associated a number—a value of the random variable—with each outcome and given attention to the number rather than to the outcome itself. Any numerically valued variable which we might associate with an uncertain outcome would also be a random variable—whether it be a quantity of oranges, the time required to complete a task, or the price quoted for General Motors' common stock. *The variable* (whatever it is) *is random in the sense that its taking on each of its possible values is subject to chance.*

It should be recognized that in making the transformation from the outcome array to the payoff table for any decision problem, we are simply associating a single number with each outcome. This number, the decision maker's utility for the outcome, may then be treated as a random variable if the outcomes are not certain.

EXPECTED UTILITY

The arithmetic average of a group of numbers is a familiar concept to most of us. If, for example, one is to determine the average test score of a class of five in which three people made a score of 90 and two made 80, the calculation of the average score would be

$$\frac{3(90) + 2(80)}{5} = 86$$

Written a little differently, this is

$$\tfrac{3}{5}(90) + \tfrac{2}{5}(80) = 86$$

where $\frac{3}{5}$ is the relative frequency of occurrence of the score 90 and $\frac{2}{5}$ is the relative frequency of the score 80.

The idea of the *expectation of a random variable* is similar to that of an average. If the random variable X takes on the values x_1, $x_2, \ldots, x_i, \ldots, x_n$ with probabilities $p_1, p_2 \ldots, p_i, \ldots, p_n$ respectively, the expectation of the variable is written $E(X)$ and defined as

$$E(X) = \sum_{i=1}^{n} p_i x_i \qquad\qquad (3\text{-}1)$$

In the case of test scores, if we know that the probability of an individual making a grade of 90 is $\frac{3}{5}$ and the probability of making an 80 is $\frac{2}{5}$ (no other grade being possible), the expectation of the random variable "test score" is

$(\frac{3}{5})90 + (\frac{2}{5})80 = 86$

It should be noted that this is the same calculation as made previously for averages. In this case the $\frac{3}{5}$ is the long-run relative frequency of occurrence of a score of 90, this relative frequency being taken as the probability. In the case of the average score, the $\frac{3}{5}$ is the relative frequency associated with five *particular* individuals who took the test. Although the numerical calculations are the same and the concepts related to the "average" and the expectation are analogous, the two ideas are not identical. Hence, while we motivate the reader's concept of an expectation by referring to an average, we need to recognize that the expectation concept is a general one. In particular we shall be concerned with expectations of utilities, costs, profits, and other quantities which describe outcomes of marketing decision problems.

Consider, for example, being faced with a decision problem involving two new products which require the same initial investment and whose returns are predicted to be those given by Table 3-8.

TABLE 3-8   INVESTMENT RETURNS

Product A			
Return	$90	$300	$ 600
Probability	$\frac{1}{3}$	$\frac{1}{3}$	$\frac{1}{3}$
Product B			
Return	$ 0	$200	$1,600
Probability	$\frac{1}{8}$	$\frac{3}{4}$	$\frac{1}{8}$

The expected value of the random variable "return on investment" for product A is

$\frac{1}{3}(\$90) + \frac{1}{3}(\$300) + \frac{1}{3}(\$600) = \$330$

For product B the expectation of return is

$$\tfrac{1}{8}(\$0) + \tfrac{3}{4}(\$200) + \tfrac{1}{8}(\$1,600) = \$350$$

The potential product planner *might* interpret these expectations as indicating the superiority of product B since its expected return is the higher of the two. Note that the word "might" is used because we have already argued that he should make this evaluation on the basis of utility rather than dollars. We shall investigate the implications of this further in the remainder of this chapter. For the moment the significant point is that we may calculate expectations of wide ranges of random variables and that these expectations may be thought of as the average we can anticipate over a long series of future repetitions of the underlying event. In this case, if we were faced with a long sequence of identical choice situations, we might reasonably expect that the long-run average resulting from choices of B would be about \$350 whereas the average from A would be about \$330. With this background in probabilities and their use in calculating expectations, we return to the building of the conceptual decision making framework.

Decisions under Certainty

When the payoff table for a decision problem has been constructed, the determination of a "best" strategy from among the set of alternatives may proceed according to any one of three paths as determined by the real-world problem, the level of information which the analyst has concerning the problem, and the analyst's desire for "accuracy" of representation. We shall first discuss the three paths and then subsequently, we shall take up the question of representing a real-world problem in one of these formats.

In the case called *certainty* the existing state of nature is known to the decision maker; i.e., each strategy leads with certainty to one outcome of a known set of outcomes. The payoff table in this situation has only a single column, that corresponding to the existing state of nature, as shown in Table 3-9.

TABLE 3-9   DECISION PROBLEM UNDER CERTAINTY

Strategy	State of nature, N_0
S_1	O_{10}
S_2	O_{20}
.	.
.	.
.	.
S_n	O_{n0}

For example, suppose that two units of a product are available in two warehouses, one in each, and that orders have been received for both of the units. If no other items are available, the decision problem involves a matching of buyers and warehouses, deciding which of the two items to ship to each buyer.

The unit profits for the various combinations of warehouses and buyers might be those given by Table 3-10. Since the shipment

TABLE 3-10   UNIT PROFITS

Warehouse	Buyer	
	A	B
1	$40	$25
2	$50	$30

of a unit from warehouse 1 to buyer *A* makes it necessary to supply buyer *B* from warehouse 2, the decision maker has only two alternative strategies.

S_1: ship from 1 to *A* and 2 to *B*
S_2: ship from 1 to *B* and 2 to *A*

If all of the vagaries of nature and competitors are wrapped up in our knowledge of the profit array, the existing state of nature is known. The outcome array then may be constructed, as shown in Table 3-11.

TABLE 3-11   OUTCOME ARRAY

Strategy	State of nature, N_0
S_1	Total profit = $40 + $30 = $70
S_2	Total profit = $50 + $25 = $75

If the symbol *c* represents total profit, the utility function for profit[3] (arbitrarily chosen for illustrative purposes) may be

$$U = 3c + 10$$

The utilities associated with the two outcomes made up of $70 and $75 total profit are then 220 units and 235 units respectively. The payoff table is shown as Table 3-12.

Modern decision theory argues that in such a case *the rational decision maker will act so as to achieve maximum utility.* In this

---

[3] The question of a utility function for monetary quantities will be discussed later in this chapter.

TABLE 3-12   PAYOFF TABLE

Strategy	State of nature, N_0
S_1	220
S_2	235

case the rational decision maker should choose S_2 rather than S_1 since by doing so, he achieves the greatest utility. This statement as to the criterion which should be used in the case of decision making under certainty is often taken to be a part of the scientific definition of a *rational decision maker*. Indeed, in any sense, this is a reasonable criterion, for the utility concept is defined in personalistic and subjective terms so as to include *all* aspects of the decision maker's desires, hopes, and needs. To do other than to maximize such a comprehensive measure of satisfaction would indeed appear to be irrational.

Two additional aspects of decision making in the case of certainty should be noted. First, the definition of certainty—as the case in which each strategy leads with certainty to one outcome of a known set—does not imply that there is necessarily a one-to-one correspondence between strategies and outcomes. Two different strategies could well lead to the same outcome, for example. In this simple case, the outcomes, as represented by total profit, could well be identical for the two strategies.

Secondly, the apparent simplicity of the case of certainty is deceptive. If the shipment example involved five warehouses and buyers, 120 strategies would have to be considered. If ten warehouses and buyers were involved, over 3½ million strategies would need to be evaluated.[4] It is clear that this relatively small problem can quickly explode beyond the limits of practicability. Even in terms of the microseconds required for operations by electronic digital computers, the scale of the problem of enumerating strategies soon becomes impractical. Hence, most real decision problems under certainty are not at all so simple as our example might imply.

The quick-witted skeptic might retort that it is our normative decision framework that is at fault, for it has specified that an enumeration of alternative strategies is a necessary component of rational decision making. Note, however, that we have emphasized that the procedure presented in terms of outcome arrays, utilities, and payoff tables is a *conceptual* basis for decision making. One

---

[4] In general, if there are n warehouses and buyers, there will be n factorial strategies. The quantity n factorial is written $n!$ and is calculated as $n! = n \cdot (n-1) \cdot (n-2) \cdots 2 \cdot 1$.

may not in all cases actually *enumerate* all alternatives and develop the tables. Indeed, much of the methodology to be presented in the subsequent chapters obviates the necessity for doing so. The conceptual foundation laid here, however, must form the basis for all of quantitative marketing analysis and the alternatives must always be considered even if not enumerated. The search for an efficient process for sifting through such large numbers of alternatives and their associated outcomes constitutes a large portion of the research effort which has been expended in the case of decision making under certainty.

Decisions under Risk

In the case of decision making under *risk*, the decision maker knows the likelihood that each of the various states of nature will occur; i.e., each action leads to one outcome of a known set of outcomes, each with a known probability.

The payoff table in this case has more than one column. With each state of nature N_j is associated a known probability p_j. The states of nature should be defined—since *it is the province of the decision analyst to define the strategies, states of nature, and competitive actions which will be considered*—to be disjoint and exhaustive so that

$$\sum_{j=1}^{m} p_j = 1 \qquad 0 \le p_j \le 1 \qquad \text{for all } j$$

For example, suppose that a salesman must decide whether to spend the next day in an urban or a rural area within his territory. He knows that the sales results he can expect to achieve depend on what sort of weather will be encountered, for on a sunny day the farmers will be off in their fields, making it difficult to contact them, but on a rainy day they will be working near the farmhouse, where they can be contacted easily. On the other hand, the city traffic will be so congested on a rainy day as to prevent the salesman from making many calls. The payoff table (in terms of utilities), Table 3-13, describes the situation for the various outcomes.

TABLE 3-13 PAYOFF TABLE

Strategy	State of nature	
	N_1 (sunny)	N_2 (rain)
S_1: city	20	10
S_2: farm	15	25

The weather bureau has predicted that the chances of rain tomorrow are 2 in 10; i.e., that the probability associated with N_1 is $8/10$ and with N_2 it is $2/10$. The *expected utility* associated with each course of action is

$$E(S_1) = 8/10(20) + 2/10(10) = 18$$
$$E(S_2) = 8/10(15) + 2/10(25) = 17$$

The rational decision maker under risk will choose an alternative which maximizes his expected utility. In this case the expected utility associated with S_1 is the maximum; hence S_1 is the better strategy.

The idea of expected utility can be readily extended to larger numbers of states of nature. The expected utility of strategy S_i is simply

$$E(S_i) = \sum_{j=1}^{m} p_j U(O_{ij}) \tag{3-2}$$

where the summation indicates that the products of utilities and probabilities are being summed over all states of nature. The decision criterion of maximum expected utility can then be expressed as requiring the selection of a strategy from the set of available strategies in a fashion which maximizes expected utility; i.e., "choose the strategy which has greatest expected utility." This criterion is complementary to the criterion of maximum utility in the case of certainty. As such, it represents another portion of the scientific definition of rational decision making.

Just as in the case of certainty, there is no need that outcomes be uniquely identified with strategies; one outcome may result from any number of different strategies. Indeed, there is no unique correspondence between outcomes and states of nature; e.g., S_1 and N_2 may lead to an outcome identical to that resulting from S_3 and N_7.

One of the common arguments which the uninitiated might present against the use of this criterion can be used to gain further understanding of the criterion. It may be illustrated using the payoff table shown as Table 3-14.

TABLE 3-14 PAYOFF TABLE

Strategy	State of nature	
	N_1	N_2
S_1	10	100
S_2	40	80

If the probabilities associated with N_1 and N_2 are $\frac{1}{4}$ and $\frac{3}{4}$ respectively, the expected utilities are

$$E(S_1) = \frac{1}{4}(10) + \frac{3}{4}(100) = 77.5$$
$$E(S_2) = \frac{1}{4}(40) + \frac{3}{4}(80) = 70$$

Strategy S_1 has the maximum expected utility and should, according to the criterion, be chosen.

The skeptic, however, might reason as follows: "If I choose S_1, I can end up getting very little (if N_1 occurs). So, should I not be willing to sacrifice a few units of expected utility and choose S_2, where I am sure of getting at least 40?" At first glance this appears to be a plausible argument. However, the recognition that the numbers in the payoff table represent *utilities*, and not numbers of dollars, pineapples, blond chorus girls, or dry martinis, leads to the realization that although this is a perfectly logical argument with any of these units of measure, it does not hold for utilities. The utilities sum up *all* of the decision maker's desires and needs; everything which needs to be considered *is within* the utility numbers. To argue that the expectation criterion is invalid, then, is really to argue that the utility measures are incorrect. In any real situation, when utility numbers must be estimated, this is, of course, possible. Conceptually, however, if the utility numbers in the payoff table are valid, this argument is invalid; for if the decision maker really felt what he has expressed, he would have evaluated the *outcome* associated with S_1 and N_1 at less than 10 utility units. In short, his evaluation of the outcome as being worth 10 units of utility has compacted *all* of his preferences, needs, and desires into a single number—his utility for that outcome.

If the payoff table were really an outcome table in terms of dollars, as shown in Table 3-15, and the decision maker owed

TABLE 3-15   OUTCOME TABLE

Strategy	State of nature	
	N_1	N_2
S_1	\$10	\$100
S_2	\$40	\$ 80

an auto payment of \$40 to the finance company, for example, he might have great utility for \$40 (since repossession of his auto would prevent his further selling) and little additional utility for amounts above \$40. His utility function for these outcomes might

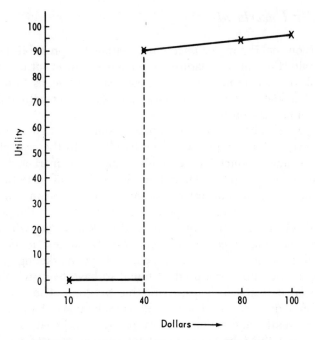

FIGURE 3-4   SALESMAN'S UTILITY FUNCTION FOR MONEY

be that shown as Figure 3-4. A payoff (utility) table corresponding to this outcome table would then be like Table 3-16. The expected utilities would be

$$E(S_1) = \tfrac{1}{4}(0) + \tfrac{3}{4}(96) = 72$$
$$E(S_2) = \tfrac{1}{4}(90) + \tfrac{3}{4}(94) = 93$$

and S_2 would be the best strategy. In this case an argument against the maximization of the expected number of *dollars* would be a valid one, since expected dollars are maximized by choosing S_1 (with an expected return of 77.5 *dollars* versus 70 *dollars* for S_2), but expected utility is maximized by a choice of S_2.

TABLE 3-16   PAYOFF TABLE

Strategy	State of nature	
	N_1	N_2
S_1	0	96
S_2	90	94

Decisions under Uncertainty

In decision making under *uncertainty*, either the probabilities associated with the states of nature are unknown or it is not meaningful
to speak of such probabilities; i.e., each strategy leads to one outcome of a known set of outcomes, each occurring with unknown
(or meaningless) probability.

The payoff table in this case appears similar to that in the case
of risk except for the absence of probabilities. In the case of introducing a new product into a highly competitive market, the relevant strategies might involve the alternatives of extensive test
marketing S_1, minimum test marketing S_2, or no test marketing S_3.
If extensive test marketing is performed, good data on the potential
demand will be obtained, but time will be lost during which competitors may introduce similar products. If no test marketing is
undertaken, time will be saved, but a large marketing program
may be carried out for a product found to have flaws which could
have been detected in a test marketing program. The generalized
states of nature may be taken to be combinations of the competitor's test marketing actions of his newly developed competitive
product and the inherent demand for the category of product in
question:

N_1: competitor tests; demand high
N_2: competitor does not test; demand high
N_3: competitor tests; demand low
N_4: competitor does not test; demand low

The payoff table might be Table 3-17. The utilities in the payoff
table account for sales revenue, the cost of test marketing, and
other costs involved in the testing and marketing operations.

TABLE 3-17   PAYOFF TABLE

Strategy	State of Nature			
	High demand		Low demand	
	Competitor tests N_1	No competitive tests N_2	Competitor tests N_3	No competitive tests N_4
S_1: extensive test marketing	50	30	20	5
S_2: minimum test marketing	80	40	10	15
S_3: no test marketing	120	50	5	0

If no sales experience with similar products is available to use as a basis for estimates of the likelihood of occurrence of high or low demand and no information concerning the action of competitors is available, the payoff table describes a case of decision making under uncertainty. In the cases of certainty and risk the maximization of utility and expected utility, respectively, are universally accepted decision criteria, but *there is no generally accepted criterion for choice under uncertainty.*

A number of criteria for choice under uncertainty have been proposed, each with its own rationale and proponents. Each, in turn, has been subjected to criticism on logical grounds. The net result is an apparent inability on the part of scientists to arrive at a single comprehensive criterion. This leaves the choice of a criterion as part of the analysis of every decision problem under uncertainty. In effect, the problem under uncertainty is twofold. First, which criterion is to be used? Secondly, which is the best strategy in terms of that criterion?

THE MAXIMIN CRITERION

In the face of so large a degree of uncertainty, the decision maker might choose to act pessimistically by deciding that he should maximize his *security level*. He might reason that if nature were to be perverse and attempt to give him as small a payoff as possible, he should in turn act to maximize his return, consistent with this malevolent intention of nature. In effect, he could identify the worst outcome (smallest utility) which can result from each strategy and *choose the strategy which gives him the best of these worst outcomes.*

In the test marketing problem, the decision maker determines that if he chooses S_1, the worst outcome is the five utility units associated with O_{14}. If he chooses S_2, the worst is O_{23} with its payoff of ten. Under S_3 the worst is the zero utility units of O_{34}. These worst outcomes are security levels for each of the strategies in that the decision maker can be confident that no worse can happen under that strategy.

To make the best of a potentially bad situation, the decision maker may act to ensure himself of the best of this set of worst outcomes. He can do this by choosing S_2. The worst outcome under S_2 has utility of ten, but lower utilities *could* result under both S_1 and S_3. The ten units of utility represent the *maximum security level* which he can obtain. By choosing S_2, he can be certain of getting no less than ten utility units, and he cannot be certain of as much as ten under any other strategy.

This idea of a "best" strategy in the case of uncertainty, the one which *maximizes the minimum utility* which can be achieved, is appropriately termed a *maximin strategy*. To select a maximin strategy, one determines the worst outcome (minimum payoff) for each strategy and selects the strategy with the largest value of this minimum. The decision maker is guaranteed at least the maximin payoff, since whatever state of nature is actually realized, no lesser payoff can be obtained by using the maximin strategy. Operationally, one determines the maximin strategy simply by finding the lowest payoff for each strategy (the smallest entry in each row of the payoff table) and then choosing the largest of these lowest payoffs. The strategy associated with this "largest of the smallests" is the maximin strategy.

THE MINIMAX CRITERION

One can also construct payoff tables in terms of disutilities rather than utilities. Disutilities are simply measures of *dissatisfaction* associated with negatively stated outcomes, e.g., losses. If one applies the maximin reasoning to such situations, he arrives at a *minimax* criterion—the minimization of maximum loss—since worst outcomes have associated with them the largest measure of disutility rather than the smallest measure of utility.

To determine a minimax strategy from a table of disutilities, one simply chooses the largest disutility in each row and then the smallest of the largest disutilities. The associated strategy is a minimax strategy. All other arguments are the same as for the maximin criterion. One must only remember that to ensure the best security level, one "plays" *maximin when dealing with gains* (utilities) and *minimax when dealing with losses* (disutilities).

THE MAXIMAX CRITERION

If the decision maker chooses to be completely optimistic rather than pessimistic, he might use a *maximax* criterion, i.e., choose the strategy which makes the best of the best which can occur. In this case he would choose the largest utility for each strategy and then choose the strategy which is associated with the largest of these; in effect, he would simply determine the greatest payoff in the table and select the strategy which *might* result in this greatest payoff. The optimistic decision maker in this case decides to go for broke. In the example of Table 3-17 the largest payoff is the 120 utility units associated with S_2N_1; hence the maximax strategy is S_3.

It should be noted that this criterion completely ignores the relative sizes of the various payoffs for each strategy. The choice of

S_3 would still be dictated by the maximax criterion if the least of the other payoffs under S_3 were 119, 0 or —1 billion. The maximin criterion has the same failing since it completely ignores all but the smallest payoff associated with each strategy.[5]

THE HURWICZ ALPHA CRITERION

The oversimplification of a complex problem which is inherent in the complete optimism of the maximax criterion and in the pessimism of the maximin criterion seems to preclude their use by any but the wildest speculator or the eternal pessimist. This realization led Leonid Hurwicz to propose a combination of the two criteria, termed the *Hurwicz alpha criterion.*

This criterion assumes that the decision maker possesses some *degree of optimism* which may be measured by a *coefficient of optimism* α. This coefficient is a number between 0 and 1 which allows the decision maker to be completely rational about any feeling of luck he may have. If the decision maker is completely optimistic, he will specify his optimism by a coefficient of 1, and if he is completely pessimistic, by a coefficient of 0. In any other case he will indicate the degree of his optimism by some decimal between 0 and 1.

This optimism coefficient, in conjunction with its complement $1 - \alpha$, is then used to calculate a decision index d_i for each strategy. The index is calculated as

$$d_i = \alpha M_i + (1 - \alpha)m_i \qquad (3\text{-}3)$$

where M_i is the best payoff associated with any of the outcomes to which strategy S_i might lead, and m_i is the worst payoff associated with S_i. *The "best" strategy is then taken to be the one which has a maximum value of this index.*

Let us assume a coefficient of optimism of 0.8, i.e., that the decision maker is fairly optimistic. In the test marketing example of Table 3-17, the best payoff associated with S_1 is 50 and the worst is 5. He therefore calculates his criterion index for S_1 to be

$$d_1 = 0.8(50) + (1 - 0.8)5 = 0.8(50) + 0.2(5) = 41$$

For S_2, the calculation is

$$d_2 = 0.8(80) + 0.2(10) = 66$$

and for S_3, it is

$$d_3 = 0.8(120) + 0.2(0) = 96$$

---

[5] The minimax criterion has the same deficiency since it is really just the maximin criterion applied to losses.

According to the Hurwicz alpha criterion, the decision maker should act to maximize this index. He can do so here by choosing S_3 with its associated index of 96.

Thus, the Hurwicz alpha criterion obviates one of the difficulties inherent in the maximin and maximax criteria. The decision maker in this case takes account of *both* the largest and the smallest payoffs for each strategy, rather than considering only the smallest, as in the case of maximin, or only the largest, as in the case of maximax. If there are more than two possible states of nature, however, all of the other payoffs are still omitted from consideration.

It is interesting to note that what his criterion actually does is to convert a decision problem under uncertainty to one under risk involving only two states of nature:

N_1: the best outcome for each strategy
N_2: the worst outcome for each strategy

The "probabilities" for each of these states are determined by the coefficient of optimism α and its complement $1 - \alpha$.

If the decision maker is completely optimistic, his optimism coefficient will be unity and the Hurwicz alpha criterion reduces to the maximax criterion. In this case the best outcome for each strategy is multiplied by 1 and the worst by 0 in the calculation of the index. In the case of complete pessimism a coefficient of 0 multiplies the payoff for the best outcome, and consequently the worst payoff is multiplied by 1—the maximin criterion.

THE LAPLACE CRITERION

Another criterion, worthy of note if only because of its historical significance, is the Laplace criterion. The decision maker using this criterion, in the absence of any information concerning the likelihood of occurrence of the various states of nature, simply assumes them to be equally likely.

Since the m states of nature are disjoint and exhaustive, the assumption of equal likelihood imputes probabilities of $1/m$ to each. The expected utilities may be calculated by using these assumed probabilities and a strategy chosen which maximizes expected utility. In the test marketing example of Table 3-17 there are four states of nature which would be assumed to be equally likely, each with probability $\frac{1}{4}$. The expected utilities are

$$E(S_1) = \frac{1}{4}(50) + \frac{1}{4}(30) + \frac{1}{4}(20) + \frac{1}{4}(5) = 26\frac{1}{4}$$
$$E(S_2) = \frac{1}{4}(80) + \frac{1}{4}(40) + \frac{1}{4}(10) + \frac{1}{4}(15) = 36\frac{1}{4}$$
$$E(S_3) = \frac{1}{4}(120) + \frac{1}{4}(50) + \frac{1}{4}(5) + \frac{1}{4}(0) = 43\frac{3}{4}$$

The strategy which maximizes expected utility is S_3 with its expectation of 43¾.

The Laplace criterion simply converts the decision problem under uncertainty to one of risk through the assumption of equal likelihood. This assumption is based on the *principle of insufficient reason*, which, when applied to this sort of situation, simply states that if we know of no reason why one state of nature should be more likely to occur than another, we should assume them to be equally likely. Debates concerning the use and misuse of this principle have raged for centuries. Suffice it to say here that the states of nature are precisely defined by the decision maker or analyst. He may do so in any way which appears to be simple enough to handle, yet useful, as long as they remain disjoint and exhaustive. Under the Laplace criterion the probabilities attached to the states of nature depend on the number of states and hence on the precise definition of the states. As the decision analyst ponders various definitions of the states of nature, the probabilities to be assigned under this criterion are manipulated with such abandon as to terrify most scientists.

This is not to say, however, that the Laplace criterion is little used. Whether they recognize the name Laplace or the principle of insufficient reason which forms the basis of the criterion, many analysts decide that for lack of a better way, they should assume equal likelihoods for the states of nature in their problems. Especially in situations involving problems of uncertainty in which the states of nature are more or less fixed in advance, who is to say that the approach is not a good one?

THE REGRET CRITERION

Savage (see the references) has proposed a criterion for decision making under uncertainty which is based upon a common psychological quirk of most humans. Most of us, when confronted with the outcome of a choice situation, apply our 20/20 hindsight vision and view the outcome which we *could* have obtained had we known in advance that the realized state of nature would occur. Savage uses the term *regret* to describe the dissatisfaction associated with not having fared as well as one might, had the state of nature been known. In effect, the decision maker views the past and regrets his choice.

A measure of such regret for any outcome might be the difference in utility between the payoff for the outcome and the largest payoff which could have been obtained under the corresponding state of nature. In the test marketing payoff table, the regret associated with O_{11} is 70, the difference between the 50 utility units realized and the 120 which could have been realized had one known in advance that N_1 would occur. The remaining regrets are given in Table 3-18, a *regret table* corresponding to the payoff array of Table 3-17. It should be noted that there is at least one outcome which has no regret for each state of nature: the outcome which has the highest payoff for that state of nature. If one chooses the strategy which leads to the highest payoff for the state of nature which is actually realized, he experiences no regret at not having chosen a better strategy.

TABLE 3-18   REGRET TABLE

Strategy	State of nature			
	N_1	N_2	N_3	N_4
S_1	70	20	0	10
S_2	40	10	10	0
S_3	0	0	15	15

To utilize the regret table in a decision criterion, one must recognize that the regrets are of the nature of losses. Hence the pessimistic decision maker might decide to use a strategy which will minimize his maximum regret, a *minimax regret* criterion. In this case the maximum regret for each strategy is 70, 40, and 15 respectively.[6] The strategy which minimizes maximum regret is S_3, with a security level of regret of 15.

The appeal of this criterion is diminished for many by a logical flaw which appears to be inherent in it. If we add a fourth strategy S_4 to the basic payoffs of Table 3-17, which has utilities 60, 40, 30, and 60 respectively for the four states of nature, the regret table becomes Table 3-19. Here, the minimax regret payoff is 45 and the minimax regret strategy is S_2.

TABLE 3-19   REVISED REGRET TABLE

Strategy	State of nature			
	N_1	N_2	N_3	N_4
S_1	70	20	10	55
S_2	40	10	20	45
S_3	0	0	25	60
S_4	60	10	0	0

[6] Note that the security level of regret for S_3 is the same for two states of nature.

Consider now what has resulted from the introduction of S_4 into the decision problem. Although S_4 is not chosen as "best," its consideration has shifted the preference from S_3 to S_2 *within the set of previously considered strategies.* That this is illogical is perhaps best illustrated by a simple example. If one is offered a choice of an apple or an orange and he chooses the apple, it is assumed that in a choice between an apple, an orange, and a peach he will choose either the apple or peach, but not the orange, since one has already expressed a preference for the apple rather than the orange. The introduction of a new alternative, the peach, should not cause a shift in preference between the apple and orange. This is exactly what has occurred here. The introduction of S_4 has caused a shift in preference from S_3 to S_2. Logically, if S_3 were the "best" strategy of the set consisting of S_1, S_2, and S_3, the introduction of S_4 should not have resulted in a switch in preference from S_3 to S_2. This logical deficiency, termed the *independence of irrelevant alternatives,* forms the basis for much of the criticism of the regret idea. To some, however, the cases in which the invalidity of the criterion can be questioned on these grounds seem to be pathological. To them the criterion is a good one "for most practical purposes."

Sensitivity Analysis

The current inability of decision analysts to present a universally valid criterion for decision making under uncertainty appears to be a major deficiency of the conceptual framework. Indeed, there is a great deal of uncertainty associated with decision problems under uncertainty!

However, the operational difficulties are not so great as the conceptual ones. To illustrate this, we need first to recognize that each of the criteria we have discussed for uncertainty essentially represents an attempt to convert a decision problem under uncertainty to one under risk or certainty. By means of the maximin, minimax, or maximax criterion, the problem under uncertainty is converted to one under certainty, the "single" state of nature being the one leading to the best (or worst) outcome for each strategy. The Hurwicz alpha criterion converts the decision problem to one under risk involving the states of nature that lead to the best *and* worst outcomes for each strategy. The "probabilities" attached to these states are simply α and $1 - \alpha$. Similarly, the Laplace criterion directly converts the problem to one under risk by simply assigning equal probabilities to each state of nature. The regret criterion involves a change in the basic payoff measure—from utility gain to utility regret—and the application of the minimax criterion, and so it is equivalent to the conversion to certainty inherent in the minimax criterion.

The simplest way to convert a decision problem under uncertainty into one under risk, however, is not so complex as any of these. If we make subjective probability estimates for the likelihood of each of the states of nature in a problem under uncertainty, we have immediately transformed it to one involving risk. Indeed, *any information concerning the relative likelihood of occurrence of any of the states of nature precludes our formulating the problem as one under uncertainty.*

In most real-problem circumstances, some information, however vague, is available concerning such likelihoods. Only the most obscure circumstances which one can hypothesize, such as "war versus peace," do not involve some information as to the likelihood of the states of nature on which probability estimates may be based. When such information is available, the problem becomes one under risk rather than under uncertainty.

Indeed, the uncertainties in such situations are still great. If probability estimates are based on meager information and these estimates are used to make numerical calculations, what degree of faith is one to have in the validity of the "best" strategy which is delineated? The answer to this question lies in two areas. First, there is the question of the inherent validity of the probability estimate, the degree of expertise of those making the estimates, the reliability of the information on which they are based, etc. All of these aspects must be evaluated by the decision maker in the individual circumstance. No general conclusions are possible.

Second, and perhaps more important, is the idea of *analyzing the sensitivity of the solution to the probability estimates.* In conducting this sort of *sensitivity analysis,* we are taking the strategy which has been determined to be best using the numerical probability estimates and asking the question: "How far wrong would the probability estimates need to be to make some other strategy be best?" We can easily illustrate this with an example. The illustration, represented by Table 3-20, is extremely simple; yet it demonstrates the basic idea of conducting a sensitivity analysis. If gross

TABLE 3-20   PAYOFF TABLE

	N_1	N_2
S_1	80	40
S_2	82	30
S_3	20	41

probability estimates of 0.10 and 0.90 are made for N_1 and N_2 respectively, the expected utilities are

$E(S_1) = 0.10(80) + 0.90(40) = 44.0$
$E(S_2) = 0.10(82) + 0.90(30) = 35.2$
$E(S_3) = 0.10(20) + 0.90(41) = 38.9$

and S_1 is judged to be best on the basis of maximum expected utility.

Now suppose that we have serious questions as to the validity of the probability estimates, as we might have in a problem of great uncertainty which we had forced into the risk framework. How far off would our probability estimates need to be to make S_2 superior to S_1, for example? We can easily determine this by writing the expected utility expressions for S_1 and S_2 in terms of the probability estimate for N_1, which we shall call p, and recognizing that S_1 is superior to S_2 as long as its expected utility is greater than the expected utility for S_2. In algebraic terms S_1 is better than S_2 as long as

$$80p + 40(1 - p) > 82p + 30(1 - p)$$

which is equivalent to

$$40p + 40 > 52p + 30$$

or

$$10 > 12p$$

which is

$$^{10}\!/_{12} > p$$

Hence, if p (the probability of N_1) is less than $^{10}\!/_{12}$, S_1 is superior to S_2. If p is greater than $^{10}\!/_{12}$, S_2 is superior to S_1. Thus, in the example in question, the true probability p would need to be over eight times that which we have estimated for S_2 to be preferred to S_1. In most instances we could probably conclude that our estimate is unlikely to be that bad, and we would consequently have great confidence in the superiority of S_1 over S_2 *even though we may have only little confidence in the probability estimate itself.*

If we make the same calculation for S_3 in Table 3-20, we find that the 0.10 estimate would need to represent a true value of less than 0.02 for S_3 to be superior to S_1. Depending on circumstances, we may or may not feel confident that S_1 is superior to S_3 on that basis.

In any case, the concept demonstrated here gives us an idea of the reliability which we can place in optimal strategies developed on the basis of unreliable probability estimates. In some cases delineations of optimal strategies on which we can place a great deal of reliance can be based on relatively unreliable probability estimates. In other cases no such reliability is possible. In the latter situation, however, the sensitivity analysis which we can conduct tells us this, and in this way it serves to reduce our uncertainty.

Some Difficulties and Implicit Assumptions

Like any tidy conceptual construct, the decision theory framework has a number of loose ends and practical difficulties inherent in it. In addition, some assumptions which are implicit in the framework and some of the common ways of attacking decision problems need to be pointed out. These practical difficulties and the implicit assumptions are often of no great significance, and so the conceptual framework may be directly applied. They should, however, always be kept in mind by the analyst, for it is only a fool that becomes the victim of a creature of his own making.

UTILITIES AND STRATEGIES

One assumption implicit in the conceptual framework is that the decision maker has included all preferences which he may have for strategies (as opposed to "pure" outcomes) in the utility measure for each outcome. For instance, an advertising manager with a predilection for television might prefer a profit of $100,000 and market share of 10 percent achieved by using TV advertising to the same profit and market share achieved by using only newspaper advertising. If this is so, his preferences should be reflected in the utility measure which represents each outcome in the payoff table, for the conceptual framework does not allow for preferences outside the realm of the utility numbers.

INDIVIDUAL, GROUP, AND ORGANIZATIONAL UTILITIES

Of the difficulties which may arise in constructing the payoff table for a marketing decision problem, not the least irksome is the decision as to exactly whose utilities should be used.

If the decision maker is actually a group of people, for example, the concept of a group's utility based upon some nebulous idea of a group consensus is not at all a simple thing. A group is made up of individuals, each with different utilities. The problem of even comparing the utilities of different individuals, much less arriving at a group utility, is a complex one, for my values and preferences are mine, yours are yours, and the two may not be reconcilable.

If the decision maker is an individual, the situation appears to be much simpler. However, the realization that there are at least two utility systems wrapped up *within* an individual serves notice that even the utilities of individuals are not simple phenomena. Each individual first has a set of preferences and values which are his own: his favorite color is green, he likes redheads, and a Corvette is just his speed. In addition, he has a set of values which emanate from the organization within which his business decisions are made. He would be less than human if both value systems did not play a part in some of his organizational decisions. It is not uncommon, for example, for most of us to consider the possible influence of our organizational actions on our own future promotions, remuneration, etc., as well as their influence on the purely organizational outcomes.

There is no clear-cut method for resolving such difficulties. When we speak of the utilities associated with outcomes of marketing decision problems, we shall be referring to organizational utilities, often without reference to the facts that most measurement techniques apply to individuals and that the utilities generated may or may not truly represent the organization's preferences. The rationale for such an approach lies solely in the unavailability of something better.

VALUEWISE INDEPENDENCE

Another implicit assumption often made in the analysis of decision problems concerns the value (utility)-wise independence of the outcomes. For example, consider the problem involving the assignment of two salesmen to two sales territories. If Table 3-21 gives

TABLE 3-21   UTILITIES FOR SALESMAN–TERRITORY COMBINATIONS

Salesman	Area A	Area B
1	20	30
2	25	32

the utilities associated with each salesman-territory combination, the problem is one of decision making under certainty. Only two strategies are available:

S_1: salesman 1 to A; salesman 2 to B
S_2: salesman 1 to B; salesman 2 to A

The payoff table might be taken to be the simple one shown as Table 3-22, where 52 is the sum of the utility associated with

TABLE 3-22   PAYOFF TABLE

Strategy	State of nature N_1
S_1	52
S_2	55

S_1, i.e., assigning salesman 1 to area A (20) and assigning salesman 2 to B (32), and 55 represents the same computation for S_2.

It seems clear that the best strategy is S_2 with its utility value of 55. Note, however, that the conceptual framework dealt with the *utilities of outcomes, not with utilities for components of outcomes*. In this example an outcome is the result generated by the assignment of *both* salesmen to territories, one to each. The salesman-area table (3-21) gives utilities associated with component parts of the possible outcomes—with the assignment of *one salesman to one territory*. *The operation by which the utilities associated with outcome components were added to develop the utilities of outcomes assumed the outcome components to be valuewise independent.*

A definition of valuewise independence should make this more clear. The necessary understanding can be attained by defining this concept in a qualitative fashion in terms of a number of states $A, B, C, \ldots, N$ and the absence of those states $\bar{A}, \bar{B}, \ldots, \bar{N}$. *The states are value (utility)-wise independent if and only if the utility associated with any outcome composed of the simultaneous occurrence of states and the absences of those states in any combination is the sum of the utilities associated with the component states; e.g.,*

$$U(AB \cdots N) = U(A) + U(B) + \cdots + U(N)$$
$$U(A\bar{B}C \cdots \bar{N}) = U(A) + U(\bar{B}) + U(C) + \cdots + U(\bar{N}) \qquad (3\text{-}4)$$

etc., for all such possible combinations.

The above expressions say that the utility associated with the outcome which encompasses the simultaneous occurrence of A, $B, \ldots,$ and N is equivalent to the sum of the utilities associated with the individual component states $A, B, \ldots,$ and N, and the utility associated with the intersection of A with the absence of B, and C, etc., is a similar sum. This idea of valuewise independence is often termed *additivity of values* for obvious reasons. The analogy to the idea of probabilistic independence (which involves products rather than sums) is apparent. In the case of the assignment of salesmen, we are using this idea in determining that the utility associated with the results produced by salesman 1 in territory *A* and salesman 2 in *B* is equivalent to the utility associated with the results produced by 1 in *A* added to the utility related to the results produced by 2 in *B*.

There are obvious cases where such an assumption is a poor one. In everyday life, for example, if the states *A* and *B* are respectively "drink a cup of black coffee" and "drink an ounce of cream," for the ardent creamed-coffee drinker the utility associated with the simultaneous occurrence of *A* and *B* is undoubtedly *not* the sum of the utilities associated with *A* and *B* in the abstract, for the worth of the coffee with cream is significantly greater than the worth of the coffee plus the worth of the cream. The value of the outcome "*A* and *B*" is a function not only of its components *but also of their interaction.* The whole is worth more than the sum of its parts.

This, then, is another of those difficulties which may arise in some applications of the conceptual framework, for using an assumption in a situation where it is not warranted can lead to unhappy results. It should be noted, however, that this difficulty is not an intrinsic part of the conceptual framework. It is simply one which often arises from misapplications of the concepts to real problems. If the assumption of valuewise independence is one which would simplify the analysis of a particular problem situation, one should be careful to question the likelihood that it is a valid assumption.

UTILITIES AND DOLLARS

After this enumeration of difficulties one might be led to think that at least the simple case of an individual with no utility for strategies, with valuewise independent composition of outcomes, making a personal choice involving only one measure of the attainment of his single personal objective, money, would be rather clear-cut. After all, since everyone values money, why would it be necessary to consider a utility for money? The uninitiated might say, "After all, everyone can evaluate and compare various dollar amounts. Hence, the imposition of a measure of utility is a needless complexity."

The answer to this logical query is that normally it is a valid assumption that anyone can *order* outcomes expressed in terms of dollars. For example, for the decision problem involving the salesman who must make an auto payment of $40, the outcome table is repeated as Table 3-23. Everyone would agree that the salesman would prefer $100 to $80, $80 to $40, and $40 to $10, or at least that he would not prefer a lesser sum to a greater

TABLE 3-23   OUTCOME TABLE

Strategy	State of nature	
	N_1	N_2
S_1	$10	$100
S_2	$40	$80

one except in the most pathological of circumstances. And indeed, if these were outcomes in a decision problem under certainty, this realization would itself lead to the selection of a best strategy.[7]

In the case of risk or uncertainty, however, a simple preference for higher dollar amounts is not sufficient, since the choice is between strategies from which *any one of a number of outcomes may occur*. In the salesman's problem the likelihoods associated with N_1 and N_2 are $\frac{1}{4}$ and $\frac{3}{4}$ respectively, so that the *expected monetary values* are $77.50 and $70 for S_1 and S_2 respectively.

To argue that S_1 is preferable to S_2 on the basis of expected monetary value is to forget the salesman's need for $40 to make a car payment. Although S_1 has a higher expected monetary value, it also leaves the salesman with the possibility of obtaining only $10 ($O_{11}$). In the case of S_2 his expected monetary value is lower, but he is assured of getting at least $40. It is apparent that this decision maker might choose S_2 rather than the strategy which maximizes his expected monetary value—S_1. The explanation for this is that he is basing his decision on his personal preferences or on some more basic measure of his needs than dollars. But *this is exactly the basis for the utility concept which we have previously discussed*. So, even in the case of outcomes expressed only in dollars, some decision makers (in fact, as we shall later argue, *most* decision makers) will (and should) base their decisions on *their utility for money* rather than the amount of the money alone.

Utility Functions for Money

A great many hypotheses have been put forth concerning the form of a typical individual's utility function for money. Of course, each individual has his own utility function, but it seems logical to think in terms of a functional form (shape) which is common to many individuals.

One characteristic which this form should undoubtedly have is

[7] This can be seen by constructing a table such as Table 3-9, in which the outcomes are $100, $80, $40, and $10. Everyone would obviously prefer the strategy which resulted in the $100 outcome.

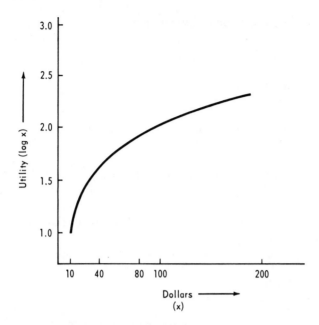

FIGURE 3-5   LOGARITHMIC UTILITY FUNCTION

that the function be nondecreasing, i.e., that an increasing quantity of dollars should not have decreasing utility. It seems illogical that in any but the most pathological circumstance, anyone would have a greater utility for $100 than he has for $125, for instance. Secondly, the function probably should have a decreasing slope: the more money one has, the less the additional utility added by an additional dollar. To one who is penniless, the dollar which represents the difference between nothing and $1 in his purse adds greater utility than does the dollar which represents the difference between $99 and $100, for the first enables him to satisfy some basic needs and the latter simply provides him with a little greater store of resources to sustain his requirements.

Of course, many functional forms satisfy these two basic requirements. In the eighteenth century Bernoulli proposed that the utility of dollars could be represented by the logarithm of the number of dollars.[8] The general shape of this function is shown in Figure 3-5. It is apparent from this sketch that the logarithmic function does satisfy the two general requirements noted above.

[8] The logarithm (to the positive base b) of a positive number M is the exponent x to which the base must be raised in order to equal M. This is written $\log_b M = x$, which says that $b^x = M$. Typically, the base chosen is 10 and we may simply write $\log M = x$, the base being implied to be 10. We shall use logarithms to the base 10 in each instance concerning logarithmic utilities in this book.

TABLE 3-24   PAYOFF TABLE USING LOGARITHMIC UTILITIES

Strategy	State of nature (probability)	
	N_1 (¼)	N_2 (¾)
S_1	1.0000	2.0000
S_2	1.6021	1.9031

With this utility function the outcome array presented previously (Table 3-23) in terms of dollars may be converted to the payoff table (in terms of utilities) shown as Table 3-24. These entries simply indicate that the logarithm of 10 is 1, the logarithm of 40 is 1.6021, etc.[9]

The expected utilities associated with S_1 and S_2 are

$$E(S_1) = ¼(1) + ¾(2) = 1.7500$$
$$E(S_2) = ¼(1.6021) + ¾(1.9031) = 1.8279$$

Hence, S_2 is the best strategy on the basis of expected logarithmic utilities, whereas S_1 is the strategy which maximized expected monetary value.

In using the logarithmic utility function here without regard to the assets of the decision maker, we have oversimplified the situation drastically. The rationale for the logarithmic function is best applied to total assets. Hence, we could construct a similar table in which the dollar quantities express this measure. Here, we have implicitly assumed that the salesman's other assets are negligible.

There is, of course, nothing sacred about a logarithmic functional form for the utility of money. Because of its elementary form and extensive tabulation, it is a most simple and convenient form to use as an approximation for an individual's utility for money.

Estimating Utilities

As an alternative to the use of a logical but arbitrary utility function for the outcomes of a decision problem, such as logarithmic utility for monetary amounts, we might wish to estimate empirically the utilities of a decision maker. One technique which has proved use-

---

[9] In using the logarithmic function as a utility function, one is faced with the difficulty that the logarithm of 1 is zero and the logarithm of zero is minus infinity. To circumvent this difficulty, we may simply use $U = \log(1 + x)$ where x is the positive dollar quantity and U represents utility. Hence, the utility of zero dollars becomes zero.

ful in accomplishing this is the *standard gamble* of Von Neumann and Morgenstern.[10]

In the standard gamble approach, the decision maker who faces a real-world decision problem is presented with a series of fictitious simple choice situations involving outcomes of his real decision problem. These choice situations are constructed in a way to be easily comprehended. On the basis of the expressions of preference which the decision maker is asked to make concerning these choice situations, the numerical utilities which are applicable to the outcomes of the real decision problem may be developed.

Let us assume that the decision maker is faced with a decision problem under certainty, risk, or uncertainty, which involves 100 distinct outcomes. *The first step in the standard gamble approach requires that the decision maker identify the most preferred and least preferred of these outcomes.* Let us assume that he can do this and that we number the outcomes so that O_1 is the least preferred and O_{100} is the most preferred, with the other 98 outcomes numbered arbitrarily. By virtue of this determination of the best and least preferred of all of the outcomes, we know that O_1 has the smallest utility of any of the 100 outcomes and O_{100} has the greatest utility.

If the decision problem is one under certainty, all of the outcomes will be related to one state of nature. This simple information is sufficient to determine which of the available strategies is best in this case, since all that we need to do is choose the strategy which leads with certainty to the most preferred outcome (that having the greatest utility). If the decision problem is under risk or uncertainty, however, we must develop numerical utility estimates for each of the 100 outcomes.

To do this, we construct a choice situation involving three of the outcomes—the most preferred O_{100}, the least preferred O_1, and any other, say O_2. The decision maker is presented with a fictitious decision problem involving two strategies whose outcomes are defined as follows:

Outcome for S_1: O_2
Outcome for S_2: O_{100} with probability p
$\qquad\qquad\qquad O_1$ with probability $1 - p$

In other words, the decision maker is asked to choose between a strategy S_1, which results in the certain outcome O_2, and a strategy S_2, which results in either the best or worst of the outcomes of the real decision problem, the best occurring with probability p and the worst with probability $1 - p$.

[10] J. Von Neumann and O. Morgenstern, *Theory of Games and Economic Behavior,* Princeton University Press, Princeton, N.J., 1944.

To begin the utility estimation procedure, the numerical value of p is set equal to 1, making the numerical value of $1 - p$ equal to 0. The decision maker is asked to choose between S_1 and S_2 in this fictitious decision problem. To be consistent, he must choose S_2, since with p equal to unity, the certain outcome for S_2 is simply the most preferred of the 100 outcomes, O_{100}.

Next, p is set equal to 0.99, 0.999, or 0.90, depending upon the situation, and the decision maker is again asked to express a choice between S_1 and S_2. In the case of $p = 0.999$, he is likely still to prefer S_2, since it still leads to O_{100} with virtual certainty. If he does still prefer S_2, the numerical value of p is lowered and he is again asked to choose. This process is continued until at some point his preference shifts from S_2 to S_1. To demonstrate that this shift in preference *must* eventually occur, consider that p is lowered to the value 0. Then S_2 results in outcome O_1 with certainty. Since O_1 is the least preferred of all 100 outcomes, the decision maker must, to be consistent, prefer S_1 to S_2; i.e., he must prefer the certainty of O_2 to the certainty of O_1. Since he begins by preferring S_2 when p is 1 and ends by preferring S_1 when p is lowered to 0, there must be some numerical value of p between 1 and 0 at which he is indifferent between S_1 and S_2. Let us assume that this indifference occurs at $p = 0.80$.

In terms of his (unknown) utilities for the outcomes, the decision maker is implying that

$$1.0U(O_2) = 0.80U(O_{100}) + 0.20U(O_1)$$

which simply says that his expected utility for the certain outcome O_2 is equal to the expected utility resulting from S_2.[11] It is important to note that this equality holds only at the point of indifference, in this case at $p = 0.80$.

We may arbitrarily assign numerical utilities to any two outcomes of any decision problem. In doing so, we fix the scale of utility to be used in terms of its zero point and unit of measure. If we were not able to do this, it would be implied that there is some absolute measure of utility, i.e., that there is some unique zero point of the utility scale and some unique unit of measure. Of course, we have already argued that utilities are subjective and personalistic—mine are mine and yours are yours. Also, we intuitively recognize that there is probably nothing which has zero utility for anyone.

To aid us in estimating utilities, it is simplest to choose arbitrary utility numbers for the best and least preferred of the outcomes

---

[11] Since S_1 achieves outcome O_2 with certainty, its expected utility is simply 1 times the utility of O_2.

of a decision problem. Further, since the utility scale is arbitrary, we may as well use simple numbers, such as 0 and 1, for these two utilities. If we do so, we will be arbitrarily determining that

$$U(O_1) = 0.0$$
$$U(O_{100}) = 1.0$$

Using these utility numbers in the expression of the equality of the expected utilities at the point of indifference, we obtain

$$1.0U(O_2) = 0.8(1.0) + 0.2(0.0)$$

or

$$U(O_2) = 0.8$$

This value for the utility of O_2 is *relative* to the arbitrarily chosen utilities of 1 and 0 for the most and least preferred outcomes respectively. Since there are no absolute utility measures, but only relative utility measures, this presents no difficulty. The numerical utility measures 0.0, 0.8, and 1.0 may then be inserted into the payoff table for the real decision problem in the positions corresponding to O_1, O_2, and O_{100} respectively, and the criterion of maximum expected utility (in the case of risk) or some other criterion (in the case of uncertainty) may be applied.

Of course, each of the other 97 outcomes first needs to be evaluated in the same fashion. We would simply sequentially repeat the same procedure, using O_1, O_{100}, and each of the other outcomes. Although the numerical utilities which are attached to O_1 and O_{100} are arbitrary, once they are chosen for a particular problem, the same values must be used throughout. In doing this, *we numerically evaluate all outcomes relative to the best and worst.*

One of the advantages to this utility estimation technique is that it may equally well be applied to qualitatively described outcomes. This may be illustrated by reviewing the procedure and observing that we ask the decision maker to select only the best and worst outcomes and to express preferences between outcomes obtained with certainty and outcomes obtained subject to chance. In no way does the technique depend upon the outcomes being described in terms of dollars or any other quantitative measure. Of course, if the qualitative outcomes utilized are complex, the decision maker may be unable to comprehend them.

The primary difficulty which is experienced in applications of this utility estimation procedure evolves from the inconsistencies which may arise in the decision maker's expressions of preference. Often the difficulty is the result of actual inconsistencies of the decision maker or his inability to comprehend complex outcomes.

In any event, a procedure involving the illustration to the decision maker of the inconsistencies which may arise will usually enable us to arrive at reasonable approximations for his utilities.

One other difficulty, which will not normally arise in marketing situations, should be mentioned. If one of the outcomes is overwhelmingly bad for the decision maker, say bankruptcy or death, the technique may collapse. A little reflection will make the reason quite clear. If he is faced with a choice between any outcome and a lottery involving his own death as a possible outcome, however small the probability, he is unlikely to be able to express a preference for the lottery. Many people would choose the lottery with probability 1 for best outcome over the certainty of the mediocre outcome, but would choose the mediocre outcome for any other value of p. This would result in the assignment of the same numerical utility to all outcomes except the one involving death. In effect, whereas small differences in the value of p may actually represent large utility differences, the individual is unable to express these differences because of the tremendous magnitude of the worst outcome. In this instance it would indeed be pompous to certify him as irrational, and so we shall simply remember that this difficulty can possibly arise and therefore guard against making utility estimates which do not adequately reflect an individual's preferences.

Monetary Values

When outcomes can be expressed in monetary amounts, a simple approximation to the utility for money is just the number of dollars each outcome represents. This is clearly more convenient than either the standard gamble or a logarithmic utility since it requires no transformation to go from the outcome array to the payoff table.[12]

This approximation is implicitly assumed in the great majority of marketing decision problem formulations. When profits are maximized or costs minimized, the implicit assumption is that the amount of profit or cost (dollars) approximates the organization's utility for these monetary values.

Like any approximation, this one is sometimes valid and sometimes invalid. It is possible, however, to apply a simple test for its validity. To do this, we offer the decision maker a series of fictitious choice situations exactly like those used in the standard gamble. If we observe that he expresses preferences which are

[12] To be perfectly correct, we must remove the dollar sign from the outcomes to convert them to utilities in this instance.

consistent with the criterion of expected monetary value in those choice situations, then we may feel confident that monetary value is an adequate approximation for utility in the real decision problem.

To demonstrate this, consider that the best monetary outcome of a decision situation is B dollars and the worst is W dollars. If S_1 leads to M dollars with certainty and S_2 to B dollars with probability p or to W dollars with probability $1 - p$, the decision maker will be expressing preferences which are consistent with the expected monetary value criterion if he chooses S_1 when

$$\$M > p\$B + (1 - p)\$W$$

and S_2 when

$$\$M < p\$B + (1 - p)\$W \tag{3-5}$$

He should have no preference between S_1 and S_2 if a similar equality holds. Since the choice of the zero point and unit of a utility scale are arbitrary, we may always choose a scale such that the number B (without the dollar sign) represents the utility of the best outcome ($\$B$), and the number W (without the dollar sign) represents the utility of the worst outcome ($\$W$). On such a scale the expected utility for the lottery involved in the fictitious choice situation is, by definition,

$$E(U) = pB + (1 - p)W \tag{3-6}$$

When presented with a series of choice situations such as those used in the standard gamble involving different values for M (when $W < M < B$) and p, the decision maker may express preferences which are consistent with the maximization of expected monetary value; in this case we may conclude that monetary value is a good approximation to utility *within the range of outcomes for the decision problem in question*.

This last statement should be emphasized, for it is all too easy to draw sweeping inferences when only limited ones apply. The utility approximation of monetary value which is tested by using this approach is applicable only for the range of outcomes between B dollars and W dollars and only for the decision problem in question. A little reflection, along with a rereading of the comments relative to the overwhelmingly bad outcome discussed in the previous section, will verify the former point. The latter is operationally sound since we cannot hide from the decision maker what we are trying to accomplish. Consequently, when we ask him to express a preference in the standard gamble variety of choice situations, the outcomes are framed in the terms of his real-world decision problem rather than in the abstract. As a result, it is unlikely that the decision maker can express preferences for monetary amounts which are independent of the real-world decision problem which will produce the outcomes.

The skeptic might respond to this procedure by pointing out that to verify that monetary values are "good enough" approximations to utilities, we must, in effect, apply the standard gamble utility estimation technique; i.e., to show that we need not estimate utilities, we need to estimate them. This is valid up to a point. However, when one recognizes that the actual number of outcomes in a decision problem may be very large and that we need to select only some of them to judge adequately the worth of monetary value measure as an approximation to utility, the saving in effort becomes apparent. Moreover, since the monetary value approximation is in such wide use, the procedure used here represents a basis for testing the approximation *when there is some doubt as to its validity*.

It is a well-advised marketing manager who constantly bears these points in mind when he analyzes a decision problem in terms of dollars alone. First of all, he should be aware that the conceptual framework on which he bases his analysis is in terms of utilities and that his treatment in terms of dollars is an *approximation*. Secondly, he should recognize that it is possible for a monetary approximation to the utility criterion to yield a "best" strategy different from that of the underlying utility criterion. Most of all, however, he should recognize that, like any approximation, the use of monetary values instead of utility is sometimes valid and sometimes invalid. It is this determination by the analyst of the circumstances in which the approximation is appropriate that is so important; for the conclusion arrived at may be much more sensitive to the approximation than to the controllable variables of the decision problem.

Summary

This chapter outlines the conceptual framework which can be used to logically evaluate and compare the alternative actions available to the marketing decision maker.

The basic controllable and uncontrollable elements of a problem situation—the strategies, states of nature, and competitive actions—are related to the states of affairs which may result from a decision—the outcomes. The outcome array is the vehicle used to analyze the interaction of these elements.

The concept of utility is used to facilitate the transformation of the outcome array to a payoff array. The payoff array, each element of which is a single utility number which incorporates all of the salient aspects of the organization's objectives and their degree of satisfaction, may then be analyzed according to the likelihood that information which is available places the problem in one of the classes termed *certainty, risk,* or *uncertainty.*

Of course, we should realize that problems do not exist in the real world as one of these classes. Most real problems "look as if" they are highly uncertain. The available information concerning likelihoods may naturally place a problem into one of these categories, but more frequently the analyst must fictionalize the real-world situation and formulate the problem in one of these categories for ease of analysis. The worth of this fictionalization process will be discussed in Chapter 4.

In problem situations which do not fit neatly into either the risk or the uncertainty category, sensitivity analysis may be used to provide the decision maker with some confidence in the analytic solution which has been developed for a rather ill-defined likelihood situation.

The significance of the concept of utility, even in situations in which every outcome is adequately described in dollar terms, can easily be demonstrated by illustrating that the maximization of expected monetary value and the maximization of expected utility can lead to different problem solutions ("best" strategies). The *standard gamble* approach of Von Neumann and Morgenstern is a basic method of estimating an individual's utilities, and yet it is awkward and cumbersome in problems involving many possible outcomes. A method for approximately determining whether monetary value can be used as a proxy for utility can be developed by using a variation of the standard gamble approach.

In most analyses of marketing decision problems, monetary values are used as approximations to utility. One reason for this is the likelihood that money represents a good approximation to utility in decision problems which do not involve a significant proportion of the total resources of an organization. Of course, it is also true that relatively less is known about utilities than about monetary values, and one is led to suspect that the use of monetary value as a proxy for utility may often be motivated by this lack of knowledge.

To use the conceptual framework which has been developed here, it is necessary to have predictions of a wide range of possible outcomes for a decision problem. The devices used to do this—i.e., to predict the elements of the outcome array—are termed "models." The construction, use, and solution of models is taken up in Chapter 4.

EXERCISES

1. In a decision problem four distinct competitive actions and three states of nature have been defined. How many generalized states of nature need to be considered?
2. Consider the outcome array presented below.

	N_1	N_2	N_3
S_1	$10	$20	$ 8
S_2	$19	$40	$11
S_3	$ 5	$19	$ 4

Can you determine which strategy is best? How did you do this? Is this a decision problem under certainty, risk, or uncertainty?
3. How might you attempt to translate the outcome array of Table 3-1 into one which has more meaningful outcome descriptions? In what way is your new outcome table more meaningful?
4. How might you transform Table 3-2 into a more meaningful outcome array?
5. In the revised outcome array which you developed in exercise 4, how can preferences which might exist for strategies be taken into account? What is the most obvious measure of such preferences?
6. Consider yourself as a marketing manager who must express a preference between the outcomes below. What would be your preference? Try to describe the mental process which you went through to arrive at this conclusion.

	Sales revenue	Total costs	Employee morale level
O_1	$2.0 million	$1.5 million	Low
O_2	1.5 million	1.0 million	Medium
O_3	1.5 million	0.9 million	High
O_4	2.0 million	1.6 million	High
O_5	1.4 million	0.9 million	Medium
O_6	1.4 million	0.8 million	Low

7. Attempt to construct the payoff table for the competitive bidding problem of Table 3-2, using the utility function in Figure 3-2. How would the strategy preferences found in exercise 5 enter into your determination of the payoff table?
8. What general shape do you think is reasonable to assume for the utility function of General Motors for its share of the auto market? The government's economic planners might have a utility function for General Motors' returned earnings (those left after taxes, dividends to stockholders, etc.). Why? Can you hypothesize what general shape this might be? (Hint: Consider the effect of earnings on the wage demands of labor.)
9. A particular strategy may result in any one of four possible outcomes. The probability that O_1 or O_2 will occur if this strategy

is chosen is ½ and ¼ respectively. Outcomes O_3 or O_4 are so vaguely defined that we find it difficult to get estimates of their relative likelihood of occurrence under the strategy in question. What might we do to facilitate the application of the conceptual framework?

10. If many of the same outcomes may result for various strategies in a decision problem under risk, the form of presentation of the probability table (3-6) may be adopted. If the utilities for O_1, O_2, and O_3 in the problem depicted in that table are respectively one, two, and five units, what is the best strategy in this decision problem?

11. A salesman wishes to travel to each of four cities, beginning at his home office, which is not in any of the cities, in a manner which will minimize the total distance traveled. What are his available strategies? Why is this a decision problem under certainty?

12. A purchasing agent must purchase 1,000 units of an item which is to be processed and delivered during the next year to fulfill a contract. What is the agent's decision problem? What are his available strategies? Why is the decision problem one under certainty?

13. Suppose that dollars are being used as an approximation to utilities and that an individual is faced with the decision situation under uncertainty described below.

	N_1	N_2	N_3
S_1	$ 5	$10	$15
S_2	$10	$20	$30
S_3	$ 5	$30	$10

If the entries in this array are *losses*, what is the best strategy using the minimax criterion?

14. (From Luce and Raiffa, *Games and Decisions*, p. 283) Consider the payoff table under uncertainty shown below.

	N_1	N_2
S_1	1	0
S_2	x	x

Suppose the decision maker has an (unknown) coefficient of optimism k. If the decision maker expresses an indifference for S_1 or S_2 if x is ⅜, what is the coefficient k?

15. (From Luce and Raiffa, *Games and Decisions*, p. 283) Consider the decision problem under uncertainty given below.

	N_1	N_2	$N_3 \cdots N_j \cdots N_{100}$
S_1	0	1	1 $\cdots$ 1 $\cdots$ 1
S_2	1	0	0 $\cdots$ 0 $\cdots$ 0

What is the best strategy according to the Hurwicz alpha criterion? Is this confirmed by intuition?

16. What is the minimax regret strategy in the decision problem situations below? Compare your answers and explain the comparison.

	N_1	N_2
S_1	0	100
S_2	1	1

	N_1	N_2
S_1	$0 + a$	$100 + b$
S_2	$1 + a$	$1 + b$

17. Each of the n individuals who handle a product along its path from the manufacturer to the consumer may be thought of as adding value to the product. If it were possible to evaluate numerically the utility of each individual's contribution in terms of the states:

E_i = individual i performs his task correctly and does not ruin the product

$\bar{E}^i$ = individual i ruins the product so that it is valueless

do you think that the utility of a finished product would be adequately described by a valuewise independence assumption relative to these states?

18. Outline the standard gamble procedure for the estimation of the payoff table for the decision problems described by the outcome arrays below. In a use 0 and 100 as the utilities for the worst and best outcomes; in b use 5 and 50 for these utilities, and in c use 0 and 65.

(a)

	N_0
S_1	$ 5
S_2	$10
S_3	$20
S_4	$15

(b)

	N_1	N_2	N_3	N_4
S_1	$100	$200	$100	$500
S_2	$500	$600	$400	$100
S_3	$100	$200	$500	$600

(c)

	N_1	N_2
S_1	$10	$20
S_2	$30	$40

19. You will note that no statement was made in exercise 18 concerning the decision maker's expression of preference for the best and worst

outcomes. Why is it possible for you to completely outline the standard gamble without this information?

20. Might the utilities which are estimated in exercise 18a and c be used directly in the development of the payoff table for the decision problem described by the outcome array shown below? Why or why not?

	N_1	N_2	N_3
S_1	$ 5	$20	$30
S_2	$10	$15	$40

21. Reoutline the standard gamble procedure for exercise 18c using 0 and 100 as the utilities for the best and worst outcomes. (These are the same as those used in 18a.) May the utilities estimated in this fashion be used in conjunction with those in 18a to develop the payoff table for exercise 20? Why or why not?

22. A decision maker expresses the following preferences in a series of choice situations involving only two strategies:

Outcome resulting from preferred strategy	Outcome resulting from nonpreferred strategy
$10	$5 with probability $\frac{1}{2}$ or $14 with probability $\frac{1}{2}$
$5 with probability $\frac{1}{2}$ or $14 with probability $\frac{1}{2}$	$8
$5 with probability $\frac{1}{2}$ or $14 with probability $\frac{1}{2}$	$9.75
$5 with probability $\frac{1}{2}$ or $14 with probability $\frac{1}{2}$	$9.25

Would monetary value be a good approximation to utilities in a real decision problem in which this decision maker was faced with outcomes between $5 and $14?

23. Describe the "equivalent" decision problems under certainty or risk to which the various decision criteria under certainty convert the decision problem of Table 3-17. Do this by constructing the "equivalent" payoff table (under certainty or risk) for this problem for each criterion.

24. Show that p would need to be less than 0.02 in the decision problem of Table 3-20 before S_3 would be a better strategy than S_1. (This statement is made a few paragraphs subsequent to the table.)

REFERENCES

Feller, W.: *An Introduction to Probability Theory and Its Applications,* John Wiley & Sons, Inc., New York, 1957.

Fishburn, P. C.: *Decision and Value Theory*, John Wiley & Sons, Inc., New York, 1964.

Hurwicz, Leonid: *Optimality Criteria for Decision Making under Ignorance*, Cowles Commission Discussion Paper, Statistics, no. 370, 1951 (mimeographed); cited in Luce and Raiffa (see below).

Luce, R. D., and H. Raiffa, *Games and Decisions*, John Wiley & Sons, Inc., New York, 1957.

Miller, D. W., and M. K. Starr: *Executive Decisions and Operations Research*, Prentice-Hall, Inc., Englewood Cliffs, N.J., 1960.

Mosteller, R., R. E. K. Rourke, and G. B. Thomas, Jr.: *Probability with Statistical Applications*, Addison-Wesley Publishing Company, Inc., Reading, Mass., 1961.

Savage, L. J.: *The Foundations of Statistics*, John Wiley & Sons, Inc., New York, 1954.

Schlaifer, R.: *Probability and Statistics for Business Decisions*, McGraw-Hill Book Company, New York, 1959.

4

Marketing Models

The use of *models* in solving marketing decision problems is basic to the application of the conceptual framework developed in the previous chapter. In this chapter we shall discuss the scientific meaning of the term "model" and illustrate the usefulness of models in marketing analysis.

Models as System Representations

The layman's understanding of models is simultaneously helpful and injurious to understanding the scientific use of the term. Discounting the female fashion variety, if the proverbial man in the street were asked to react to the word "model," he would be likely to respond with an example such as a child's model airplane. Indeed, this kind of model is familiar to all of us. In this context a model is simply a scaled-down *representation* of a real-world system—the full-size airplane. Each exterior dimension of the real airplane is accurately represented in miniature on the model. But many features of the real airplane are completely excluded from the model; e.g., the model is often of solid construction whereas the airplane's interior is a maze of electronic gear and cables. This feature of the child's model is an intrinsic part of the scientific model. Namely, in both kinds of models *some aspects of the real system are included* (such as exterior dimensions, color, markings, etc.), *and some are excluded* (interior configuration, materials, etc.). This is consistent with the scientist's idea of a model as an *abstraction of reality.*

Other models of the same system might abstract different elements of the system. For example, the ground training devices used for pilots are little more than movable enclosures which incorporate all of the interior makeup of the cockpit, but little else. The pilot seats himself in an exact representation of a portion of an airplane's interior and is closed off from his surroundings. By reference to the instruments in the cockpit, he proceeds to "fly" the training device while sitting in a room with other trainers. The trainer's response to the pilot's actions is quite similar to that of the real airplane in actual flight.

Such training devices are also models of airplanes. The elements of the real system which they incorporate are very different, however, from those accounted for in the child's model. In a trainer no attention is paid to exterior detail and dimensions since the pilot cannot see outside. The interior instruments and dimensions and the responses of the controls to actions of the pilot are faithful duplicates of the real system, however, because the pilot is expected to carry over what he has learned in the trainer to the operation of the real airplane.

These two kinds of airplane models illustrate, in their similarity, the applicability of the layman's view of a model as a "representation of something else" to the scientific concept of a model. Although they are different models of the same system, their similarity lies in their *inclusion of important aspects of the real system and exclusion of unimportant aspects of the real system.* The determination of which factors are important and which are unimportant clearly depends on the *use to which the model is put.* In the case of the model to be used to decorate a child's room, the exterior configuration, color, and markings are important and the portions which are not visible are unimportant. In the training case, the aesthetic value is insignificant, but the interior design and control responses are of utmost importance. Since both these models involve changes in physical structure, however, their applicability to the more abstract scientific use of the term "model" is limited. To illustrate these essential differences, we shall first need to give a scientific definition of the term.

Models Defined

A *model,* in the scientific sense, *is a representation of a system which is used to predict the effect of changes in certain aspects of the system on the performance of the system.* The applicability of the first phrase of this definition—"a representation of a system"—to the layman's use of the term "model" has already been illustrated. The essential distinction lies in the subsequent phrase—"used to predict the effect of changes in certain aspects of the system on the performance of the system." Clearly, most commonly known models are not used in this way. To understand the definition better, it is simpler to consider the ways in which models may be used in marketing analysis.

Using Models in Marketing

One of the essential differences between the well-recognized physical sciences and the more recent scientific investigations in business environments is that in business one is more frequently dealing with systems which cannot be duplicated in the laboratory. Moreover, in many instances the scientist's penchant for experimenting on the system in question cannot be satisfied in the business environment.

In marketing, for instance, the researcher recognizes that to make decisions regarding advertising expenditures, he must first learn something about the basic relationship governing sales response to advertising. In other words, how many sales dollars are produced by each advertising dollar? To discover this, the physical scientist might attempt to vary advertising expenditures in different areas in predefined ways and to measure the resulting sales changes. To do so, however, is often impractical simply because sales response to advertising is so little understood. Executives are generally hesitant to lower advertising expenditures because the potential resulting sales decrease might be impossible to recover. Consequently the vicious circle of meager knowledge motivating experimentation and at the same time precluding it means that the traditional approach of the physical scientist is impractical in the advertising context.[1]

Such points of conflict between the necessary practicalities of business and the traditional methods of science are not at all atypical. As a result, there is a great need in the analysis of business decision problems for some device to facilitate drawing the kind of conclusion which could be arrived at through experimentation. A model is such a device. In using a model, the scientist makes assertions which express the relationship of various elements of the system with each other and, in turn, their effect on the perfor-

[1] It must be noted that advertising experiments have been carried out by a number of companies. The example here might be thought of as a typical reaction by executives to the advertising-sales response measurement problem.

mance of the system. In doing so, he creates an entity, the model, which he can use in lieu of the actual system. He can then *experiment on the model* and on this basis make his predictions of the effects which changes in the system have on the performance of the system.

A model airplane might be used in the same way. If we were to place the model in a wind tunnel and vary the angle of sweep of the wings·to predict the effect of various angles on the performance of the airplane, we would have a representation of the airplane on which experimentation was being conducted. Hence, the model airplane would be a model in the scientific sense. The key distinction between the common garden-variety model and a scientific model is in the *use to which the model is put.*

In the advertising context, if we had a way of relating advertising expenditures to sales in a mathematical equation, say

$$R = 2A \qquad\qquad\qquad (4\text{-}1)$$

where R is sales dollars and A is advertising dollars, we would have a representation of the advertising system which would enable us to predict that \$100 in advertising would produce \$200 in sales, \$200 in advertising would produce \$400 in sales, etc.

Two things should be made quite clear regarding the simple advertising model of expression (4-1). First, it is a model which is obviously a poor description of reality. To imply that one can indefinitely continue to increase advertising expenditures and expect to obtain sales in the two-to-one ratio is beyond the wildest dreams of the most optimistic advertising manager. This implies that although this simple expression constitutes a model of the advertising system, it is not a very good model. We shall take up the question of building *good models* in a later section.

Secondly, one recognizes that the kind of model involving only changes in physical dimensions, such as the model airplane, and the kind described by expression (4-1) are vastly different. To understand models better, it is useful to view these essential differences by developing a simple classification of models.

A Taxonomy of Models

A useful categorization of models is based upon their basic characteristics of being a *transformation* of the real-world system which they represent. In the simplest model type, called *iconic,* the scale of the real-world system is transformed. The model airplane is clearly of this kind because the important aspect of the real system—the exterior configuration—is related to the corresponding feature of the model by a simple scale transformation; e.g., 1 inch equals 1 foot. One may think of such iconic models and the real systems which they represent as being "look alikes."

A more abstract kind of model is the *analog,* in which properties are transformed; i.e., one property is used to represent another. A graph in which a unit distance along a line is used to represent a unit of time, speed, market share, or sales is an example of such a model. Topographic maps in which the property color is substituted for height above sea level are other common analog models.

The kind of model represented by the advertising–sales response relation of expression (4-1) is even more abstract than the analog. This kind of model, termed *symbolic,* is one in which symbols are substituted for properties; e.g., the symbol A is substituted for advertising expenditure. This is the sort of model we try to develop for marketing analysis because it is both the most general and the easiest to manipulate. The simple model $R = 2A$ might just as well be used to relate the number of machine hours available, A, to total production quantity, R, if two units of a product may be produced per hour; or if A is allowed to represent the quantity sold and \$2 is the unit selling price, the same model could be a relationship between sales revenue, R, and the quantity sold, A. In this sense a symbolic model such as $R = 2A$ is of a general nature. Such a model is easy to manipulate in the sense that all of the tools of logic and mathematics may be brought to bear, whereas no such storehouse of rigorous methods is available in manipulating and experimenting on iconic and analog models.

Of course, the difficulty in constructing symbolic models is the area in which we must pay for this generality and ease of manipulation. Symbolic models are generally much more difficult to construct than either of the other kinds. In the succeeding chapters we shall deal primarily with symbolic models. Analog models, particularly graphs, will usually be made use of for pedagogic purposes rather than out of the necessity forced by our limited knowledge.

Models and the Conceptual Framework

To understand the role of symbolic models in the conceptual framework as developed in the previous chapter, it is necessary to recognize the time dimension which is inherent in most decision problems. In general, the point in time at which a strategy is selected by the decision maker is prior to that at which the outcome is realized. This implies that at the time the alternatives are being evaluated, the outcome descriptors in the outcome array are *predic-*

tions of future states which will result from the various strategy–state of nature interactions.

We have defined scientific models as representations of systems
which may be used in *predicting* performance. If we recognize
that the outcome descriptors which we must choose are of the
system performance type, one possible use for models becomes
apparent. We need to have outcome (performance) predictions
in order to apply the conceptual framework, and models may be
used as devices to predict performance. Hence, *models may be
used to generate the predicted states of affairs associated with
each strategy–state of nature combination in the outcome array.*
This is one simple way of viewing the role of models in the conceptual framework—as devices for predicting the performance level
associated with each outcome.

The *performance* levels may be expressed in a number of different ways. In one common situation there are a number of descriptors of the "pure" performance of the system as defined by the
decision maker's objectives. If, for example, the objectives relate
to revenue, market penetration, and image, the descriptors of "pure"
outcomes might be "dollars of revenue," "market share," and some
verbal description of the image level. An appropriate model would
be one which would permit the prediction of these performance
measures.

Often the decision maker's objectives are expressed in terms
which make it appropriate to view the performance of the system
as something other than a number of outcome descriptors. In a
problem involving the selection of a new product to be marketed,
natural descriptors might be the revenues to accrue to the organization in the first, second, third, etc., years after introduction. Hence,
if a product will produce revenues for 10 years, one might consider
10 outcome descriptor variables—"revenue in first year," "revenue
in second year," etc. If declining revenues are causing the organization to be viewed unfavorably by Wall Street, the decision maker
might have objectives solely with respect to immediate revenue;
in this case only the first of the descriptors would be meaningful.
Or, if total revenue is of major concern, he might wish to choose
the product with greatest potential for total revenue; in this case
the 10 descriptors could be summed to obtain the single performance measure, total revenue.

In other instances measures which are simultaneously associated
with the strategies and objectives can be combined into a single
measure. We have given such illustrations in the previous chapter
involving the revenue derived from a contract award and the costs
which may be associated with various strategies. If the strategies

are simply levels of advertising expenditure and the sole objective is short-run profit, the cost of each strategy (the advertising expenditure) must be incorporated into the overall performance measure, profit.

The two basic situations which may arise should now be clear. In one case outcomes may be described simply by a number of performance measures. In the other, outcomes are composed of both "pure" performance measures and other measures relevant to the decision maker's objectives. Often these several measures may be combined into a single measure, such as total cost or profit, which incorporates both aspects.

The basic and most important point is that *it is the objectives of the decision maker which dictate the measure(s) of performance.* If the decision maker wishes to increase market share, and to have a good image, and to make a profit, *and,* etc., a number of outcome descriptors, each with its own dimension, are undoubtedly the best choice. If "to make money" is the primary objective, a single measure—"profit"—may adequately incorporate both pure outcome descriptors, such as "revenue," and measures associated with the strategies, such as the cost of utilizing each strategy.

Of course, it is sometimes possible to have a situation which appears to be of the former kind—one requiring a number of performance measures—and to treat it as being of the latter kind—one having only a single performance measure. This may be accomplished if one aspect of the decision maker's objectives appears to be dominant, as might well be the case with the portion of his objectives related to profit, for example. In such an instance the other aspects of his objectives may be ignored and the profit measure used as an approximation. Or it may be possible to express other aspects of his total objectives as *constraints* on the basic objective; e.g., one might maximize profit subject to the constraint that a given image level must be maintained, rather than attempt the simultaneous maximization of both profit and image level.

A MODEL UNDER CERTAINTY

Let us revise our almost trivial advertising model to be more descriptive of reality and use it to demonstrate the usefulness of models in predicting performance. We may do this by first describing outcomes in terms of a single performance measure which incorporates both "pure" performance and the cost of each strategy—the difference between sales revenue and advertising expenditure. We shall not concern ourselves with costs other than advertising costs since the simple decision problem we shall deal with

involves only advertising costs as controllables. The difference between revenue and advertising expenditure may be written as

$$Q = R - A \qquad (4\text{-}2)$$

where R is sales revenue and A is advertising expenditure. The
quantity Q becomes our single performance measure.

Another way of viewing this might be in terms of two outcome
descriptor variables—one "pure" (sales revenue) and one associated with the strategies (advertising expenditure). It is more
convenient, however, simply to combine them into a single measure
since they both have the same dimension, dollars, and the combined
measure Q is easily interpreted by both analysts and executives.

Secondly, we may revise our sales response function to be more
descriptive of reality. We believe, for example, that sales response
to advertising, for most products, has the characteristic of decreasing marginal return; i.e., each additional dollar spent on advertising
produces smaller and smaller additional quantities of sales dollars
in return. Let us account for this by selecting the advertising–sales
response model to be that given as expression (4-3).

$$R = 20\sqrt{A} \qquad (4\text{-}3)$$

Expression (4-3) is a simple model which says that the quantity
of advertising dollars A is related to sales revenue by a factor
of 20 times the square root of A. This arbitrarily chosen model
has the property of declining marginal return and implies that
a $100 advertising expenditure will return $200 in revenue, $400
spent on advertising will return $400 in revenue, etc.

Let us consider the decision problem under certainty involving
a choice from the advertising expenditure levels $50, $100, $200, and
$400.[2] These advertising expenditures represent the strategies, and
the decision problem is one under certainty since we assume that
we know the single outcome to be achieved by each strategy. The
outcome table in terms of revenue R is given as Table 4-1.

TABLE 4-1   OUTCOME TABLE (R)

Strategy	State of nature
S_1: spend $50	$R = 20\sqrt{\$50} = \$141.40$
S_2: spend $100	$R = 20\sqrt{\$100} = \$200$
S_3: spend $200	$R = 20\sqrt{\$200} = \$282.80$
S_4: spend $400	$R = 20\sqrt{\$400} = \$400$

[2] We might have only these alternatives if $50, $100, $200, and $400 represented the costs of four different TV commercials which might be used.

In terms of $Q = R - A$ the outcome table is depicted as Table 4-2.

TABLE 4-2   OUTCOME TABLE (Q)

Strategy	State of nature
S_1: spend \$50	$Q = \$91.40$
S_2: spend \$100	$Q = \$100$
S_3: spend \$200	$Q = \$82.20$
S_4: spend \$400	$Q = \$0$

The payoff table for this decision problem would be one in which the individual's utility for the various values of Q would appear in an array similar to Table 4-2. Since this is a simple decision problem under certainty involving only dollars as an outcome descriptor, we recognize that the utility for the largest amount of dollars will be greatest. Hence, S_2 can immediately be determined to be the best strategy. It is important to note that our ability to arrive at this conclusion without resorting to a payoff table rests on the peculiar circumstances that we have only one state of nature (certainty) and one simple outcome descriptor (dollars) and that we hypothesize a nondecreasing utility function for dollar amounts.

In generating the outcomes of Table 4-1, we have used the basic advertising–sales response model of expression (4-3). In generating the outcomes of Table 4-2, we have used a more complex model obtained by incorporating expressions (4-2) and (4-3). This demonstrates our ability, in this simple case, to use a model in predicting outcomes which are expressed either in terms of sales revenue R or as the difference between revenue and advertising expenditure Q.

A MODEL UNDER RISK

To gain further insight into the use of models in predicting outcomes, we may consider a more realistic decision problem under risk. Suppose that we have only the four alternative advertising expenditures which were previously used and that we are unsure about the degree of sales response to advertising. Suppose further that we may summarize all that we know about sales response into the simple statement that sales revenue is either related to advertising expenditure as 20 or 25 times the square root of the number of advertising dollars and that the likelihoods of these two states of nature are assessed to be 0.8 and 0.2 respectively.

It is apparent that we are severely oversimplifying the situation by making these statements concerning our knowledge of sales response to advertising; nonetheless such gross abstractions are sufficient for illustrative purposes.

In effect, we are using a more general advertising–sales response model

$$R = b \sqrt{A} \qquad (4\text{-}4)$$

where b is a sales response coefficient which in this case may take on either the value 20 or the value 25.

The outcome array, in terms of R, for this problem is given as Table 4-3.

TABLE 4-3   OUTCOME ARRAY (R)

Strategy	State of nature	
	N_1 $(b = 20)$	N_2 $(b = 25)$
S_1: spend $50	$R = 20\sqrt{\$50} = \$141.40$	$R = 25\sqrt{\$50} = \$176.75$
S_2: spend $100	$R = 20\sqrt{\$100} = \$200$	$R = 25\sqrt{\$100} = \$250$
S_3: spend $200	$R = 20\sqrt{\$200} = \$282.20$	$R = 25\sqrt{\$200} = \$353.50$
S_4: spend $400	$R = 20\sqrt{\$400} = \$400$	$R = 25\sqrt{\$400} = \$500$

In terms of $Q = R - A$ the outcome array is given as Table 4-4. These arrays demonstrate our ability to use the basic adver-

TABLE 4-4   OUTCOME ARRAY (Q)

Strategy	State of nature	
	N_1 $(b = 20)$	N_2 $(b = 25)$
S_1: spend $50	$91.40	$126.75
S_2: spend $100	$100	$150
S_3: spend $200	$82.20	$153.50
S_4: spend $400	$0	$100

tising–sales response model of expression (4-4) or the revised model obtained by incorporating expressions (4-2) and (4-4) in predicting outcomes in the case of risk.

We may continue and select the best strategy in this problem on the basis of the arbitrary utility function

$$U = \log(1 + Q) \qquad (4\text{-}5)$$

TABLE 4-5   PAYOFF TABLE

Strategy	State of nature (probability)	
	N_1 (0.8)	N_2 (0.2)
S_1	1.97	2.11
S_2	2.00	2.18
S_3	1.92	2.19
S_4	0.00	2.00

which appears to be reasonable since Q is a dollar quantity.[3] The payoff table in this case would be that shown as Table 4-5.[4] The expected utilities are then

$$E(S_1) = 0.8(1.97) + 0.2(2.11) = 1.998$$
$$E(S_2) = 0.8(2.00) + 0.2(2.18) = 2.036$$
$$E(S_3) = 0.8(1.92) + 0.2(2.19) = 1.974$$
$$E(S_4) = 0.8(0.00) + 0.2(2.00) = 0.400$$

Strategy S_2 is the best strategy since it is the one which maximizes expected utility.

Optimization Models

One kind of symbolic model of tremendous usefulness in applying the conceptual framework is called the *optimization model*. Such a model explains the operation of a system in terms of the controllable and uncontrollable variables which influence the performance of the system. This may be put into general symbolic terms as

$$P = f\{C_i, V_j\} \tag{4-6}$$

which says that the system's performance P is some function f of two *sets* of variables—the set of *controllable variables* $\{C_i\}$ and the set of *uncontrollable variables* $\{V_j\}$.

Now, since an outcome is the product of the interaction of a strategy, made up of controllables, and a state of nature, composed of uncontrollables, we can view the model of expression (4-6) in terms of the outcome array for a decision problem. Each outcome is described by variables which are related to the objectives of the decision maker. Depending on his objectives, these outcome

---

[3] Note that we are using a form of the logarithmic utility function for monetary amounts which was discussed in the previous chapter.
[4] In this table the logarithms are rounded to two decimal places for simplicity.

descriptors may all be predicted from models of the form of expression (4-6), or only some of them may. We have already discussed this point in Chapter 3; we repeat it here for clarity.

A good illustration of the several situations which may arise is given by the simple advertising model discussed previously in this chapter. In the case of the model relating advertising expenditure A to sales revenue R

$$R = \sqrt{20}\, A$$

we are thinking of revenue as descriptive of system performance. Depending on the decision maker's objectives, the outcomes in such a situation might be described in various ways. If the decision maker has objectives related only to revenue, R itself may be an adequate outcome description. In such a case the outcome is being predicted directly by the model.

If the decision maker's objectives relate to both revenue and the cost of each strategy, the best outcome description might be in terms of both R *and* A. In this case A would appear both as a predictor in the model and as a variable related to the strategy in the outcome description.

On the other hand, the relevant aspects of performance and the cost of each strategy may be directly incorporated into the model, as they were in the model

$$Q = \sqrt{20}\, A - A$$

where Q is simply the difference between revenue and advertising expenditure. If no other measures related to the decision maker's objectives are necessary, Q may be the single outcome descriptor which is used.

Thus, the performance predicted by using a model of the form of expression (4-6) and the outcome description may not always be identical. The most common difference between the two is that *the performance predicted by the model may be complemented by other variables in the outcome description.*

We should be careful to realize that we mean expression (4-6) to be a very general symbolic representation of a *class* of models, any one of which may be quite complex and/or appear to be very little like this simple expression. For example, performance P may itself involve a number of variables like sales revenue, market share, profit. Moreover, the relation itself may or may not actually be in the form of an equality. It is likely, for example, that we can describe the relationships of certain aspects of a system best in terms of inequalities such as "not less than" or "not more than." In addition, we may well have more than one symbolic expression in our model.

The payoff table expresses the decision maker's utility as a function of each outcome. Since each outcome is described by some variables predicted from a model and others which are related to the strategies and/or states of nature, we may think of the payoff table as being equivalent to the application of the decision maker's utility function to the predicted outcomes. Symbolically the payoff table is $U(O_{ij})$ for all of the strategies (values which all controllables may take on) and all of the states of nature (values which the uncontrollables may take on).

The basic problem of selecting a best strategy may be thought of in these terms as one of *selecting values of the controllables* $\{C_i\}$, *given some knowledge of the uncontrollables* $\{V_j\}$, *in a fashion which maximizes the (expected) utility.* The "knowledge" of the uncontrollables which we need to have may be deterministic (as in the case of certainty), probabilistic (as in the case of risk), or simply a capability of enumerating all possible combinations of the uncontrollables (as in the case of uncertainty).

If we can use mathematical or logical operations on a symbolic model to arrive at the set of controllables which are predicted to maximize utility, it may not always be necessary actually to construct the outcome array or the payoff table. We may be able to *optimize* the values of the controllables, i.e., to select a best strategy, using an *optimization model.* In this case, the outcome array and payoff table exist only in our minds as parts of the framework to remind us of the steps we must take and operations we must perform to arrive at a best strategy. If we can find a more efficient way to perform the necessary steps, however, it is to our advantage to do so, for the number of strategies and states of nature may be so great as to preclude their actual enumeration. An optimization model is the basis of an efficient technique for this procedure.

A SIMPLE OPTIMIZATION MODEL

To illustrate this, let us consider again the advertising expenditure decision problem under certainty. The basic model relating advertising expenditure A to sales revenue R was expression (4-3), i.e., $R = 20\sqrt{A}$. The single performance measure (outcome descriptor) which was used was the difference between revenue and advertising expenditure, $Q = R - A$. The single controllable variable is advertising expenditure, and the underlying uncontrollable is consumer demand, the level of which determines that the response coefficient in the model is 20 rather than some other number.

If we assume that instead of being restricted to the four strategies previously considered, we may choose any positive value of the controllable variable *A* as our advertising expenditure, we have an infinity of available strategies from which to choose. Since it is impractical to enumerate an infinite number of strategies, we need to make use of an optimization model and the mathematical devices which permit us to sift through the strategies conceptually to determine one which is optimal.

We may use the basic advertising–sales response model to *write the expression for the performance measure Q in terms of the controllable variable A.*

$$Q = R - A = 20 \sqrt{A} - A \qquad (4\text{-}7)$$

If, for simplicity, we define our utility function for *Q* as

$$U = Q \qquad (4\text{-}8)$$

i.e., that each dollar of *Q* is identically worth one unit of utility to us, we may write our utility as a function of *A*, as in expression (4-9).

$$U = 20 \sqrt{A} - A \qquad (4\text{-}9)$$

We wish to maximize utility by a choice of a value of the controllable variable *A*, of which the available strategies are composed. One simple way to do this is to plot the graph of equation (4-9)

TABLE 4-6   ADVERTISING EXPENDITURES AND UTILITIES FOR RESULTING OUTCOMES

Advertising expenditure A	Utility $U = 20 \sqrt{A} - A$
0	0
4	36
16	64
25	75
64	96
100	100
144	96
225	75
256	64
324	36
400	0

and view the way in which *U* varies with *A*. To do this, we simply select various values of *A* and calculate the corresponding values of *U* from expression (4-9). Some of these combinations are shown as Table 4-6. Plotting the graph represented by the pairs in Table 4-6, we arrive at Figure 4-1.

Note in Figure 4-1 that utility increases as more money is expended in advertising *up to a point* at which utility begins to

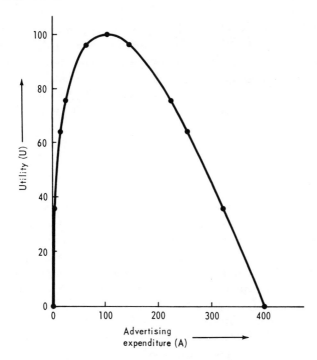

FIGURE 4-1 UTILITY–ADVERTISING EXPENDITURE GRAPH

decrease. At this point, which appears to be at about $A = \$100$ on this rough graph, the maximum utility is realized. This, then, appears to be the best of the infinite number of strategies available to us, i.e., spend $100 on advertising and obtain a payoff of 100 utility units. Any other strategy results in a lower payoff in utility units.[5]

Using this simple optimization model in conjunction with our knowledge of graphs, we have chosen the best strategy of all those which are available to us. Of course, we should investigate the behavior of the value of expression (4-9) at points close to $A = \$100$ to assure ourselves that we have not missed the maximum value of U simply because it was not one we chose to include in Table 4-6. In other words, we should find the general shape of the function and the approximate optimal value of A by a rough graph and then investigate the region near the approximate optimal to ensure discovery of the true optimal. We might, for example, plot only the portion of the graph near $A = \$100$ in 1-cent, $1 or $10 increments of A, whichever is appropriate, to determine whether $A = \$100$, and not $A = \$99$ or $A = \$99.96$, is the true optimal.

[5] Recall that the advertising cost of using each strategy is incorporated into our outcome descriptor Q and hence into our utility for each outcome.

Of course, we usually need not resort to a graph for simple optimization models. The techniques of marginal analysis and the differential calculus may be utilized for discovering the optimal value of the controllable variables in such models. We shall introduce these techniques in later chapters in the context of the problem areas of marketing management in which they prove most useful.

Descriptive Models

The lack of a basic understanding of the underlying structure which governs the operation of many aspects of the marketing system often leads to the necessity for using a "lower-level" model in marketing analysis. This kind of model, termed *descriptive*, may be used either as an alternative or as a complement to an optimization model in any particular analysis. A descriptive model, as the name implies, is one which describes how a system *does* work rather than explaining its operation and suggesting how it *ought* to be manipulated to achieve some objective, as does an optimization model. In a sense the descriptive model is behavioral in nature.[6]

The usefulness of a descriptive model is best understood in terms of the little understanding which anyone possesses of the phenomena which regulate the response of the marketing system to the various promotional devices used by marketing organizations. For example, little is known about the reaction of consumers to TV advertising, sales promotion, etc. Even less is known about the degree of generality of the few tidbits of information which are available. Does a basic feature of the sales response to a TV commercial for soap flakes apply as well to a newspaper advertisement? Moreover, does the same phenomenon apply to TV advertisement for cleanser? No definitive answers to such questions are available.

In such instances the analyst is initially forced to aim at something less than an optimization model of the decision problem. He usually begins with an attempt to understand the system better. First, he lists factors which might be causally related to the performance of the system; e.g., the weather and advertising expenditure might be believed to be causally related to sales. He then defines variables which appear to be measures or indicators of these factors. In the case of weather, for example, he might use average temperature, a humidity index, and mean annual rainfall as measurable variables.

---

[6] We do not wish to imply that the optimization-descriptive dichotomy is either disjoint or exhaustive. It is simply a useful way of viewing a spectrum of models, many of which have characteristics of both of these types.

Using these variables as representations of the underlying factors, he then attempts to *describe* the effect of the variables on system performance or, more typically, on some single aspect of system performance. Usually this is accomplished by reference to the *history* of the system; e.g., he might observe the effect which various combinations of weather have had on past sales and attempt to describe symbolically the relationship between weather and sales by using the variables sales revenue, average temperature, a humidity index, and mean annual rainfall. The techniques which may be used to accomplish this description range variously from the completely subjective to advanced statistical methods. We shall introduce some of these specific techniques in later chapters.

As an example of a very simple descriptive model, consider the relationship of gross sales of food products S to the average per capita disposable income I. We might hypothesize the form of the relationship to be

$$S = aI + b \tag{4-10}$$

which might be interpreted as indicating that individuals must spend, on the average, at least a fixed amount annually for food—b dollars—and that their additional expenditures are a fixed percentage of their disposable income, as represented by the first term in expression (4-10). If the analyst went into historical records and found that 0.05 and 225 were good estimates for a and b, the descriptive model would be

$$S = 0.05I + 225 \tag{4-11}$$

which says that, on the average, an individual spends \$225 plus 5 percent of his disposable income for food products.

This simple model illustrates an additional important aspect of descriptive models. Often, either because the analyst does not understand how the system under study may be manipulated or because there is no history of changes in controllable variables on which to base estimates of their effect on system performance, descriptive models may contain only uncontrollable variables. In the language of expression (4-6), such a model would be represented as

$$P = f\{V_j\} \tag{4-12}$$

i.e., system performance expressed as a function of a set of uncontrollable variables.

This is the kind of model which is often dealt with by economists who study the workings of the national economy. For example, there is no known relationship between the research and development expenditures of government and business and the gross national product. Yet many economists hypothesize that such a relationship exists. Consequently the use of a descriptive model involving these uncontrollable expenditures may serve to cast light on this aspect of the underlying structure of the economy.

In marketing, the effect of important controllables is often either unknown or unrecognizable. Descriptive models may then be used to sneak up on a decision problem solution. In one such situation an oil company used a descriptive model to relate uncontrollables such as traffic flow, car ownership, income, and number of competitive service stations to the gasoline sales volume in an area. The *pattern* of traffic flow through intersections was then studied from data gathered at similar service stations. The sales contribution of each of the twelve possible paths through an intersection, consisting of the three alternatives right turn, left turn, or straight ahead for each of four approach directions, was incorporated into another descriptive model. Experiments aimed at determining the reason for significant variations in the sales contribution of the twelve paths were then proposed on the basis of service stations with various combinations of number of pumps, pump location, number of entryways and exit ramps, and location of entry and exit points. In this manner the influence of controllable variables was introduced into later models.

In such circumstances it is often possible to apply this stepwise procedure to the eventual construction of a model incorporating both controllable and uncontrollable variables and their effect on performance. Sometimes enough is learned to construct an optimization model, for such is the goal of the analyst. In the meantime, however, the knowledge of the *direction and magnitude of the causal relations underlying the system which* is gained by using descriptive models is often in itself a significant aid to decision making.

Model Construction

The processes involved in constructing a model, particularly a symbolic one, are often baffling to the novice. The mathematical expressions comprising a particular model may be simple enough to understand, after the fact, but how does one create a model from nothing and decide which is best of all of the possible models which might be used?

THE MAKEUP OF MATHEMATICAL MODELS

To begin with, we may consider the *sources of the relationships which may be incorporated into mathematical models.* For these, as for most such listings which one can produce in marketing, no pretense to uniqueness or exhaustiveness is made. The sources to be enumerated here are some which have proved useful to analysts in the past and which may therefore reasonably be expected to be useful to the novice model builder.[7]

First, one may consider a class of *definitional* relationships which may bear on a problem situation. Profit expressed as sales revenue minus costs and sales revenue expressed as unit price times the number of units sold are two simple examples of marketing relationships which are definitionally correct.

Further, expressions which describe physical processes or physically invariant laws may be termed *technological* in nature. The product of a department's production rate and the time to be devoted to production expressing the total production quantity would be an example of such a relationship.

Behavioral relations, which describe the behavior of human beings, are another possible input for a mathematical model. The descriptive model of expression (4–11), which describes an aspect of economic behavior, is itself such a relationship.

Policy restrictions which are placed on decision problems may also be incorporated into models. A constraint requiring that at least 10 percent of the work force be women or a policy requiring the immediate filling of open orders can usually be translated into symbolic terms rather easily and included in a model of the situation in question.

More significant than the simple question of the potential sources of relationships for use in mathematical models is the question of which particular combination of such relationships constitutes a good model of a situation—the model selection problem.

THE MODEL SELECTION PROBLEM

One of the typical reactions on the part of the nonscientist to an explanation of the use of a model in solving a decision problem may be summed up in the expression, "Well, that's all well and good, but you've left out *x*." In that statement *x* is anything—the weather, the number of tiny tea leaves per pound, or the attitude of the Kremlin's leaders—which in the opinion of the nonscientist bears on the problem and does not explicitly appear in the model.

The scientist's retort to this would most likely be that this may be a virtue of the model rather than a defect in it. A good model accounts for all the significant factors which bear on the perfor-

[7] This categorization is based on a similar one given by R. S. Weinberg, *An Analytical Approach to Advertising Expenditures*, Association of National Advertisers, Inc., 1960. (Reprinted in the Bass et al. reference.)

mance of the system and omits all those which are unimportant.[8] As such, the abstraction of reality represented by a model is subject to comprehension and manipulation by using the tools of logic and mathematics. The complexities of the real world may prevent us from interacting in this way with anything but the model.

The determination of which factors are important enough to be included and which are unimportant enough to be excluded from the model is of central importance, of course. As we pointed out earlier, this question depends on the use to which the model is to be put. This determination forms the basis of the model selection problem.

As an illustration of the *model selection decision problem*—for, indeed, the analyst is typically faced with a decision problem involving alternative models to use in representing a system—consider a situation in which there is available a complex model which quite accurately depicts the operation of a system. Most frequently such a model will be costly to manipulate. In fact, often a very complex model may be completely beyond the scope of one's mathematical capabilities, so that no solution can be achieved by using it. Conversely, overly simple models, such as those which have been used for illustrations in this chapter, are easily manipulated and solved. Such models, however, are usually rather inaccurate representations of the systems in question.

The basic decision problem before the analyst is one of *balancing complexity and accuracy* or, more precisely, *the cost of complexity versus the cost of errors*. From the available models he should choose one which is sufficiently detailed to represent accurately the essential characteristics of the system, and yet simple enough to be solved and understood. To accomplish this, one would like a perfectly accurate model which would require only trivial solution techniques and negligible time for solution. Of course, no such model is ever available to the analyst. He must satisfy himself with a model he considers the best of those imperfect ones which are available.

The importance of a model's being simple enough to be understood—in its basic structure, if not in every mathematical detail—cannot be overemphasized. The analyst constructs and solves a model as an input to the decision maker in whom the authority and responsibility for the decision is vested. The decision maker will undoubtedly value this input in direct proportion to his understanding of the way it was obtained. No successful business execu-

---

[8] In practice, of course, we are often forced to leave out factors because we are simply unable to measure or manipulate them. If we make these falsifications explicit, however, we have not necessarily detracted from the value of the model.

tive is likely to cede his responsibility unquestioningly to some mystical mathematical construct. In order that the model and its solution have an impact on the real decision process—the one going on in the mind of the decision maker—a belief in the validity of the model must be shared by the decision maker. Such a belief can only be the product of *understanding*. On this point, a leader in the management field has said that "most managers would rather live with a problem they can't solve than use a solution they don't understand."

There is, then, no inherent goodness in a model which depends only on its accuracy and simplicity, for a "good" model, in this restricted sense, can have no impact upon real-world decision problems if it is discarded because of a lack of understanding on the part of the decision maker. In reality, aside from the abstract value gained by those who participated in the construction and solution, such a model is a totally bad one.

SOLVING THE MODEL SELECTION PROBLEM

No precise formalization or methodology is available for determining the best model to use in a particular situation. To a large extent the resolution of this problem is dependent upon the analyst's ability to estimate the degree of accuracy and costs involved. Since we do not have a perfectly accurate representation of the marketing system, we do not even have a standard against which to judge the accuracy of representation of a model. We are therefore forced to compare two models which have different degrees of abstraction, and so we have only relative indications of accuracy. Moreover, since many problems cannot be approached with a ready-made model whose solution is already known in the form of a decision rule, the analyst must be able to estimate the difficulty which will be encountered in manipulating and solving a model. It is apparent that this is not an easy task.

The analyst may generally approach this problem in one of two ways, both of which are based upon the common situation in which the available models which might be used to represent a system are variations of one another, each being constructed by adding or deleting factors from a basic system representation. He can begin with a complex, and presumably accurate, model and successively simplify it until he obtains a model which is manageable, yet still sufficiently accurate. In accomplishing this simplification, he may omit variables which appear to have little relevance to the outcomes or he may change the nature of variables and relationships to make them more manageable. For example, the use of a continuous variable to approximate a discrete-valued one or the approximation of a curvilinear function by a linear one over some restricted range usually permits one to operate on the model by using mathematical techniques which are more familiar and better understood.

More frequently, however, the analyst will begin with the simplest model that seems at all likely to be an adequate representation of the system. He can then successively introduce factors which contribute to accuracy, but only until the addition of a factor contributes less to accuracy than to the unwarranted complexity of the model. This approach of sneaking up on a good model of the system permits one to make a rather easy initial determination of the direction and gross magnitude of the effect of each falsification of the real system. Very early in the process one can often use a quite simple model to make an order-of-magnitude estimate of the potential results which will be obtained. Sometimes it is possible to determine in this fashion that the decision problem in question does not warrant the cost of extensive analysis. In low-level tactical problems, for example, often the significant costs of analysts, computers, and the like may be compared with the estimated difference in return between the best strategy and the "good" strategy to which the experienced decision maker might evolve without formal analysis. In fact, it is possible that the difference in return from the best strategy and worst strategy may not be adequate to cover the cost of analysis. In such situations the expending of analytic resources is unwise. The analyst who "analyzes himself out of a job" in this case has performed a service to the organization, himself, and his profession, for there is no better way to degrade the analytic approach in the eyes of decision makers than to proceed with complex and costly quantitative analyses of simple problems which could be approached adequately on a completely subjective basis.

In most strategic problems, however, the cost of analysis is relatively insignificant when compared to the potential returns from selecting a best strategy. In product selection and advertising, for example, the amounts of money to be gained or lost are usually so large that even small percentage gains more than warrant large analytic expenditures.

Models in the Analytic Process

In the first chapter the role of analysis in the marketing decision process was outlined. Chapter 2 included a description of the elements of the marketing decision process itself. In this section we shall make precise the role of models in the analytic process. To do this, it is helpful to view the steps which an analyst must logically take to formulate and solve a decision problem. One possible sequence of steps is given below.[9]

1. Analyze the organization and environment.
2. Formulate the problem.
3. Construct a model.
4. Solve the model.
5. Test the solution.
6. Develop controls for the solution.
7. Make recommendation and/or implement the solution.

ANALYSIS OF THE ORGANIZATION AND ENVIRONMENT

Since a part of the analysis of major marketing decision problems is often performed by persons other than the decision maker, the first step in the analytic process is to become thoroughly familiar with the organization and its environment. One valuable way for the analyst to do this is through a series of interviews which reveal how the organization operates. Each of the persons interviewed might be asked, "With whom do you communicate daily?" "When event x occurs, what do you do?" "Whom do you see if y goes wrong?" etc. The information gained may be summarized in terms of a flow diagram on which the nodes represent people or functional units and the connecting lines represent flows of information, material, etc. A portion of such a diagram is given as Figure 4-2.

[9] This formulation is similar to that given in the Churchman, Ackoff, and Arnoff reference, Chapter 1.

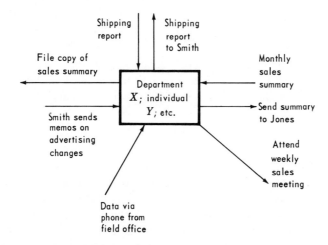

FIGURE 4-2 FLOW DIAGRAM

PROBLEM FORMULATION

A problem which is formulated is one in which all of the elements of a decision problem, as discussed in Chapter 2, are precisely defined and delineated. The identification and definition of the decision maker, the alternative courses of action which are available, objectives, and the other components of a problem result, in part, from the analysis of the organization and environment. Some of the practical difficulties involved in formulating problems precisely have already been discussed in Chapter 2.

MODEL CONSTRUCTION

After the decision problem has been formulated as precisely as possible, the model construction phase is begun and the model selection decision problem is approached by the analyst. He must identify relevant factors which bear on the performance of the system under study and determine measurable variables which represent those factors. Then he determines the relationship of the variables and the measures of performance and attempts to estimate the utility function of the decision maker. The choice of an overall best model to represent a system and the development of utility measures represent the climax of this model-building phase.

MODEL SOLUTION

After the best model has been chosen, the analyst is faced with developing a solution to the model. In the case of an optimization model this involves the use of mathematical or numerical techniques for sifting through the relevant strategies and determining a best strategy. In the case of a descriptive model the word "solution" must be interpreted more generally. In this case we might mean the manipulation of the model to put it in a form which conveys the most information in the most meaningful way.

SOLUTION TESTING

After a best strategy has been developed for the model, some test of the strategy's effect in the real world should be made, *for the selected strategy is "best" relative to the model used* and if one has made some gross error in any of the previous steps, it may not be "best" relative to the real world. Early recognition of this will permit revisions in the model and its solution.

Some degree of assurance that the selected strategy is truly best relative to the real world, as it is relative to the model, is the objective of the testing phase. Assurance may be obtained either retrospectively or prospectively. We might go back into history and attempt to calculate what the performance results *would have been* if the strategy indicated to be best by the model, rather than the one actually used, had been in effect. In doing this, we must be careful that we do not fall prey to one of the common mistakes of the novice—testing a solution to a model by using data which were used in developing the model itself.

If, for example, data from the last five years were used in estimating profit margins, demand levels, or any parameters of the model, we cannot logically take the solution to the model and test it against these same data, for it would be a poor model indeed which did not survive a test composed of circular reasoning. *We must test the solution against independent data.* Such data might be generated during a different time period, for example. One commonly used procedure is to test the solution against operational data generated during the model construction and solution phases. Often, because the quantity of such data is not adequate, we can develop a procedure which splits the available historical data into two distinct parts, one of which is used in constructing the model and one in testing the solution. Whatever the procedure for retrospective testing of the model's solution, one should ensure that some independent data base is used as a standard against which the solution may be evaluated.

On a prospective testing basis, the solution may be implemented in a limited fashion. The procedure is similar to a field test or experiment to gain insight into the value of a solution. Such a test might be conducted in a limited number of market areas, for example. In some cases the procedure called *test marketing*, which we shall discuss in detail later, may provide the basis for the prospective testing of the solution.

CONTROLLING THE SOLUTION

The analyst's satisfaction with a model and its solution should be tempered by the significant realization that a good model and solution will likely not be good forever. The model and its solution have been based upon a particular structure for the system and upon particular numerical values of relevant parameters of the system. Thus, the model often does not explicitly incorporate *the effect which changes in the structure or in parameter values might have on the system.*

If an unexpected government decision to lower import duties

which might affect the choice of a new product or a price to charge has not been incorporated into a model, some procedure should be available for identifying the potential effect of such a structural change on the model and its solution. At the very least an automatic red flag should be run up to identify changes which might reduce the validity of the model and its solution. In such a case the problem may need to be reformulated and a new model constructed and solved. In the instance of a change in labor costs, a procedure for altering the cost parameter in the model and developing an up-to-date solution should be an integral part of the model package.

These control aspects of the solution to a model are magnified in importance through the realization that model solutions in the form of decision rules are often repetitively applied by someone other than the analysts who have been responsible for the original construction and solution. To permit the model application to attain its greatest usefulness, control procedures which indicate the need for reanalysis must be provided to the people who will use the model, since one who is not intimately involved in building and solving a model cannot be expected to have the detailed understanding necessary to determine the significance of changes in the system after the original solution.

IMPLEMENTING THE SOLUTION

After a best solution to the decision problem has been determined and validated and adequate procedures have been developed for controlling the model's use over time, the solution should be recommended to the decision maker; i.e., the output of the analysis should become the input to his subjective decision process.

To ensure that this input achieves its maximum value—for indeed, it is the concise resultant of all of the factors which have been incorporated into the model and, therefore, a potentially invaluable entity—it should be supplemented with a clear understanding, on the part of the decision maker, of the model, its assumptions, and its omissions. When the decision maker has this understanding, together with the recommended strategy, he may weigh the solution along with the relevant omissions of the analysis and arrive at a strategy to be implemented. This last phase is, of course, the culmination of the analytic process, for analysis is valuable only when it has an impact on the operation of the organization.[10]

[10] A significant impact can occur in another sense, of course, if analysis prevents the use of a bad strategy. In this case the optimal strategy may be considered to be "do nothing."

This last phase vividly illustrates that the steps enumerated here are neither necessarily disjoint nor presented in chronological order. One can easily see how the decision maker's consideration of some previously overlooked factor in the last phase could lead to reformulations of the problem and the construction of models which incorporate the newly discovered factor. If one were to step at random into a marketing problem analysis which is in process, it would not always be possible to fit the observed analytic activity into one of these neat steps. Where, for instance, does the analysis of the organization end and problem formulation begin? In addition, the order of the steps is not invariant. One may constantly use information gained in one stage to aid in reworking a previous stage, so that many stages may be occurring simultaneously. The delineation of steps used here should be viewed as a simple and logical way of considering the role of models in the analytic process rather than a dogmatic order of rigid steps which one must follow.

Summary

Models of marketing systems may be used to predict the outcomes to which various combinations of the controllable and uncontrollable elements of a decision problem may lead. If an adequate model of a system is available, we may "experiment on the model" by changing the levels of its component controllables and uncontrollables. Since we are often prevented from experimenting on the marketing system itself, the use of models in this fashion offers a viable scientific alternative.

Models can be classified in various ways according to their composition and/or the use to which they are put. In all cases the categorizations are somewhat inexact, since their primary utility is for organizing one's thinking and not as precise dictionary definitions.

The process of building models often appears to be rather mysterious. Although there is still a great deal of art left in this aspect of science, a number of theoretical and pragmatic guidelines can be followed in constructing models and selecting a "best" model from the several which may be available.

In the succeeding seven chapters the application of the conceptual framework and models in various marketing problem situations will be discussed. Since the first decisions to be made in marketing a product are those related to the product itself, product decisions will be treated in Chapter 5. Then, market analysis, test marketing, purchasing decisions, pricing decisions, advertising decisions, other promotional decisions, and distribution decisions will be taken up.

EXERCISES

1. Which kind of model is each of the following?
   a. a road map
   b. a pilot plant
   c. a flow diagram
   d. a world globe
2. If a decision maker must consider four prices and five brand name possibilities for a new product, how many strategies are available to him?
3. What is the value of having several outcome descriptors, each with the same quantitative dimension?
4. What kind of information would be necessary to achieve the same value as in exercise 3 if we were faced with two outcome descriptors, each having a different quantitative dimension?
5. Consider that the two relevant outcome descriptors in a decision problem have the dimensions dollars and hours and that I feel that I know (through some mystical process) that the utility functions of the organization for money and time are $U_1(\$)$ and $U_2(t)$ respectively. One particular outcome in the decision problem involves $100 and four hours. May I simply evaluate the utility of the outcome as $U_1(100) + U_2(4)$? Why or why not?

REFERENCES

Ackoff, R. L.: *Scientific Method—Optimizing Applied Research Decisions,* John Wiley & Sons, Inc., New York, 1962.

Arrow, K. J.: *Social Choice and Individual Values,* John Wiley & Sons, Inc., New York, 1951.

Bass, F. M., et al. (eds.): *Mathematical Models and Methods in Marketing,* Richard D. Irwin, Inc., Homewood, Ill., 1961.

Churchman, C. W., R. L. Ackoff, and E. L. Arnoff: *Introduction to Operations Research,* John Wiley & Sons, Inc., New York, 1957.

Roberts, H. V.: "The Role of Research in Marketing Management," *Journal of Marketing,* vol. 22, no. 1 (July, 1957), pp. 21–32.

Weinberg, R. S.: "The Uses and Limitations of Mathematical Models for Market Planning," in *An Analytical Approach to Advertising Expenditures,* Association of National Advertisers, Inc., 1960.

CHAPTER 5

Product Decisions

In a dynamic economy such as that which we have enjoyed for the past several decades, the design, development, and marketing of new products is of tremendous importance. It is the sales dollars generated by the stainless-steel razor blades, diet colas, and cold-water detergents—items not even conceived of only a few years ago—which provide the impetus necessary to sustain the continued expansion of our economy.

The importance of new products is vividly realized when one considers that it is not atypical for a consumer product manufacturer to be deriving over 50 percent of his gross sales receipts from products marketed within the past 10 years. The dependence on new products, coupled with the fantastic failure rates for new products introduced since World War II—80 to 98 percent in some industries[1]—serves to illustrate the extent to which new product decisions pervade the marketing manager's function. Indeed, it is the unusual company which does not continually have dozens of new product proposals in various stages of evaluation.

[1] Peter Hilton, "New Product Introduction for Small Business Owners," Small Business Management Series, no. 17, U.S. Government Printing Office, Washington, D.C. The economic magnitude of these failures is estimated by Batten, Barton, Durstine & Osborn to be over $4 billion annually for ". . . ideas that did not develop or . . . products that failed . . . ." (quoted in *Saturday Review*, March 11, 1967, p. 128).

Why Introduce New Products?

To understand the product decisions which the marketing manager faces, one must understand the organizational objectives and motivations relevant to the introduction of new products.

In one popular-priced beer introduction made recently, the stated objectives were:

1. To make use of capacity not utilized by current production levels of the major product, a premium beer
2. To enable exclusive wholesalers who were distributing only the company's premium brand to participate in a larger share of the total beer market
3. To have an entry in the large and growing popular-priced beer market
4. Long-term profit

The order of these objectives is the one most frequently used by the executives who were queried by an analyst. It is interesting to note that profit was mentioned almost as an afterthought. Even then it was always noted that profit expectations were long-term; no immediate profits were anticipated. The influence of the wholesalers, who felt that the distribution of only a premium-priced brand did not give them access to the popular market, is also of significant importance. Here, some of the extraorganizational elements in the marketing system were expressing opinions with direct impact on the actions of the manufacturer.[2]

In similar instances when management was questioned concerning their motivations in introducing new products into the marketplace, the answers often incorporated the familiar words "survival," "growth," and "diversification." The latter term is undoubtedly one of the most frequently heard words in the boardrooms of large American corporations. The typically successful U.S. business has capital to invest in profit-making ventures, the problem being to choose investments which show promise of profit and the ability to sustain the company as an operating organization. What better investment opportunity than a new product which will produce profits and provide the organization with a job to perform, the managers with something to manage, and the president with something to preside over?

---

[2] We shall find that this is a pervasive feature of marketing decision problems. In many situations the decision maker does not merely need to consider what the impact of his decisions will be on the other elements of the environment but also finds that the objectives of other organizations must be taken into account in the initial stages of formulating the decision problem for analysis.

Further, what better kind of investment opportunity than a new product which is similar enough to current products to require allied technological and managerial skills, and yet different enough to respond differently from present products to changes in economic and social conditions? For example, a product which sells well during periods of economic slowdown is an effective complement to one whose sales are sensitive to economic recession, and a product which is oriented to a civilian market might well fit into the product mix of a defense contractor, since its sales would not be likely to react to changes in the world political situation.

This idea of temporal stability through product diversification interacts with another crucial sales characteristic of most products—the *product life cycle*. The typical consumer product will have a sales pattern much like that shown in Figure 5-1—a period of slow establishment in the market, a period of rapid sales increase followed by a peaking, and a long gradual decline in sales.

In most multiproduct companies the desire for a relatively stable or steadily increasing total sales volume is fulfilled by the introduction of new products, so that the total sales resulting from products in various stages of their life cycles achieve some desired level as time progresses. In the situation in Figure 5-2 as shown for 1973, product *A* is in the long declining portion of its life and product *B* is in the early stages of a short cycle. Total sales revenue

FIGURE 5-1   PRODUCT LIFE CYCLE

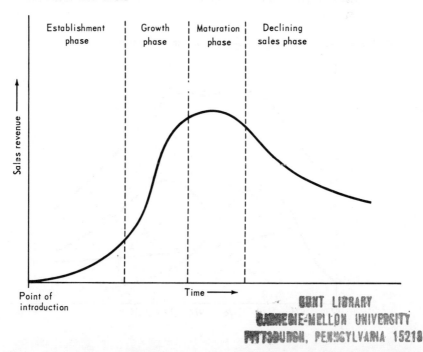

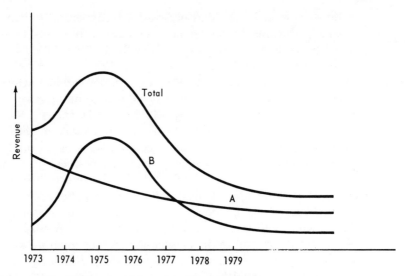

FIGURE 5-2   SALES REVENUE DERIVED FROM TWO PRODUCTS

for the two-product company is represented by the sum of the heights of the two curves. It is clear that total sales for this company will begin to fall during 1975 if no third product is introduced prior to that time. If a new product, product *C*, is developed and introduced early in that year, the situation might be that shown in Figure 5-3. In this case the introduction of a new product, whose fast-growing sales period overlaps the period of decline in sales

FIGURE 5-3   SALES REVENUE DERIVED FROM THREE PRODUCTS

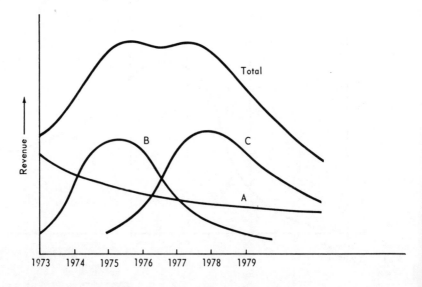

for the other two products, serves to stabilize total sales for several additional years. Another new product would need to be introduced in 1978 to maintain sales levels in the future.

Product Planning

The introduction of new products requires significant advance planning because large capital expenditures are involved in their design, development, and introduction into the marketplace and because the time lag between the generation of an idea for a new product and the marketing of the product is so great; a period of two to five years is frequently required before an idea reaches fruition in a salable consumer product.

One can distinguish three phases of product planning which must be the concern of the marketing manager:

1. Product idea generation
2. Product search
3. Product evaluation

In the first of these phases, ideas for potential new products are generated. These ideas provide the basis for the choice of a new product. Secondly, in the product search phase, these ideas are "filtered" and given preliminary evaluation to determine their gross suitability. Typically this phase centers on intangible considerations which must be met by an acceptable product idea. Finally, in the product evaluation stage, stringent economic analysis is performed and a "best" product alternative is selected from those which have survived the search phase.

Of course, these phases are not always either distinct or exhaustive. Sometimes it may be difficult to decide which of the latter two phases is represented by a particular analysis, since both can involve economic considerations. Such difficulties are of no real consequence, for it is irrelevant what label we attach to the phases. Their enumeration here serves only to illustrate the logical process which should be followed in new product decisions.

The differing nature of the three phases makes it apparent that the analytic approach to each must be carried out at a different level. The generation of ideas is a creative function which requires little analysis. In fact, the formal techniques which are sometimes called upon to aid in this creative process often explicitly preclude any analysis. Product search is largely descriptive. However, the models used in this phase need not always be of the purely descriptive kind. In effect, this phase is often carried on, as was the oil company's analysis of service stations described in the previous chapter, both to learn more about the structure of the system and to lead to the development of optimization models. Such models may then be used in the final product evaluation phase.

THE GENERATION OF PRODUCT IDEAS

Although no formal techniques are available to obviate the need for creativity in the generation of new product ideas,[3] the tremendous problems inherent in this phase of product planning should be recognized. In one company, for example, the estimate was made that forty seriously considered new product designs and ideas are required on the average to produce one marketable new product. In another study[4] it was reported that 2,100 ideas resulted in fourteen marketable products of which only two achieved a sales performance which would be rated as successful. Such fantastic rejection rates for product ideas and high failure rates for newly introduced products illustrate the tremendous quantity of ideas which is required in a typical product planning process. This is the first portion of the investment in a new product, for most of the valuable time invested in creating ideas is spent on products which never reach fruition.

It should be emphasized that these ideas are not necessarily of a very specific kind. Sometimes the "mad inventor" from the organization's laboratory might propose a specific new product which he has designed and carried along to the prototype stage. Sometimes an idea will take the form of a formal proposal concerning a rather specific consumer demand which has been detected. Quite often, however, the idea will represent a large *class* of products rather than a specific one: "Let's go into the frozen food products business" rather than "Let's freeze and sell broccoli spears in supermarkets." Often, of course, specific new product ideas may emanate from a source outside the organization. Not infrequently the marketing manager must deal with a proposal such as the one in which a major producer of beverages was asked to form a partnership with a street-corner purveyor of a tangy Latin-American pastry. The argument used by the small businessman—that the beverage and pastry complemented one another and could make use of the same distribution facilities—was one which would have done justice to a professional analyst. Some companies rely heavily on such extraorganizational idea sources. Wham-O Manu-

---

[3] The composition of poetry and the suggestion of brand names such as CITGO by computers perhaps forecast the day when the fantastic speed of the computer may be used to enumerate combinations of factors which constitute new product "ideas."

[4] These data are reported by Stillson and Arnoff (see references).

facturing Company, the firm which first marketed the Frisbie flying saucer, the hula hoop, and the Super-Ball, is an example of an organization whose major successful products have originated in the ideas of amateurs.

In fact, the utilization of inventions as a source of new products has been institutionalized to the degree that an annual International Inventors and New Products Exhibition is held at the New York Coliseum, where hundreds of inventors display their ideas to representatives from large corporations seeking marketable product ideas. Although the likelihood of finding a profitable product in this way might be rather small, it is recognized, e.g., by one new product manager quoted by *Business Week:* "Don't forget that Xerox and Polaroid grew from inventions made in someone's basement."[5] The degree of success achieved in bringing together feasible product ideas with potential producers is illustrated by the approximately $20 million in sales of goods and patents, licensing agreements, distribution rights, and other long-term arrangements which result from this single "new product fair." Moreover, there are a number of companies which are in the new product idea business. Their methods of operation are varied, but several restrict themselves to generating and developing an idea to the point where it can be licensed to another company to manufacture and market.

Rather than placing total reliance on outside ideas, however, most firms have some formal or informal group in which the responsibility for new product ideas is vested. In many organizations, of course, laboratories are constantly pursuing applied and basic research which can eventually lead to new products. However, the kind of creativity involved in producing a new durable plastic is different from that involved in foreseeing use of the plastic in the manufacture of shoes; although it may be the scientist who frequently displays the former kind of creativity, it is not necessarily he who possesses the latter.

One way of aiding the nonscientific creative process which is so necessary for the generation of new products is a technique called *brainstorming*. In a brainstorming session a group is called together and encouraged to discuss possible new products or fruitful areas for new products. Blue-sky thinking aloud is encouraged. No critical analysis of ideas is permitted, although it is hoped that each wild idea may lead to another and eventually, perhaps, to a radical idea which has merit. The interaction of the individual members of the group is believed to have a stimulating effect and often, during later analysis, when most of the recorded ideas put forth in the brainstorming session are discarded as impractical,

[5] *Business Week,* Sept. 17, 1966, p. 117.

a few ideas which merit study are uncovered. Some who have participated in apparently fruitless sessions have felt that, at the very least, the sessions served to open new avenues of thought in the minds of participants who might later produce meritorious new product ideas.[6]

The organization should not view the concept of a new product idea so narrowly as to preclude specific ideas for new products that may actually only encompass *changes in existing products.* The "creation" of new products, through changes in style, appearance, packaging, or brand name serves to define "new" products from the marketing point of view, since the appeal of the transformed product may be to a different segment of the market. Typically the creation of new products in this manner is less costly than the development of a radically different or innovative product. Manufacturers of women's apparel, who have made a way of life of product style changes, give evidence of the potential returns of this sort of thinking.

Similarly changes in the design of existing products which encompass more significant departures than simple style or brand name, such as power brakes rather than standard auto brakes, are analogous to the introduction of a new product from the viewpoint of the marketing manager, since they may require different promotional emphasis, distribution techniques, etc.

The cranberry industry offers a good example of product changes of this kind. The cranberry was previously a seasonal delicacy, consumed primarily during the Thanksgiving and Christmas seasons. Now, by extensively promoting such derivatives as "cranapple" juice, "cramwiches," cranberry juice cocktail, etc., the industry has increased sales volume and reduced the seasonal sales pattern.

In fact, one may go so far as to describe a drastic change in the attitude of the consuming public toward a product as being equivalent to the introduction of a new product. One good illustration is the change in attitude toward pancakes resulting from an "educational" campaign which stressed novel recipes and comple-

---

[6] Pseudosophisticates may respond to this emphasis on brainstorming with the comment that it is old hat. Nonetheless the pragmatist will find that the basic idea is both useful and used (although the term "brainstorming" may not always be applied). A good illustration of its utility is the important role brainstorming plays in the operations of Van Dyck Corp., a company specializing in devising new products for clients such as Olin Mathieson, J. C. Penney, and Textron. *Business Week* (July 2, 1966, pp. 52–54) describes Van Dyck's brainstorming sessions: "the staff bats around ideas, and scrawls them down on scraps of paper, which are tossed into a huge fishbowl. . . . Any idea goes in if it has aroused even a glimmer of response from the group. Later on, a two-man team—always one engineer and one industrial designer—cull out the most promising candidates."

mentary foods for pancakes at any meal. The concept of pancakes as a versatile food rather than a traditional breakfast food resulted in new sales life for the product.

PRODUCT SEARCH—A BASIC APPROACH

The search for a product which is best in terms of the organization's objectives begins by focusing on the general acceptability of each of the ideas which have been put forth. Typically such a preliminary analysis concentrates on intangible considerations, such as general compatibility with the existing production and distribution facilities, together with gross economic factors, such as the potential competitive nature of price, etc. Initial insight into such factors may often be gained through a series of questions asked by the analyst of himself, the decision maker, and experts in the various areas of concern. Some questions which might be asked about a new product idea are:

1. Can the present production facilities be utilized?
2. Can present raw materials be utilized?
3. Can the product be distributed through the present organization?
4. Does it make use of the know-how of our present organization?
5. Will it be feasible to produce it in a relatively small number of sizes, grades, etc.?
6. Can it be priced competitively?
7. Is there an existing demand?
8. Will the demand continue into the future?

There are no simple good or bad answers to such questions. In general, positive responses might appear to be favorable for most companies. Products which fit into the existing raw material, production, and distribution scheme would appear to be preferable to those which do not. Hence, an oil company might consider products which use oil as a raw material, require refining, and may be sold through service stations as preferable to products not meeting these criteria. In fact, however, the list of potential products meeting all of these criteria is rather limited. As a result several major oil companies have broadened the limited view of raw materials, production facilities, and the distribution organization by including their clerical and managerial resources and correspondence with credit-card customers. One result has been the marketing of travel insurance policies. This new product utilizes many of the human resources already available to the companies as "raw material" and "production capacity." In addition, the sale of these policies through existing postal correspondence with credit-card customers represents a broad view of promotion and distribution facilities. The possibility of direct insurance marketing through the standard outlets—the service stations which sell the company's brand of gasoline and oil products—has also not been omitted from consideration.

This example illustrates how a broader view of the function of an organization can lead to feasible new product ideas. The oil company thinks of itself as in the business of travel rather than the oil business. The result is travel insurance and arrangements with food processors to distribute hot meals to weary travelers at their service stops. The insurance company, on the other hand, thinks of itself as being in the business of finance rather than the insurance business and begins to invest its capital in new ventures and to develop new insurance policies to meet the changing needs of an affluent society.

In any case the answers to questions concerning the feasibility of new product ideas, such as those posed, should be interpreted in two ways. First, *are the definitions of the present organization broad enough?* Secondly, *do positive responses to the questions really reflect preferences which are consistent with the organization's objectives?* If diversification is a primary organizational goal, positive answers to questions concerning the compatibility of a new product with the existing system might reflect negatively on the product idea. The primary overall consideration must always be, *What are the objectives of the organization and how compatible is the product idea with them?* rather than the more restricted view of always finding new products which fit in well with the existing system. Of course, in a great many situations the feasibility of new product ideas will be enhanced if they are compatible with the existing system, for the objectives of many companies in new product searches are of this nature rather than diversification for its own sake.

The answers to questions relating to the objectives of the organization and the compatibility of a product idea with those objectives can themselves often serve as a gross-level filtering device for the preliminary evaluation of ideas. A qualitative delineation of a set of ideas which are clearly feasible may well be the result of the questioning process.

In general, however, one of the central difficulties of marketing decision making precludes total reliance on this simple process. That difficulty—*the problem of qualitative outcome descriptors with no easily discernible utility function*—is of primary importance to product planning. Consider, for example, the product idea which is determined to fit well into the existing production and raw material procurement setup but poorly into the distribution organization, as contrasted with one which may be easily distributed but requires new sources of raw materials and new production facilities. Here we may think of the ability to fit into the production, raw materials, and distribution setups as three qualitative outcome descriptors in a decision problem under certainty. The outcome table, based on only these two product ideas as the available alternatives, might be that shown as Table 5-1.

TABLE 5-1   OUTCOME ARRAY

Strategy	*State of nature*	
Product idea *A*	Production	yes
	Raw materials	yes
	Distribution	no
Product idea *B*	Production	no
	Raw materials	no
	Distribution	yes

The difficulty in choosing between these two alternatives emanates from the difficulty in estimating a utility function which will permit the construction of the corresponding payoff table. If we could estimate the utility of the qualitatively described outcomes "yes-yes-no" and "no-no-yes," in this case, we could easily determine which of the two ideas is preferable. It should be apparent that this is no trivial task.

To resolve the dilemma, it is necessary to develop a method for approximating organizational utilities in situations involving intangibles such as those incorporated in the questions discussed earlier. Since intangibles are involved, any such method must rely heavily on the judgments of the people who are involved—the decision maker, other executives, and technical experts. However, their judgments should not be garnered in such a fashion that they are simply being asked to choose a best product alternative from an outcome array. To do this would be to defeat the purpose of the conceptual framework.

In reality this is a basic point which is appropriate to the operational use of the conceptual framework. The framework is objective and logical in itself, and yet it incorporates subjective judgment in the utilities which it employs. An obvious way of estimating utilities is, therefore, to ask individuals for their judgments. To do so, however, one must be cautious that the judgments asked for are not so complex as to subsume the decision problem itself. After all, we might simply ask the decision maker to view the outcome array and adjudge the best alternative. This would obviate the need for the construction of the payoff table. But it would also be a foolish course, for we have already argued that logical expressions of preference are difficult, if not impossible, for a human being to make when he is presented with a number of levels of satisfaction in a number of dimensions. To be useful in product search, a utility estimation device must therefore involve judgments concerning preferences and future courses of events which are made at an elementary level.

Indeed, the opportunity to utilize such judgments is one of the strong points of objective analysis. Since only elementary judgments are involved, various individuals who are experts in the several areas may be called upon to make judgments related to the same decision problem. Nonscientific decision making usually requires that one person make overall judgments involving economic, sociological, technical, and other elements in which he has no expertise. Also, the making of elemental judgments serves to distinguish value judgments clearly from predictions of future happenings—something not so easy to do when overall judgments are being made.

DEVELOPING UTILITY MEASURES FOR PRODUCT SEARCH

The previous example illustrates some of the difficulties involved in the consideration of intangibles in the product search phase. The questions which are asked often have no simple *yes* or *no* answers. In any case the relative importance of each of the factors treated in the questions is usually unknown. For example, is a product which fits into the production setup but not into the distribution network more valuable than one which fits into the distribution network but cannot be produced with existing facilities? Many difficulties would be overcome if a comprehensive measure of worth could be developed on the basis of these intangible considerations. Such a measure would be a kind of utility measure in that it would express the organization's preference for products.

To formulate such a measure, it is necessary to call upon executives and technical experts for expressions of their opinion and judgment, for it is they who best exemplify the organization's preferences and it is they who are the best available sources of information concerning the organization's values and the course of the future. Indeed, we should always remember that we are dealing with product ideas which exist only in the minds of men and that the outcomes which we need to consider are the predictions of the states of affairs which would result *if* the products were produced and marketed.

To facilitate the rendering of meaningful opinions about *the states of affairs which will occur* and *what the organization's prefer-*

ences are for those states, the individuals to be queried should always be presented with oversimplified situations about which to make judgments. It is preferable that each of them be presented with a series of one-dimensional evaluative and comparative situations rather than a single multidimensional one, for example.

Most product search situations may be considered in terms of risk situations—those in which it is meaningful to attach likelihoods to the various outcomes which may occur. In this section we shall illustrate the development of a utility estimate for product search which incorporates general intangibles in a risk environment. It will be beneficial to accomplish this illustration by using an example of rather wide applicability.

Suppose that general factors which we shall call marketability, durability, and productive ability are chosen as those basic intangibles which are relevant to the organization's objectives and are therefore to be used as a basis for preliminary filtering of new product ideas.[7] The precise determination of the factors involved should be the first step in attempting to determine the utility to attribute to an outcome or the expected utility to attribute to a product idea. In the area of marketability, for example, one might be concerned with the relationship to present distribution channels, the competitive quality-price relation, and the effect on the sales of existing products. The sales durability of a product might be evaluated in terms of the size of the potential market, the resistance of sales to economic fluctuations, and the life cycle. Productive ability might be defined as the equipment, personnel, and raw materials which will be necessary to produce the product.

Having made this specific delineation of the factors which relate to the organization's objectives, we may measure the degree to which each satisfies the objectives in terms of qualitative levels such as "good," "very good," "poor," etc. While we are still dealing with qualitative terms like the *yes* or *no* answers sought in the basic approach, the difference here is that we shall then proceed to describe, as precisely as possible, the conditions which warrant each level. Normally a subjective approach using two levels of attainment would not be at all precise.

Descriptions of these levels for the marketability factors might be those in Table 5-2. The third row in that table describes the effect which the new product might have on the sale of existing products in ordinal terms, i.e., "significantly positive," "small posi-

---

[7] An example along these lines is presented in J. T. O'Meara, Jr., "Selecting Profitable Products," *Harvard Business Review*, vol. 39, no. 1 (January–February, 1961), pp. 83–89. Some of the presentation in this section is adapted from his paper. The papers by Kline and Richman (see references) also follow the same logical lines.

TABLE 5-2   MARKETABILITY FACTORS AND LEVELS

Factor	Level				
	Very good	Good	Average	Poor	Very poor
Relation to present distribution channels	Uses present channels exclusively	Some new channels required	Equal distribution between new and existing channels	Only a few existing channels used	Entirely new distribution channels necessary
Competitive quality-price relation	Can be priced below all competing products of similar quality	Priced below most competing products of similar quality	Priced same as competitive products of similar quality	Priced above many competitive products of similar quality	Priced above all competing products of similar quality
Effect on sales of existing products	Will significantly aid sales of existing products	Possible small positive effect on sales of existing products	No effect	Possible small negative effect on sales of existing products	Will significantly reduce sales of existing product

tive," "none," "small negative," and "significantly negative." While the terms are still rather nebulous, they serve to define the various levels more precisely than do the levels "good," "average," etc. The levels of some other factors can be described more precisely. The "competitive quality-price relation" levels and those of the "relation to present distribution channels" involve numerical counts of one form or another, for example.

Descriptions of factors relating to durability and productive ability may be similarly displayed as in Tables 5-3 and 5-4. There again the objective is to specify the levels in terms more precise than the "good," "average," etc., descriptors.

Once the various levels for each of the factors have been defined, they should be numerically scaled in an arbitrary, but consistent, fashion which indicates their relative worth. We might begin by assigning a weight of 10 to the "very good" level of the first marketability factor—the relation to present distribution channels. Next, we would ask the relevant individuals about the relative worth of the "good" and "very good" levels and attempt to arrive at a consensus numerical weight for the "good" level. This weight must be lower than 10 since the "good" level is preferentially inferior to the "very good" level. If the use of a few new channels

TABLE 5-3 DURABILITY FACTORS AND LEVELS

Factor	Level				
	Very good	*Good*	*Average*	*Poor*	*Very poor*
Size of potential market	Large national and foreign market	Wide national market	Regional markets only	A single regional market	A very specialized narrow market
Resistance to economic fluctuations	Will sell well in inflation or depression	Moderate effect of economic cycles	Sales will follow economic trends generally	Heavy dependence on economic conditions	Total dependence on economic conditions
Life cycle	Can expect to sell product indefinitely	Longer life cycle than competitor's products	Average life cycle for industry	Shorter life than competitor's products	Will rapidly become obsolete

TABLE 5-4 PRODUCTIVE ABILITY FACTORS AND LEVELS

Factor	Level				
	Very good	*Good*	*Average*	*Poor*	*Very poor*
Equipment necessary	Utilizes excess equipment capacity	Utilizes present equipment	Some additional equipment necessary	Mostly new equipment but some existing equipment	All new equipment necessary
Personnel necessary	Utilizes present personnel exclusively	Few new people needed at low levels	Many new people except high-level key people from present organization	Mostly new people except existing low-level personnel	All new personnel necessary
Raw materials necessary	Same raw materials; can be purchased from present best supplier	Same raw materials; must obtain new suppliers	New raw materials; have contacts necessary for efficient procurement	New raw material; some experience available	New raw material from unknown suppliers

("good") is not deemed much worse than the sole use of existing channels ("very good"), the number 9 might be assigned to the "good" level of this factor. Similarly we should obtain opinions as to the relative worth of the other levels and assign weights to each, say 5, 3, and 1 for the "average," "poor," and "very poor" levels.

The other factors should be similarly weighted, beginning with 10 units for the "very good" level of each. One simple way to accomplish this might be to define the levels of all factors in such a way that, as nearly as possible, they have the same relative worth, i.e., so that the same weight applies to the "very good" level of all factors, the same weight to the "good" level of all factors, etc. In many cases it will not be practical to do this, of course, and it will be necessary to assign different weights to the various levels of each factor.

It should be emphasized that *these weights should be independent of the worth of the factor itself.* In assigning them, we should concern ourselves only with the *relative worth within the factor*, regardless of the importance of the factor itself. The overall utility measure must also account for the relative importance of the factors, of course.

To weight the factors numerically, we might simply divide 100 units of worth among the factors according to the decision maker's estimate of their relative importance to the organization's objectives. The factors described by the previous tables might be weighted as in Table 5-5. In this weighting of factors it is apparent that relatively little emphasis is being put on the existing equipment and raw material resources, a bit more on personnel resources, and primary emphasis in the area of marketability. The organization in question might be one which is seeking profitable diversification rather than compatibility.

TABLE 5-5   FACTOR WEIGHTINGS

Factor	Weight
Relation to present distribution channels	5
Competitive quality-price relation	10
Effect on sales of existing products	15
Size of potential market	20
Resistance to economic fluctuations	10
Life cycle	20
Equipment necessary	5
Personnel necessary	10
Raw materials necessary	5
Total	100

The difference between these two weightings should be noted. The former is an expression of relative preference for various levels of attainment for the various factors which is *independent* of the importance of the factor. To obtain this, we ask ourselves, "If the only criterion to be used were this factor, what would the relative worth of each level be?" The latter weighting of the factors themselves is an expression of the relative importance of the factors to the organization.

For a particular product idea to be evaluated, the likelihood of achievement of each of the levels for each of the factors must be estimated. This may be done on a subjective basis simply by questioning the concerned executives and technical experts who have specialized knowledge about the various factors.[8] These likelihood estimates may then be entered, along with the two weightings discussed previously, in a tabular form such as Table 5-6. In this table we have assumed for simplicity that 10, 8, 6, 4, and 2 relative weights hold for the levels of all factors, i.e., that the descriptions of the levels have been so made that the differences in relative preference are approximately equal for all factors. The last two columns of this table are to be discussed later and may therefore be disregarded at this point.

The probability entries in Table 5-6 are for either a particular product idea (e.g., refrigerators) or an idea concerning a class of products (e.g., major electrical appliances). In the case of a product class it is undoubtedly best to obtain the likelihood estimates from questions asked about a specific product which is considered representative of the class, for we all tend to think more logically in terms of specifics rather than generalities.

The weights which have been developed are entered at the top of Table 5-6 (for the factor levels) and at the left side (for the factors). These weights are determined from analysis of the organization's *general objectives.* The probabilities in the table on the other hand describe opinion concerning a *particular new product idea.* The table may be interpreted as saying that the likelihood that the particular product idea in question will exclusively use existing distribution channels is almost certain (0.9) and that with small probability (0.1) some new channels will be required. It is virtually impossible that the descriptions of the "average," "poor," or "very poor" levels will be applicable to this product. Similarly, the experts are certain that there will be no positive effect on the sale of existing products.

[8] It should be noted that in obtaining these judgments, we are switching from value judgments to judgments involving the future course of events. Hence, our process has served to distinguish effectively between these two vastly different judgmental processes.

TABLE 5-6　PROBABILITY TABLE

Factor	Factor weight	Level Very good (10)	Good (8)	Aver- age (6)	Poor (4)	Very poor (2)	Expected level weight	Contribu- tion to total expected utility
Relation to present dis- tribution channels	5	0.9	0.1	0.0	0.0	0.0	9.8	49.0
Competitive quality-price relation	10	0.2	0.2	0.4	0.2	0.0	6.8	68.0
Effect on sales of existing products	15	0.0	0.0	0.2	0.4	0.4	3.6	54.0
Size of poten- tial market	20	0.2	0.4	0.2	0.1	0.1	7.0	140.0
Resistance to economic fluctuations	10	0.0	0.1	0.5	0.2	0.2	5.0	50.0
Life cycle	20	0.9	0.1	0.0	0.0	0.0	9.8	196.0
Equipment necessary	5	0.8	0.1	0.1	0.0	0.0	9.4	47.0
Personnel necessary	10	0.8	0.1	0.1	0.0	0.0	9.4	94.0
Raw materials necessary	5	0.6	0.2	0.2	0.0	0.0	8.8	44.0
Total expected utility								742.0

The overall utility measure can then be derived by multiplying the factor weight, the weight attached to a particular level, and the probability of achieving that level, and summing across all factors and levels. We may view this in two steps. First, we obtain an expected level weight for each factor, which is independent of the importance of the factor, by multiplying each level weight and probability and summing across the five levels. This expected level weight is given in the next-to-last column of Table 5-6. The calculation for the first factor is

$$10(0.9) + 8(0.1) + 6(0.0) + 4(0.0) + 2(0.0) = 9.8$$

The contribution of each factor to the total expected utility may then be incorporated by multiplying its weight by the expected level weight and summing over all factors. These individual contributions are given in the last column of Table 5-6 as the factor's "contribution to total expected utility." The total expected utility for the product idea in question is the sum (742.0) at the lower right.

The total expected utility is a number which is derived from a number of subjective judgments concerning rather elementary aspects of the entire decision problem. It represents an *overall measure which may be used as a basis for the comparison of product ideas* on a gross level, for it can be considered an approximation to an underlying utility measure which logically combines expectations related to organizational objectives, values and predictions of future events. Its usefulness is amplified if one recognizes that it is derived from simple human judgments which are aggregated as an overall measure and that the errors involved in estimating utilities for product search are of such magnitude that one should not conclude that an idea which scores a total expected utility of 738.0 is necessarily inferior to the idea which we have evaluated, with a total expected utility of 742.0. It would probably be quite reasonable, however, to conclude that a product idea with a total expected utility of 525.0 is inferior to both of these.

Interpreting the Utility Measure in Terms of the Conceptual Framework. To demonstrate the place of the estimated utility measure in the conceptual framework, we may use an oversimplified two-factor–two-level situation. Table 5-7, which gives the prob-

TABLE 5-7   PROBABILITY TABLE

Factor	Factor weight	Level		Expected level weight	Contribution to total expected utility
		(10) Good	(5) Bad		
F_1	60	0.8	0.2	9.0	540.0
F_2	40	0.3	0.7	6.5	260.0
Total expected utility					800.0

abilities for a product achieving the "good" or "bad" level of each of two factors F_1 and F_2, is completely analogous to the larger and more realistic data given in Table 5-6. In this table the factors are given weights of 60 and 40, the levels of both factors are weighted 10 and 5 respectively, and the total expected utility is calculated to be 800.

In the language of the conceptual framework there are four possible *outcomes* in terms of these two factors. The four outcomes may be enumerated as

O_1: (good on F_1, good on F_2)
O_2: (good on F_1, bad on F_2)
O_3: (bad on F_1, good on F_2)
O_4: (bad on F_1, bad on F_2)

Since an outcome is simply a manifestation of a state of nature, we may view the strategy (the product idea in question) in terms of its likelihood of achieving each outcome. If the factors are independent in the probability sense, these probabilities are simply the *product* of the relevant probabilities given in Table 5-7. The probability for O_1 (good on both F_1 and F_2) is simply 0.8 times 0.3, or 0.24, the probability for O_2 (good on F_1 and bad on F_2) is the product of 0.8 and 0.7 or 0.56; and the probabilities for O_3 and O_4 are 0.06 and 0.14 respectively.

Moreover, if value (utility)-wise independence exists among the factors, the products of the weight associated with each factor and the weight for each level should be added to obtain the weight to be associated with each outcome (since each outcome is a combination of all factors at specific levels). For instance, O_1 incorporates F_1 at the "good" level, which has an associated product of 60 times 10, or 600, and F_2 at the "good" level, which has a product of 40 times 10, or 400. These products, together with the probabilities for the outcomes, are summarized in the first three columns of Table 5-8.

TABLE 5-8   EXPECTED UTILITY CALCULATED BY USING OUTCOMES

Outcome (F_1, F_2)	Probability	Products of weights for each factor–level	Sum of products	Expected sum
O_1 (good, good)	0.24	(600, 400)	1,000	240.0
O_2 (good, bad)	0.56	(600, 200)	800	448.0
O_3 (bad, good)	0.06	(300, 400)	700	42.0
O_4 (bad, bad)	0.14	(300, 200)	500	70.0
Total				800.0

The fourth column of this table expresses the valuewise independence of the factors, since it evaluates each outcome as the sum of the product weights associated with each specific factor–level included in the outcome. In the last column we simply multiply the probability times this sum and add over all possible outcomes to arrive at an overall total of 800.0—the same value as in Table 5-7.

This demonstrates, for this situation, the equivalence of the outcome-oriented approach of the conceptual framework for risk and the operational approach which we previously presented. In both cases we are evaluating a product idea on the basis of expected utility. The operational approach is framed in terms in which information may be gained from questions about simple situations. The approach fits into the conceptual framework, as demonstrated, through the use of two assumptions: that the factors are probabilistically independent and valuewise independent. This permits us to *multiply* the probabilities associated with each of the factor-level combinations in each outcome and to *add* the utilities (weights) associated with all of the factor levels comprising an outcome.

One practical virtue of the operational procedure (and hence, of making these two assumptions) may be appreciated by observing that there are a total of about 75 numerical entries in Table 5-6, which describes a realistic situation of nine factors, each at five levels. In this situation there are 5^9, or 1,953,125, possible combinations of levels for the factors, and hence this number of possible outcomes. It would be impractical not only to apply the conceptual framework directly by querying individuals concerning each outcome, for each outcome is quite complex, but also even to enumerate all of the outcomes!

Summary of the Operational Technique for Estimating Utility Measures in Product Search. We may summarize the operational technique in a series of steps to be followed. To form a basis for evaluating product ideas, we should:

1. Define intangible factors in terms of the organization's objectives (as in the first column of Tables 5-2, 5-3, and 5-4).
2. As precisely as possible, describe various degrees of worth for each factor to serve as the levels (as in Tables 5-2, 5-3, and 5-4).
3. Numerically scale the relative worth of every level for each factor, considering only attainment within the factor and not the relative worth of the factors themselves (as in the heading of Table 5-6).

4. Weight the relative importance of the factors independently of the level of attainment within the factor (as in Table 5-5).

To evaluate a particular product idea or class of ideas, we should choose a specific product (as representative of the class), and continue step by step:

5. Estimate the likelihood of achieving the various levels of each factor with the product idea in question (as in Table 5-6).
6. Find the expected level weight by summing the products of the level weights and probabilities for each factor (as in the next-to-last column of Table 5-6).
7. Find the contribution to total expected utility by multiplying the factor weight and the expected level weight (as in the last column of Table 5-6).
8. Sum these contributions to obtain the total expected utility (742.0 in Table 5-6).

We may then use the estimated expected utility as an indicator of the worth of a product idea. If we use the same basis for estimating utilities for a number of ideas, the estimate of expected utility may be used as a quantitative basis for comparing them. In this case we should be careful not to read too much accuracy into our estimate, since our sole objective in product search is to delineate ideas which appear to be promising enough to warrant detailed analysis.

PRODUCT EVALUATION

Having made the preliminary qualitative and quantitative investigations into new product ideas in the product search phase, the organization is faced with the problem of choosing the best product idea from those which have survived the filtering process. This represents the basic decision problem of product planning—one which must be approached through detailed economic analysis.

The Break-even Approach. One of the simplest approaches to the economic evaluation of a new product idea is through a *break-even analysis*. In this approach the organization attempts to determine the quantity of the product which it will have to sell in order to break even.

The analysis may be viewed graphically as in Figure 5-4, where we have chosen to view costs, revenues, and profits for the first year of sale for the product as a function of the first year's sales quantity s. The horizontal line at f dollars represents the *fixed costs* which must be invested in design and development of the product, tooling for production, advertising, and the like. These are costs which are necessary to produce and market the product. *They are incurred as a result of the decision to produce the product and are not affected by the quantity of the product which is sold.* Hence they are represented as a horizontal line on the graph; whether one unit is sold or a million units are sold, the same fixed cost f must be incurred. Since we are viewing these data on a first-year basis, any expenditures, such as for new machines and tooling, which will be "used" for more than one year should be amortized and the first year's value included in numerical calculation of f.[9]

The *variable costs of production and marketing* are those costs which vary, in total, with the quantity produced or sold. In this simple analysis, we shall consider that the quantity produced is equivalent to the quantity sold. The variable costs include material costs, labor costs, shipping costs, etc., which are essentially investments in each unit which is produced and sold. Hence, their total is dependent upon the total number of units sold. Assuming a fixed unit cost, c dollars per unit, the total variable cost is

$$TVC = cs$$

which is plotted *above* the level of fixed cost f in Figure 5-4. Because the total variable cost function is plotted above f rather than above the coordinate axis, its height represents the sum of the two kinds of costs—fixed and variable—or total cost. In symbols,

$$TC = TVC + f = cs + f \qquad (5\text{-}1)$$

gives the total costs associated with each sales quantity s.

The revenue R to be gained from the sale of a quantity s, assuming a selling price of p dollars per unit, is

$$R = ps \qquad (5\text{-}2)$$

or simply the unit price times the sales quantity. Plotting this in Figure 5-4, we note that the revenue function crosses the total cost function. This is because the unit selling price p, which is the slope of the revenue function, is taken to be greater than the unit cost c, which is the slope of the total cost function. At this point of intersection, *total cost equals revenue,*

$$TC = R \qquad (5\text{-}3)$$

[9] See the Grant and Ireson reference. The use of first-year costs is not a necessary past of the procedure; any reasonable time period might be used. Often, the time period is made the output by calculation of a *pay back period*—the duration of sales activity which will be necessary in order for cumulative expenditures to equal cumulative revenues.

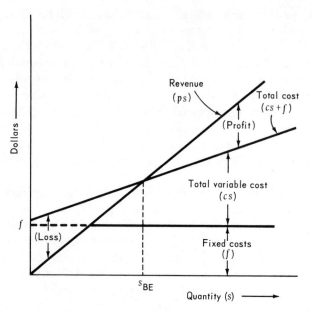

i.e., at this sales quantity, represented by the intersection of the total cost and revenue function, *all of the costs (fixed and variable) which are expended are exactly matched by revenues returned.* This point of intersection is therefore called the *break-even point* and the corresponding sales quantity is called the *break-even sales quantity.*

To determine the break-even sales quantity, we simply use expression (5-3) which, in terms of the quantity *s*, is

$$cs + f = ps$$

and solve for *s*. Since this value of *s* is the particular value at which we break even, we shall label it s_{BE}.

$$s_{BE} = \frac{f}{p - c} \tag{5-4}$$

The interpretation of expression (5-4) is simple. The quantity $p - c$ is the unit profit; hence, we may rewrite expression (5-4) as

$$(p - c)s_{BE} = f$$

This demonstrates that s_{BE} is the *number of units which must be sold in the first year at a unit profit of $p - c$, in order that the fixed costs f be recouped by the profit.* s_{BE} is the quantity which must be sold in the first year because we have included only the first year's portion of the total fixed costs in f.

An example will illustrate this point. Suppose that we plan to spend $40,000 on the first year's promotion of the product and that we must purchase a new machine for $50,000, which we expect to be useful for five years. Using a straight-line amortization policy, we would write off $10,000 per year for each of the first five years.[10] Assume further that all other fixed costs which are attributable to the product, such as heat, light, service allocations, etc., amount to $13,000 for the first year. If the unit price p is $1.50 and the unit cost is $0.80, we have f as the sum of the promotional expenditure, the first year's amortized portion of the machine cost, and the other fixed costs, as

$$f = \$40,000 + \$10,000 + \$13,000 = \$63,000$$

The equation for total cost is [from expression (5-1)]

$$TC = 0.80s + 63,000$$

The revenue calculation is [from (5-2)]

$$R = 1.50s$$

Hence, the break-even point is the solution s to expression (5-3), i.e., the solution to

$$0.80s + 63,000 = 1.50s$$

The solution is [from (5-4)]

$$s_{BE} = 63,000/0.70 = 90,000$$

which says that we must sell 90,000 units in the first year in order to break even.

To use this break-even quantity in evaluating the product idea, we must question the likelihood of achieving sales of this level. Often knowledgeable people can draw upon their experience to produce penetrating analogies. If the product in question were rather specialized and a market analysis estimated that there were only 300,000 potential single-time buyers in the nation, we might judge it to be unlikely that we would sell to almost a third of these in the first year. On that simple basis we might decide against the product idea. On the other hand, if we knew of a latent demand for 60,000 units of the product and the existence of other potential buyers, we might decide to produce the product on the basis of

---

[10] Of course, we might amortize the $50,000 over its useful life or a life chosen on some other basis.

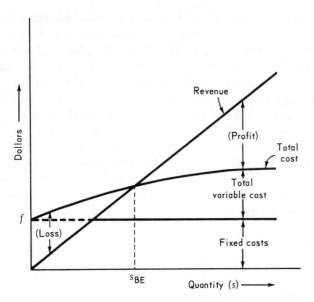

FIGURE 5-5   BREAK-EVEN ANALYSIS WITH NONLINEAR COSTS

a high likelihood of reaching the break-even sales point. The operational use which we generally make of these break-even quantities, in terms of the conceptual framework, is as a device for further filtering of the product alternatives (strategies). We may discard some ideas as totally impractical and retain others as potentially good, for organizations are usually actually looking for a set of good product alternatives rather than a single best alternative.

Returning to Figure 5-4, we note that the difference between revenue and total costs represents profit (or loss). The difference between the total cost and revenue functions, for quantities where cost exceeds revenue, is labeled "loss" and the difference between these lines in the upper right, where revenue exceeds total costs, is labeled "profit." It is not at all necessary that the total variable cost should be a straight-line function of s, of course. For example, economies of scale in production and marketing might determine unit profits which rise as a greater quantity is produced and sold. Figure 5-5 shows such a case. Note that to the right of the break-even point, the profit rises at an increasing rate (up to some point which might represent the production capacity which is available). There is no basic difference in the break-even analysis which one might conduct in this case, except that the total cost expression will be somewhat more complex.

Break-even Analysis with Fixed Rate of Return. The simplest kind of break-even analysis does not explicitly incorporate the organization's profit motive. Indeed, the business enterprise does not have an objective of simply breaking even, but of making a profit.

In terms of the basic profit relationship

$$P = R - TC \tag{5-5}$$

i.e., profit equals revenue minus total costs, one can see that the equality of revenue and total costs at the break-even point, as determined by expression (5-3), implies that *profit is zero at the break-even point.* Of course, the firm's implicit objective is *to at least break even* or *to achieve at least some minimum return.* The latter objective may easily be incorporated into a break-even analysis.

For example, if the organization set a minimum acceptable rate of return of $100z$ percent on gross sales revenue, where z is a decimal between 0 and 1, the *break-even point for minimum acceptable return* may be determined from the basic profit relation written as

$$zR = R - TC \tag{5-6}$$

i.e., total profit must equal z times R at the break-even-for-minimum-acceptable-return point. Of course, as noted previously, the organization really desires profit to be at least as great as z times R.

In terms of the sales quantity s, expression (5-12) is

$$zps = ps - (cs + f)$$

Solving this equality for s, we obtain the break-even quantity for minimum acceptable return s_{BER}.

$$s_{BER} = \frac{f}{p - c - pz} \tag{5-7}$$

Interpreting this quantity in the same fashion as we did s_{BE}, we see that the break-even point for minimum acceptable return is reached when a sufficient quantity is sold so that the gross profit minus the minimum acceptable return is sufficient to recoup the fixed costs f. This may be further clarified by writing (5-7) as

$$[(p - c) - pz]s_{BER} = f$$

in which $p - c$ is the gross *unit* profit and p times z is the minimum acceptable *unit* return.

A graph of this situation is shown in Figure 5-6. The "pure" break-even point s_{BE} is shown as in Figure 5-4. The break-even point for minimum acceptable return s_{BER} is at the intersection of the

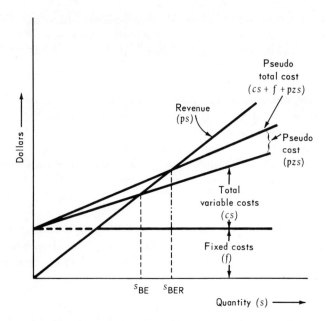

FIGURE 5-6   BREAK-EVEN ANALYSIS FOR MINIMUM ACCEPTABLE RETURN

revenue function and the *"pseudocost" function, which is composed of the fixed and variable costs and the minimum acceptable return.* This pseudocost function is simply

$$cs + f + pzs \qquad\qquad (5\text{-}8)$$

which is the sum of total costs and the minimum acceptable return. In effect, we may think of this situation as simply involving an additional cost, p times z, which is assessed against each unit. Viewing it in this way, we are then searching for the simple break-even point as discussed in the previous sections; i.e., if the variable cost were of the form $c + pz$, the total cost function would be that given as the pseudocost

$$(c + pz)s + f$$

Of course, the unit "cost" pz is really the minimum return from each unit which will be considered to be acceptable.

Multiple Decision Variables in Break-even Analysis.   Both of the previous sections discuss product evaluation techniques which involve break-even analysis. The basic assumption of such an analysis is that all of the marketing decision variables, such as price, advertising and promotional expenditures, and the like, have been fixed in advance. This is, of course, a gross abstraction of the real-world decision problem, since all of these controllable variables interact with the uncontrollables to produce the outcome of a product selection decision problem. In reality, values for all of the controllables are most often set simultaneously during pre–introductory product planning.

To introduce multiple decision variables into break-even analysis, consider a slightly more complex situation involving the determination of the unit price p and the sales commission to be paid for a new product.[11] If we symbolize the latter decision variable as d, the sales commission expressed as a percentage of sales revenue, and if c does not include commissions, total costs are

$$TC = cs + f + psd \tag{5-9}$$

where the last term is simply sales revenue ps times the commission rate which is to be determined. Hence the profit relationship may be written as

$$P = ps - (cs + f + psd) \tag{5-10}$$

If the minimum acceptable rate of return to be applied to gross sales is $100z$ percent, the break-even point in this case is determined by solving expression (5-11)

$$zps = ps - (cs + f + psd) \tag{5-11}$$

for the quantity s_{BEM}.

$$s_{BEM} = \frac{f}{p(1 - d - z) - c} \tag{5-12}$$

To illustrate the use of this expression, let us consider some numerical values for the various elements.

f = fixed costs = \$10,000
c = unit variable cost = \$0.50
z = rate of return = 0.10
p = unit price; to be determined
d = sales commission rate; to be determined

Using these values, expression (5-12) becomes

$$s_{BEM} = \frac{10,000}{0.9p - pd - 0.5} \tag{5-13}$$

Since the number of unknown quantities in expression (5-13) is too large to permit any unique solution, we may consider various

---

[11] The Stillson and Arnoff paper in the references presents a similar approach which is not stated in terms of a break-even analysis.

possible combinations of numerical values for each of the decision variables p and d, and array the break-even quantity s_{BEM} as a function of these two variables. If, for example, we consider that reasonable prices might range from \$1 to \$1.50 and that the sales commission should be between 10 and 20 percent, we might construct an array such as Table 5-9.

TABLE 5-9   BREAK-EVEN QUANTITY FOR 10 PERCENT RETURN AS A FUNCTION OF PRICE AND SALES COMMISSION

Price	Sales commission rate (d)		
(p)	0.10	0.15	0.20
\$1.00	33,000	40,000	50,000
\$1.25	20,000	22,800	26,700
\$1.50	14,300	16,000	18,200

Table 5-9 illustrates that under the ranges of the decision variables which are considered to be reasonable, the quantity of the product which must be sold in order to achieve the minimum acceptable return of 10 percent varies from 14,300 units to 50,000 units. This information, in itself, may be sufficient to determine grossly the values which should be set for p and d, since market information and expert opinion as to *likely* sales quantities may be available for comparison with Table 5-9. At the very least this display of break-even quantities should serve to narrow the range of possible values for the decision variables.

By using the break-even quantity with multiple decision variables in this fashion, we are viewing the product decision problem in a more general framework; i.e., each strategy is composed of more than just a simple product alternative. In this model the strategies might take the form

S_1: (product A, $p = \$1.00$, $d = 10\%$)
S_2: (product A, $p = \$1.25$, $d = 10\%$)
S_3: (product A, $p = \$1.25$, $d = 15\%$)
S_4: (product B, $p = \$1.00$, $d = 10\%$)

etc.

By viewing a presentation such as Table 5-9, we may often be able to eliminate large numbers of such strategies. If we determined that it was unlikely that the product in Table 5-9 could have first-year sales above 20,000 units, we could eliminate five of the nine strategies related to it, since only four of the break-even quantities in this table are not above 20,000.[12] These four strategies would be (calling the product in question product A):

S_1: (product A, $p = \$1.25$, $d = 10\%$)
S_2: (product A, $p = \$1.50$, $d = 10\%$)
S_3: (product A, $p = \$1.50$, $d = 15\%$)
S_4: (product A, $p = \$1.50$, $d = 20\%$)

Having also constructed similar tables for other product ideas, we might dismiss a large number of strategies on the basis of the break-even sales quantity.

It should be noted that there is no necessary restriction to the consideration of only two decision variables in this approach. With a greater number than two, the results are not so simple to display nor are the interactions so vividly illustrated. Nonetheless, the approach is equally applicable to any finite number of decision variables.

At best, any of these varieties of break-even analysis is a gross filtering device which may be used to eliminate obviously poor product alternatives which were not deleted from consideration in the product search phase. There are obvious deficiencies in the break-even approach—the question of the time period to be considered, the ambiguity of costs which cannot always be classified as either fixed or variable, the requirement of constant price and cost parameters over the life of the product, the omission from consideration of the time pattern of costs and revenues, and the treatment of the decision situation as one involving certainty rather than as one under risk or uncertainty. Nonetheless, break-even analysis is a useful and much used approach to gross economic evaluation of product alternatives.

Present-value Product Evaluation. Each of the product evaluation aids which incorporate break-even analysis focuses on the sales results which *must* be obtained in order that some minimum goal be achieved. The *expected* sales results are introduced as a point of comparison with the required results. The use of judgmentally desired expectations in break-even analysis implies that they are either of a qualitative or gross quantitative nature. One way of using more refined judgments in product evaluation is through consideration of the expected costs and revenues associated with a product throughout its life cycle. When this is done, consideration must also be given to the *time value of money*.

[12] We are assuming that all admissible strategies involve at least breaking even in the first year. For many products no such objective would be achievable.

Compound Interest. The principle of compound interest forms the basis for consideration of the time value of money. One dollar invested in a savings bank at an interest rate of $100r$ percent per year will, if interest is compounded annually, result in a return of r dollars at the end of one year. At a 4 percent interest rate, i.e., $100(0.04)$ percent, one will earn \$0.04 for each dollar invested. The total investment will therefore have grown to $(1 + r)$ dollars at the end of the first year for each dollar invested.

If the $(1 + r)$ dollars is left to accumulate interest during the second year, the amount earned will be $(1 + r)r$; i.e., the entire \$1.04 in the account at the end of the first year will earn interest at a 4 percent rate during the second, so that the earnings during the second year will be \$1.04 times 0.04, or \$0.0416. At the end of the second year the original dollar will have grown to the quantity

$$1 + r + (1 + r)r$$

where the terms represent the original dollar, the first year's interest r, and the second year's interest on the $(1 + r)$ dollars which the account held at the beginning of the second year.

Simplifying this quantity, we can see that it is equivalent to $1 + 2r + r^2$, which is simply $(1 + r)^2$. If we continue this process, we can see that *a single dollar invested at an interest rate of $100r$ percent will have grown to $(1 + r)^n$ dollars after n years* and in general, *if we invest A dollars at an annual rate of $100r$ percent, it will grow to*

$$A(1 + r)^n \tag{5-14}$$

dollars after n years.

The simple example of Table 5–10, involving an investment of \$10 at 4 percent, will illustrate this. The first column of this table represents the number of years from the point of initial investment (at zero years). The second column is the annual interest calcu-

TABLE 5-10   COMPOUND INTEREST CALCULATIONS

At end of year (n)	Interest during year	\$10 has grown to	$\$10(1.04)^n$
0	....	\$10.00	\$10.00
1	\$0.40	10.40	10.40
2	0.416	10.816	10.816
3	0.43264	11.24864	11.24864
4	0.4499456	11.6985856	11.6985856

lated as 4 percent of the amount in the account at the beginning of the year, which is tabulated in column three. After the lapse of no time (year zero), the $10 is still $10, as indicated by the top row of the table. The $10 earns $0.40 interest in the first year to bring the account balance to $10.40. This amount earns $0.416 interest in the second year to bring the balance to $10.816, etc.[13] The last column represents the calculation of $A(1 + r)^n$. Note that the last two columns are identical.

The Present Value of Future Monies. Since A dollars invested at $100r$ percent, compounded annually, will have grown to $A(1 + r)^n$ dollars after n years, we might well ask how one would determine the amount of money which needs to be invested to produce a fixed sum B in n years. If one expects his newborn son to enter college at a cost of $20,000 in 18 years, he might wish to know how much he needs to invest now to have $20,000 available to him then.

The answer is easily seen to be

$$\frac{B}{(1 + r)^n} \qquad (5\text{-}15)$$

since this amount invested now would produce $B/(1 + r)^n$ times $(1 + r)^n$, or B dollars n years from now; i.e., let A equal $B/(1 + r)^n$ in expression (5-14).

The quantity $B/(1 + r)^n$ is called the *present value of B, n years from now*, since it is the quantity which would need to be invested *now* to have B dollars available n years hence. The process of calculating present values is called *discounting;* i.e., we discount future values back to the present to obtain *present values*. It is necessary for us to do this because we recognize that *a dollar today and a dollar at some later time are not equivalent*. They are different by virtue of their earning power—the rate $100r$ percent. The rate r, in this context, is referred to as the *discount rate*.

Discounting Streams of Money. In marketing we are typically concerned with streams of money which involve costs and revenues at many different points in time, rather than single quantities at a single point in time. If, for example, we expected revenues of $10 at a point one year hence, $20 two years hence, and $30 three years hence, we could calculate the respective present values as $10/(1 + r)$, $20/(1 + r)^2$, and $30/(1 + r)^3$. *The present value*

---

[13] Of course, the bank rounds off these figures to two decimal places, the number of whole cents.

of the entire stream is simply the sum of the individual present values or

$$\frac{\$10}{1+r} + \frac{\$20}{(1+r)^2} + \frac{\$30}{(1+r)^3}$$

In general, *the present value of a stream involving* Q_1 *one year from now,* Q_2 *two years from now, etc., up to* Q_n *n years from now is*

$$PV = \frac{Q_1}{1+r} + \frac{Q_2}{(1+r)^2} + \cdots + \frac{Q_n}{(1+r)^n} = \sum_{i=1}^{n} \frac{Q_i}{(1+r)^i} \qquad (5\text{-}16)$$

In general, *the present value of a stream involving* Q_1 *one year* of net cash flow, given the knowledge of the revenues received and costs incurred at various points in time. If $Q_i = R_i - C_i$ is net flow at time i as the difference between revenue at time i, R_i, and total costs at time i, C_i, the present value of the net cash flow stream for n years is simply that given by expression (5-17).

$$\sum_{i=1}^{n} \frac{Q_i}{(1+r)^i} = \sum_{i=1}^{n} \frac{R_i - C_i}{(1+r)^i} \qquad (5\text{-}17)$$

This is equivalent to

$$\sum_{i=1}^{n} \frac{R_i}{(1+r)^i} - \sum_{i=1}^{n} \frac{C_i}{(1+r)_i} \qquad (5\text{-}18)$$

which represents the difference between the present values of the revenue and cost streams. This says that just as each individual net flow is a difference between the revenue and cost at that point in time, so *the present value of net cash flow is simply the difference between the present value of the revenue stream and the present value of the cost stream.* To calculate the present value of net cash flow, we may thereby simply obtain the individual net flows $Q_i = R_i - C_i$ for all points in time and obtain the present value of the Q_i stream.

To demonstrate this, consider the cost, revenue, and net flow data of Table 5-11 occurring at one, two, and three years from the current date. The corresponding table of present values, using a 5 percent discount rate, is given as Table 5-12.

Note that in Table 5-12 the difference between the present value of revenue and the present value of cost for each point in time is simply the present value of the net at the same point. Further, the present value of the stream of revenues (top row sum) minus the present value of the stream of costs (middle row sum) is just the present value of the net flow (bottom row sum).

TABLE 5-11   COST, REVENUE, AND NET FLOW DATA

	At end of 1 year	At end of 2 years	At end of 3 years
Revenue	$100	$100	$140
Cost	50	40	40
Net	50	60	100

TABLE 5-12   PRESENT VALUES

	At end of 1 year	At end of 2 years	At end of 3 years	Total
Present value of revenue	$\frac{\$100}{1.05} = \$95.22$	$\frac{\$100}{(1.05)^2} = \$90.70$	$\frac{\$140}{(1.05)^3} = \$120.94$	$306.86
Present value of cost	$\frac{\$50}{1.05} = 47.61$	$\frac{\$40}{(1.05)^2} = 36.28$	$\frac{\$40}{(1.05)^3} = 34.55$	118.44
Present value of net	$\frac{\$50}{1.05} = 47.61$	$\frac{\$60}{(1.05)^2} = 54.42$	$\frac{\$100}{(1.05)^3} = 86.39$	188.42

Of course, this procedure assures that the rate r remains constant and that the same rate is appropriate for both costs and revenues. In general, investments are best discounted at the organization's *cost of capital*—the rate which it would have to pay for externally-generated capital—and earnings at the organization's *opportunity rate of return*—the rate at which the organization has the opportunity to earn elsewhere. In the following discussion, we shall, for simplicity, assume that both rates are the same. The generalizations to situations where they are different are straightforward.

Now, let us consider the comparison of two products on the present value basis. Specifically, consider that two product ideas remain after break-even consideration and that the better of the two is to be selected. The expected cash flows for products A and B might be those given by Table 5-13.

The data of Table 5-13 determine that an initial outlay of $10,000 is required for product A and an initial outlay of $20,000 for product B. Subsequently the two products will produce the revenues and require the expenditures which are shown, the primary difference being that product A has a rather short sales life of four years and product B is expected to produce revenues for eight years. Product A produces the highest revenue and profit in any single year, but product B produces profits for a longer period.

It is clear that Table 5-13 does not present the relevant information in a form that enables us to compare the two products. Since

TABLE 5-13   CASH FLOWS FOR TWO PRODUCTS

	Years after introduction									
	0	1	2	3	4	5	6	7	8	9
Product A										
Revenue	$ 0	$2,000	$ 5,000	$20,000	$ 5,000	$ 0	$ 0	$ 0	$ 0	$0
Costs	−10,000	2,000	2,000	2,000	2,000	0	0	0	0	0
Net	−10,000	0	3,000	18,000	3,000	0	0	0	0	0
Product B										
Revenue	$ 0	$2,000	$10,000	$15,000	$15,000	$15,000	$2,000	$2,000	$2,000	$0
Costs	−20,000	3,000	3,000	4,000	4,000	4,000	1,000	1,000	1,000	0
Net	−20,000	−1,000	7,000	11,000	11,000	11,000	1,000	1,000	1,000	0

both the *amount* and *timing* of cash flows bear importantly on their value in terms of the present, we shall calculate the present value of the net cash flows, using an opportunity rate of return of 10 percent, and then use this discounted net cash flow measure as a basis for comparison of the two products.

The present value of net cash flows for product A, assuming that all flows occur at the end of the year in question, is[14]

$$PV_A = -\$10,000 + \frac{\$0}{1.1} + \frac{\$3,000}{(1.1)^2} + \frac{\$18,000}{(1.1)^3} + \frac{\$3,000}{(1.1)^4} \qquad (5\text{-}19)$$

or PV_A is about \$8,000. A similar calculation for product B determines that PV_B is about \$9,000.

To investigate the rationale of making the comparison on this basis, let us suppose that we have available three nonproduct investment alternatives in which we might make use of the initial \$10,000. Let us think of each of the three as being equivalent to depositing the \$10,000 in a bank at the (optimistic) annual interest rate of 10 percent, which is also our opportunity rate of return. The three bank investments are differentiated by our policy over the years.

In alternative C we allow the \$10,000 to accumulate interest, and at the end of n years we withdraw the total sum. The present value of cash flow for this alternative is

$$PV_C = -\$10,000 + \frac{\$10,000(1.1)^n}{(1.1)^n} = 0 \qquad (5\text{-}20)$$

[14] Note that the initial capital outlay is included without a discount factor in the present-value calculation. The absence of a discount factor implies that the investment is made immediately. In general, we may choose either to include or to exclude such investments from present-value calculations. If we exclude them, we are determining the present value of the stream of net cash flow (costs and revenues excluding initial investment). We may then simply compare this present value with the investment since both are in terms of "today's dollars."

In this calculation the $-\$10,000$ represents the initial outlay. The numerator of the second term is the amount to which the $\$10,000$ would grow in n years, as calculated by using expression (5-14). Division by $(1.1)^n$ determines the present value of that total quantity according to expression (5-15). The present value of the net cash flow is identically zero.

In alternative D we deposit the initial $\$10,000$ and then annually withdraw each year's interest for six years, at which time we also reclaim our $\$10,000$ deposit. The present value of these flows is

$$PV_D = -\$10,000 + \frac{\$10,000(0.1)}{1.1} + \frac{\$10,000(0.1)}{(1.1)^2}$$
$$+ \frac{\$10,000(0.1)}{(1.1)^3} + \frac{\$10,000(0.1)}{(1.1)^4} + \frac{\$10,000(0.1)}{(1.1)^5}$$
$$+ \frac{\$10,000(0.1)}{(1.1)^6} + \frac{\$10,000}{(1.1)^6} = 0 \qquad (5\text{-}21)$$

In this calculation the $-\$10,000$ is the initial deposit, the next six terms represent the annual interest at 10 percent discounted to the present, and the last term represents the present value of the $\$10,000$ which we shall reclaim at the end of six years. The present value of alternative C can be seen to be identically zero.

The third alternative, E, involves depositing our $\$10,000$ and withdrawing each year's interest indefinitely. Since we never reclaim our $\$10,000$, the present value of cash flow for this alternative is

$$PV_E = -\$10,000 + \frac{\$10,000(0.1)}{1.1} + \frac{\$10,000(0.1)}{(1.1)^2}$$
$$+ \frac{\$10,000(0.1)}{(1.1)^3} + \cdots \qquad (5\text{-}22)$$

where the sequence of dots indicates that successive terms are added endlessly, each term being determined in the same fashion as those shown; i.e., each term after the second is simply $\frac{1}{1.1}$ times the preceding term. The terms in this expression represent the initial deposit and the endless sequence of present values of annual interest payments, where the interest is computed at the 10 percent rate and the present value is calculated by using the same rate. To calculate the numerical value of PV_E, we must recall some basic algebra dealing with geometric progressions.

Geometric progressions. A geometric progression is a sequence of numbers such that any term after the first is obtained from the preceding term by multiplying the preceding term by a fixed number, called the common ratio. For example, the sequence

2, 4, 8, 16, 32, 64, 128, 256

is a geometric progression with the common ratio 2; the sequence

$$\tfrac{1}{3}, \tfrac{1}{9}, \tfrac{1}{27}, \tfrac{1}{81}, \tfrac{1}{243}, \tfrac{1}{729}$$

is a geometric progression with the common ratio $\tfrac{1}{3}$. Both of these progressions have a limited (finite) number of terms. The *sum* of the numbers which make up a limited geometric progression is a finite number, which can be shown to be of the form[15]

$$\frac{\text{First term} - (\text{ratio} \times \text{last term})}{1 - \text{ratio}} \qquad (5\text{-}23)$$

A geometric progression may also have an unlimited (infinite) number of terms. The sequence

$$3, \tfrac{3}{2}, \tfrac{3}{4}, \tfrac{3}{8}, \tfrac{3}{16}, \tfrac{3}{32}, \ldots$$

is meant to be of infinite length. The dots after $\tfrac{3}{32}$ indicate an unending sequence, each of whose successive terms is simply $\tfrac{1}{2}$ times the preceding term. *If the common ratio in an unlimited geometric progression is less than 1, the sum of the terms will approach a limit which can be shown to be*[16]

$$\frac{\text{First term}}{1 - \text{common ratio}} \qquad (5\text{-}24)$$

Returning to the consideration of the three bank investment opportunities, we may make use of expression (5-24) to calculate the present value of alternative E. Note that, except for the first term in expression (5-22), PV_E has the form of the sum of an unlimited geometric progression with a common ratio of $1/1.1$ and a first term of $1,000/1.1$. This sum, using (5-24), is

$$\frac{\$1,000/1.1}{1 - (1/1.1)} = \$10,000$$

[15] If a is the first term of an n-term limited geometric progression and b is the common ratio, the progression is $a, ab, ab^2, ab^3, \ldots ab^{n-1}$. The sum of these n terms is $S = a + ab + ab^2 + ab^3 + \ldots + ab^{n-1}$. If we multiply both sides of this expression by the common ratio b, we obtain $bS = ab + ab^2 + ab^3 + \ldots + ab^n$. Subtracting this expression from S, we get $S - bS = a - ab^n$, which is equivalent to $S = (a - ab^n)/(1 - b)$. This represents a simple expression for determining the sum of a limited geometric progression.
[16] The expression for the sum of an n-term limited geometric progression given in footnote 15 may be written as $S = a/(1 - b) - ab^n/(1 - b)$. If b is less than 1, b^2 is smaller, b^3 is still smaller, etc. If n is large enough, b^n is infinitely small. The limit of b^n as n increases without bound is zero. Hence, as n approaches infinity, the last term in the expression for S approaches zero and S approaches $a/(1 - b)$.

This $10,000, added to the negative $10,000 which represents the initial outlay, determines that PV_E is also identically zero.

To summarize, the three nonproduct investment alternatives are:

C: deposit $10,000 in bank; allow interest to accumulate; reclaim total after n years
D: deposit $10,000 in bank; withdraw annual interest for six years; withdraw $10,000 after six years
E: deposit $10,000 in bank; withdraw annual interest indefinitely

Each alternative has a present value of net cash flow which is equal to zero. One way of viewing our evaluation of potential products on this present-value basis is, then, that we are measuring their worth from a base (zero) which represents the present value of bank investments of the kind described by C, D, and E.

However, the present value is not an adequate basis for comparison if the product alternatives involve either very different life cycle durations or very different capital investment requirements. If the life cycles are of very different duration, as they are in Table 5-13, a simple present-value calculation does not give explicit concern to the earnings which can be generated between the end of the life cycle of the product with the shorter life cycle (A) and the end of the cycle for the product with the longer duration (B). If the initial capital investment is very different for the alternatives, the present value omits comparison of return with investment.

To obviate these difficulties, we may use a variation of the present-value idea. To do this, we do two things. First, *we may assume that investment opportunities of the variety of the product with the shorter life cycle will be available sequentially.* This permits comparison of the present values of alternatives over a fixed period of time. In the situation of Table 5-13, this would involve the comparison of the present value of two sequential opportunities of the A variety with B, since each A has a life cycle of only four years and B has a cycle of eight years' duration. Then, *we may express the discounted flows as a discounted return per dollar of discounted capital outlay.* This brings the "return on investment" idea into play.

To demonstrate this variation of a simple present-value calculation, we may take the situation of Table 5-13. If we think of two sequential opportunities of the variety of A, the first has a present value of cash flows (excluding initial investment) of $18,000. Since the second of these two would not *begin* producing revenues until four years hence, its present value of cash flows must be further discounted to the present. The present value of net cash flows of $0, $3,000, $18,000, and $3,000 beginning four years hence is about

$12,294.[17] The total present value of cash flows for the two *A*'s is therefore $30,294.

Since the outlays in the first column of Table 5-13 are presumably capital investments, the present value of investment for the first of the *A*'s is $10,000. For the second, the present value is $10,000/(1.1)^4, or $6,830, since this outlay would not be required for four years in the case of two sequential investments in *A*. The total present value of the two outlays is $10,000 + $6,830, or $16,830. Since the present value of net cash flows for the two *A*'s is $30,294, the discounted return per discounted dollar of investment for two sequential investments of the *A* type is $30,294/$16,830, or 1.80

For *B*, a similar calculation gives $29,000/$20,000, or 1.45, since the only investment required is the immediate one of $20,000 and the present value of net cash flows is $29,000. Thus, *A* is the better alternative of the two, on the basis of this criterion, and it should be chosen as best if these are the only alternatives to have survived to this stage of the analysis. Of course, Product *B* has a positive present value for the entire stream ($9,000), so it is also a good alternative. Any alternative having a negative present value for the stream (including initial investment) would be considered to be unacceptable on the basis of this evaluation.

To incorporate multiple decision variables into this analysis, we must recognize that the costs and revenues of Table 5-13 are predicated on specific values of all such variables. We may therefore use other combinations of the decision variables and hence, other revenue and cost streams to calculate the present value of net flows. These calculations might be presented in the fashion used for break-even analysis. In this manner we could make a determination of the interaction of the decision variables, and we could choose a best strategy simply by selecting the product idea, in combination with the other decision variables, which is indicated to be best.

This model gives us a basis for solution of the product selection decision problem, which is the primary objective of the product evaluation phase. Like any other model, however, this one does not consider everything. One important omission is the way in which the time pattern of revenues which are produced by the product alternatives fits into the organization's present product mix. If, for example, a product which produces revenues for eight years

[17] This may be calculated either directly as $0/(1.1)^5 + 3000/(1.1)^6 + 18,000/(1.1)^7 + 3000/(1.1)^8$ or by noting that this quantity is simply $1/(1.1)^4$ times the calculation already made for *A* which resulted in $18,000. Hence, the desired present value is $18,000/(1.1)^4$. In the latter case, we are, in effect, discounting the second of the two *A*'s back to its beginning point four years hence and then discounting this to the present.

is more compatible with the organization's desire for increasing sales revenues than one which produces revenues for only four years, product *B* may be the better choice.

Product Development

After a best product alternative has been selected, the detailed development of the product in terms of shape, color, packaging, and the myriad of other decision variables must be begun in earnest. Of course, this phase is not really distinct from the planning stage, since tentative conclusions on many of these variables are a necessary part of product evaluation. At this point, however, the organization has made a decision as to which product it will emphasize in its development program, and it must now make commitments on all of the aspects of preparing the product for production and marketing.

The decision on a best alternative is not, however, an irrevocable one. At any stage of development some new information may come to view about the product, the organization's production capability, or the potential market, which will require reevaluation of the product alternatives. In reality the phases of product planning and development, like the phases of any problem analysis, are neither disjoint nor phased in chronological order. At any point in time, aspects of product development and evaluation may be developing concurrently. In the decisions which are primarily the concern of the development phase, the effect of *timing* is of paramount importance.

THE TIMING OF PRODUCT DEVELOPMENT

To demonstrate the significance of timing in product development decisions in the context of the present-value calculation used previously, let us consider that we must decide on a target date for the introduction of a particular product, the alternatives being to market the product at the beginning of 1972 or at the beginning of any subsequent year.

Suppose that the best estimate of the level of annual net cash flows which would result from marketing the product is $10,000 per year for each year from 1972 until 20 years hence and $100,000 per year for each year thereafter. This somewhat extreme example might be the case if the product were somewhat ahead of its time in consumer acceptance, as was hybrid corn when it was first introduced to the farmer-consumer. If the initial cost of developing and introducing the product is a million dollars whenever we choose to introduce it, and if the effect of delaying the introduction for a year simply results in the loss of that year's net flow, with no effect on future flows, we may attack the timing problem on the basis of the present value of net cash flow.

The present value of net cash flow resulting from an introduction now $(t = 1)$, using a 5 percent opportunity rate, is (assuming that the current date is 1972):

$$PV_{i=1} = -\$1,000,000 + \sum_{i=1}^{20} \frac{\$10,000}{(1.05)^i} + \sum_{i=21}^{\infty} \frac{\$100,000}{(1.05)^i} \qquad (5\text{-}25)$$

in which the first term is the initial million-dollar outlay, the second term is the sum of the present values of the $10,000 net flows in each of the next 20 years, and the third is the sum of the present values of the endless $100,000 annual flows thereafter.

To evaluate this numerically, we must recognize that the first sum is the sum of a limited geometric progression whose common ratio is $1/1.05$. By using expression (5-23), its value can be found to be $125,000. The second sum is associated with an unlimited geometric progression which is determined, by using expression (5-24), to be $754,000. The entire present value of introducing the product in 1972 is therefore a negative $121,000. This indicates that the product should not be introduced then.

The next alternative which should be explored is whether it would ever be beneficial to introduce the product, i.e., to delay marketing until a later date. Consider, for example, the effect of postponing the marketing for *any* one-year period. *The loss in present value associated with a one-year postponement is the loss of the present value of that year's net flow. The gain is the savings in not having invested the million dollars during that year,* i.e., the savings on the present value of the initial expenditure. In effect, we may think of the million dollars as becoming less in present value as it is spent further out in time, since we could invest some lesser amount at our opportunity rate and have the million dollars available at a later date, that lesser amount being the present value of the million dollars.

In order that we decide to postpone the marketing of the product from year t to year $t + 1$, the gain in doing so must be greater than the loss. In symbols the gain in a year's delay from t to $t + 1$ is the difference in the present value of the million-dollar outlay,

$$PV_t(\text{million}) - PV_{t+1}(\text{million})$$

which is equal to

$$\frac{\$1,000,000}{(1.05)^t} - \frac{\$1,000,000}{(1.05)^{t+1}}$$

or

$$\frac{\$1\,000\,000}{(1.05)^t}\left(1 - \frac{1}{1.05}\right) \tag{5-26}$$

The loss in present value resulting from a delay from t to $t+1$ is the present value of the net flow for year $t+1$, which may be symbolized as

$$\frac{Q(t+1)}{(1.05)^{t+1}} \tag{5-27}$$

i.e., the net flow divided by the discount factor.

The decision to delay from t to $t+1$ is determined by the relative size of (5-26) and (5-27). If the inequality

$$\frac{\$1,000,000}{(1.05)^t}\left(1 - \frac{1}{1.05}\right) > \frac{Q(t+1)}{(1.05)^{t+1}} \tag{5-28}$$

holds, the gain is greater than the loss and we should delay from t to $t+1$. If the reverse inequality holds, development should not be delayed; i.e., it should either be begun in time for introduction in year t or the product idea should be entirely dropped, depending on whether the present value of net flows resulting from beginning sales in year t is positive or negative.

The inequality (5-28) holds for $t = 1$:

$$\frac{\$1,000,000}{1.05}\left(1 - \frac{1}{1.05}\right) > \frac{\$10,000}{(1.05)^2}$$

which tells us to delay at least until $t = 2$. The next inequality also holds:

$$\frac{\$1,000,000}{(1.05)^2}\left(1 - \frac{1}{1.05}\right) > \frac{\$10,000}{(1.05)^3}$$

telling us to delay at least until $t = 3$. Continuing this process, we find that the inequality holds until $t = 20$, i.e.,

$$\frac{\$1,000,000}{(1.05)^{20}}\left(1 - \frac{1}{1.05}\right) < \frac{Q(21)}{(1.05)^{21}}$$

where $Q(21)$ is the $100,000 net flow in the twenty-first year. This indicates that we should postpone introduction at least until 20 years hence, if at all. We need to ask two additional questions: First, do we gain anything by postponing beyond this point? Secondly, should we possibly never introduce the product?

To answer the first of these, we need only to view the gains and losses in present value from additional annual delays. We find that the losses exceed the gains in each year's postponement after 20 years. Hence, we should not postpone beyond 20 years.

The question of whether we should ever introduce the product rests on the present value of the policy consisting of introduction 20 years hence. In effect, we have decided that if we ever introduce the product, we should do so then. The question now is: Should we introduce it at all?

The present value of introducing the product at $t = 20$ is

$$PV_{t=20} = -\frac{\$1,000,000}{(1.05)^{20}} + \sum_{i=21}^{\infty} \frac{\$100,000}{(1.05)^i} \qquad (5\text{-}29)$$

where the terms simply represent the present value of the million-dollar expenditure 20 years hence and the present value of the sum of the endless stream of net flows beginning at that time. This quantity may be found to have the value $PV_{t=20} = \$376,900$. This positive present value for introduction of the product 20 years hence indicates that we should wait until then and market the product. We are assuming, of course, that nothing changes in that period which will affect the validity of our model. In reality we would reevaluate the situation at a later date in terms of any new information which may then be available.

In effect, our analysis of this timing question is in two steps. First we ask the question: When is the best time to introduce the product, if we are to introduce it at all? Having resolved this, we ask whether we should indeed ever introduce the product in question. In attacking the problem in this way, we have blended the product evaluation problem and the timing decision of product development into a single analysis. The relevant strategies in this case include the product alternatives of product evaluation and the timing alternatives of product development. Some representative strategies might be

S_1: (product A, introduce at $t = 0$)
S_2: (product A, introduce at $t = 1$)
S_3: (product B, introduce at $t = 0$)
S_4: (product B, introduce at $t = 2$)

etc.

In combining the product evaluation and product development phases in this way, we are illustrating one of the virtues of the systems approach. Since the two decision variables interact, they must be treated together rather than independently. In this fashion we are more likely to arrive at overall optimal solutions than at suboptimal ones.

PLANNING AND CONTROLLING PRODUCT DEVELOPMENT

Consideration of the timing of product development is important for many new product ideas. For many others, however, it is apparent that the product should be developed with as much speed as is consistent with some reasonable cost considerations. Since most product development projects involve many different activities which are performed by different individuals and groups, the *detailed planning* of the activities necessary to develop the product assumes great importance. Also of great significance is the *control* which is exercised over the activities as they progress. For instance, if certain activities are consuming more time than was planned, some investigation into the cause for the delay might be made and appropriate actions taken to rectify it. In the case of an activity delay which is critical to the development project, additional personnel might be assigned to the activity which is undergoing difficulty.

The Project Network.   One of the useful means of viewing a development project for purposes of planning and control is a *project network*. Such a network is simply a pictorial description of a plan for a development project. The two major elements of a network are *activities* and *events. An activity is a time-consuming task* which is a part of a development project. *An event is an accomplishment which occurs at a point in time.* For example, an activity might be described as "design the product's package" and a related event would be "package design completed."

Activities are designated by lines in the pictorial network, and events are symbolized by circles. Thus the simple network of Figure 5-7 might represent some portion of an overall product development network. Although there is no time scale in the network, there is a time *sequence*. For example, *the three lines emanating from event 2 indicate that none of the three activities "detailed design," "preliminary pricing study," and "obtain promotional ideas" is to begin until the event has occurred.* Similarly the activities "detailed design," "manufacture models," and "redesign" must occur in sequence; i.e., one must be completed before the next can begin. Since each of the events is usually defined to designate the beginning or end of an activity, this is equivalent to saying that *an activity cannot begin before its preceding event has occurred.* In general any event cannot occur until *all* of the activities preceding it have been completed. In the case of the "detailed price analysis" activity, *all* of the three preceding activities ("estimate promotional costs," "preliminary pricing study," and "redesign") must be completed before event 6 can occur and hence, before the "detailed price analysis" activity can commence.

It is easy to see that a pictorial representation such as that shown gives a clear description of the tasks necessary to achieve some end event (in this case, "complete analysis"). Such a network can

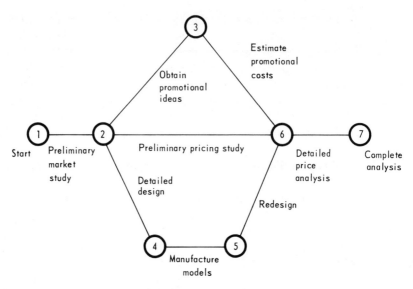

FIGURE 5-7   PRODUCT DEVELOPMENT NETWORK

be developed to represent all of the activities and interrelationships which will be necessary to develop a product. The planning value of the network is clear. A glance at a simple network vividly illustrates some gross plans which can be made concerning the resources to be devoted to each activity. For instance, the same work group would probably not be assigned to the "obtain promotional ideas" and "preliminary pricing study" activities because both can begin at the same time. The most efficient way to accomplish these activities is in parallel, as the network illustrates. The "estimate promotional costs" activity might well be assigned to the same work group which obtains promotional ideas, since the two activities are in sequence (and one would suspect that the learning acquired in the former might aid in the efficient accomplishment of the latter, since both are related to promotion).

Time Estimates.   For purposes of detailed planning and control, the manager should have estimates of the *time required to complete each of the activities.* Such estimates can be obtained from the people who are charged with responsibility for the various activities. Although there will be a large degree of uncertainty associated with most of them, such estimates are generally infinitely superior to no time estimates at all.

The difficulties involved in obtaining time estimates is amusingly illustrated in Univac's basic guidebook to network planning.

On 1 January, the Manager determines he needs a high frequency "frimbus" by 1 June. He calls George and says: "George, I've got to have this frimbus by 1 June, can you get it out by then?"

Well, George nearly faints because he knows instinctively that he cannot possibly get this thing out before 1 September. Does George tell that to his boss? Hell no, the boss would think he was some kind of nut. So, he tells his boss: "Well now, 1 June is mighty tight, I don't think I can do it by then. It will probably be closer to 1 July."

At this point the manager scratches his head, tugs his chin, and pulls out his drawer a crack to see how much fat he has in *his* estimate. Finally, the date of 15 June is negotiated which becomes the scheduled completion date.

Time passes. About 1 April, the manager remembers the "frimbus," and he gives George a call: "Say, George, how's everything going on the frimbus? Will it be out 15 June?"

By now George has had several months to think of reasons why the "frimbus" cannot be ready 15 June, so he says: "Boy, things are really in a pot. We had to redesign the I/O circuit, my best mechanical man left the company, and the vendor will be late in delivering the transformers. I don't see how we can get this out before 1 July."

The boss makes the usual comments about bearing down, but what else can he do? After all, a man cannot be fired for being just two weeks late, and 1 July becomes the new scheduled completion date.

A month later, the same thing happens, and 15 July becomes the completion date. The game goes on until the "frimbus" is ready, which will probably be about 1 September.

Sometimes this process is called "we make every extension."[18]

Of course, the point illustrated here is that the time estimate should represent an educated guess rather than a firm commitment. If it is not treated as a commitment, it is usually possible to obtain estimates which truly represent the totality of the best information which is available, and nothing more can be expected.

The Critical Path. The purpose of the time estimates is to predict activities in which delays will be critical to the total project. If the activities are defined so that they represent short controllable tasks, and if one activity encounters difficulties, some control action may be taken to hasten its completion. In particular, there are certain activities in the network which should be carefully monitored. Those *critical activities* are the *ones in which a delay will cause a delay in the completion of the entire project.* Identification of these activities is accomplished by determining the *critical*

---

[18] K. L. Dean, *Fundamentals of Network Planning and Analysis,* Univac Defense System Division, Sperry Rand Inc., St. Paul, Minn., PX 1842B, January, 1962. Quoted with permission.

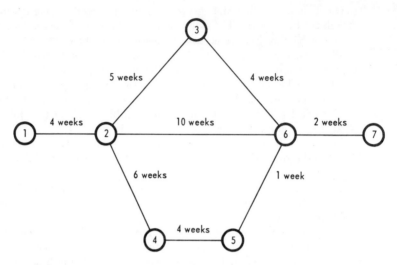

FIGURE 5-8   PRODUCT DEVELOPMENT NETWORK WITH TIME ESTIMATES

path—the path through the network which has the longest duration.

In Figure 5-8 the network previously used is reproduced, but with a number for each activity to represent the estimated time required for its completion (in weeks). The longest path through the network is the lower one (events 1, 2, 4, 5, 6, 7), which requires an estimated total of 17 weeks to proceed from event 1 to event 7. The middle path (events 1, 2, 6, 7) requires 16 weeks, and the upper path (events 1, 2, 3, 6, 7) requires 15 weeks. In this simple network there are only these three possible paths from event 1 to event 7. The lower path, the most time-consuming one, is the critical path. *Critical activities are those activities which are on the critical path.*

It is easy to see that a delay on any activity on the critical path must necessarily delay the completion of the project. If all activities except the critical activity between events 4 and 5 actually took exactly the estimated time and this activity actually required 5, rather than 4, weeks to complete, event 7 would not be accomplished until 18 weeks after event 1. However, if an activity which is not on the critical path is delayed, the completion event (7) may not be delayed. If all activities go according to schedule except the one between events 3 and 6, for instance, and that one slips by one week, the final event (7) will still be completed 17 weeks after event 1 (even though the upper path requires 16 weeks instead of 15). This is because *the final event cannot be reached until all preceding activities are completed, and this cannot take place any sooner than the time required for the longest path through the network*—the critical path.

A second important feature of the network (in addition to the critical path) is the positive *slack* associated with the noncritical paths. Slack is the extra time available along the path, i.e., the extra time the path could require before its duration would become equal to that of the critical path, and hence, before it would become an alternate critical path. The upper path in Figure 5-8 has two weeks of slack and the middle path has one week. (These periods are determined by subtracting the estimated time required for the path from the time required for the critical path.)

Management Action. The manager may use the network and critical path for a number of purposes, primary among which are planning, scheduling, and control. We have already noted how certain initial planning may be accomplished by using only the basic network. Additional planning may involve the reallocation of resources from activities which are not on the critical path to critical activities. Of course, this presumes that the time estimates are valid (or at least of the same degree of inaccuracy) and that it is possible to hasten the completion of activities on the critical path by expending greater resources on them. Clearly, if this were so, it would be advantageous to lengthen the duration of noncritical paths by reducing the resources being expended on activities on these paths and to reassign those resources to critical activities. The resulting shortening of the critical path would produce a gain (in terms of a quicker project completion date) and no loss would be incurred (in terms of the completion date) as long as the increased duration of any noncritical path was not greater than its slack.

After such initial reallocations have been made, scheduling simply involves the establishment of beginning and completion dates for the various activities as indicated by the revised activity duration estimates (revised after the initial resource allocations).

Control of the project is necessary because of changes in the activities and activity time estimates which are made as the project progresses. One way of facilitating managerial control is through the periodic preparation of a list of all paths through the project network. The list should incorporate the current revised estimates of activity durations, and the various paths should be listed *in order of increasing slack*. Thus the critical path will appear at the head of the list and each successive path will have greater "extra" time associated with it.

The manager can implement the principle of *management by exception,* which states that adequate control requires that primary attention be paid to exceptions, or in this case, extremes. The critical path is the extreme which deserves primary consideration, because any delays there must necessarily cause overall delays, and any time savings can be reflected in the completion date of the project. For example, if an activity on the critical path is being delayed, the manager may take control action by reallocating personnel just as he did during the initial planning phase. Here the action is necessitated by *changes* which occur as the project progresses, and he can "borrow" from activities on noncritical paths (as long as he does not require them to use more time than the slack in their path) and "lend" to critical activities.

These, then, are the basic uses of project networks for product development. Other sophistications can be included also. PERT[19] is one well-known variation of the basic *critical path method* (CPM) which utilizes three time estimates per activity rather than the single estimate we have used. The use of three time estimates—an "optimistic," "pessimistic," and "most likely" estimate—together with a set of assumptions dealing with estimating behavior,[20] permits the planning and control problem under certainty, as illustrated here, to be transformed into a risk situation.

Summary

Decisions involving the introduction of new products are among the most significant of an organization's problems. New products are vitally necessary to the continued growth of business enterprises and the economy.

The process of *making strategic plans* for new products may be viewed as encompassing an *idea generation* phase, a *search* phase in which the ideas are given a preliminary "filtering" as to general feasibility, and an *evaluation* phase in which stringent economic analysis is applied to the surviving ideas and a "best" product alternative is selected.

Product development involves the tactical decisions and the operations necessary to execute effectively and efficiently the strategic decisions made during the planning process. The question of timing is of paramount importance in this process, along with the detailed planning and control of the activities which are necessary to bring

---

[19] Program Evaluation and Review Technique. See the original paper by Malcolm et al. and any of the numerous manuals prepared by computer manufacturers and the armed services. (See references.)

[20] The MacCrimmon and Ryavec paper explores the basic PERT assumptions. The paper by King and Wilson presents an empirical analysis of the accuracy of CPM time estimates. (See references.)

a product idea to sales fruition. Project networks and critical path analysis may be used to effect such planning and control.

Of course, much additional information is usually necessary before a good product idea becomes a successful product. "Market analysis" is the term applied to the process of obtaining information about the market for a product. Test marketing is a pervasive method for testing both the product itself and the associated controllables of the overall product strategy—price, advertising, etc. Both market analysis and test marketing are treated in the subsequent chapter.

EXERCISES

1. A manufacturer of large construction earthmovers is considering adding new products. What are some he might consider? Why?

2. What sort of temporal pattern would you expect the annual earnings of Wham-O Manufacturing Company to display? (Wham-O is the producer of flying saucers, hula hoops, Super Balls and other novelties.)

3. Construct a chart similar to Figure 5-2 to represent what you would consider typical life cycles of Wham-O's products. What special problems are likely to exist in such a situation?

4. What special problems for the producers of women's apparel might be the direct result of their creation of "new" products via style changes?

5. Greyhound Corporation, which operates the largest U.S. bus fleet, thinks of itself in the general context of the "business of transportation." What new products do you think they might be likely to venture into?

6. Many newspapers have taken the attitude that their business is communications. An even broader viewpoint might incorporate the "greater knowledge industry." What new products might result from these two viewpoints?

7. If the level weights in Table 5-6 are 10, 9, 5, 3, and 1, rather than those given, determine the total expected utility for the same product idea.

8. Assume the availability of a product idea which is certain to achieve the "good" level of all factors in Table 5-6 except "the effect on sales of existing products," on which it is rated "very poor," and the "personnel necessary," on which it is rated "very good." Is this idea better than the one evaluated in Table 5-6?

9. Would it be meaningful to compare the total expected utility result in exercise 7 with the result in exercise 8? Why or why not?

10. What factors, other than those in Table 5-5, related to the general areas of productive ability, marketability, and durability, might be considered in the development of an expected utility for a new product idea?

11. Consider a product idea in a two-factor–two-level situation, such as that in Table 5-7 (i.e., factors F_1 and F_2, levels "good and

"bad"). If the level weights are 10 and 4 for F_1 and 10 and 2 for F_2, what is the total expected utility for the product idea which is reflected in the probabilities of that table? (Use the same factor weights as in Table 5-7.)

12. What are the implicit assumptions of a break-even approach in terms of the nature of costs, price stability, and the "learning" which occurs in production process?

13. Suppose that a firm's cost of outside capital is 10 percent and that its opportunity rate of return is 15 percent. How might an appropriate discounting procedure for future earnings and investments be developed?

14. Develop the certainty analog of the product search procedure which is developed in Table 5-6 in terms of risk.

15. What is the present value of the infinite stream of net flows given below? Use a discount rate of 5 percent.

1 year hence	2 years hence	· · ·	10 years hence	11 years hence	12 years hence	· · ·
$100	$200	· · ·	$1,000	$1,000	$1,000	· · ·

16. A new product, which will sell for $2 per unit, has a unit cost of $1.50 per unit and first-year fixed costs of $5,000. It is deemed virtually certain of first-year sales in excess of 15,000 units. What is your estimate of the likelihood that the organization will not break even with this product?

17. What is the likelihood that the organization will not achieve a minimum acceptable return of 5 percent on gross sales with the product in exercise 16?

18. The unit price of a new chocolate bar is determined by its size and the nature of the market to be either 5 or 10 cents. Suppose that a new idea for a nickel bar is under consideration. The unit cost of producing the bar is 2 cents and the first-year fixed costs are estimated to be $5,000. Proposals for a sales commission rate of 15 percent and 17 percent of gross sales are being considered along with a distributor discount of either 5 or 10 percent on gross sales. What are the break-even sales quantities for this situation?

19. Suppose that the candy manufacturer in exercise 18 desires a minimum acceptable return of 20 percent on gross sales. What are the break-even-for-minimum-acceptable-return sales quantities?

20. A product alternative requires an initial investment of $5,000 and is expected to produce net cash flows of $200, $400, $800, $1,600, and $3,200 in the next five years. Assume that the flows occur at the end of the annual periods in question and that the organization has an opportunity rate of return of 10 percent. Should the investment be made? Why?

21. Refer to the result of exercise 20 and contrast this investment opportunity with the one involving a $10,000 investment which returns $2,000 per year for 10 years.

22. In what way is the model dealing with the timing of product development an abstraction of the situation which would be likely to exist in the real world? How would these factors affect the solution to the timing problem as derived from the model?

23. Make a graph of the present value of net cash flow for the product development timing problem, using the data for the illustration in the chapter. How could one determine the best time at which to introduce the product, using only this graph?

24. The simultaneous choice of a product alternative and an introduction date for the product is an example of the systems approach to problem solving. If this is the best way to go about problem solving, why don't we also incorporate price, advertising expenditures, and the other decision variables into our strategies and evaluate them all simultaneously?

25. What two general kinds of errors may be made in the search phase of product planning? Which do you think is the more important?

26. Two products A and B are available as investment opportunities for a company whose other products average 5 percent return. The expected cash flows associated with A and B are given below. Which product would you recommend (if either)? Why? (Consider the initial investment as occurring immediately and all other flows as occurring at yearly intervals beginning in one year.) What additional information might you require before you would make a recommendation to top management in this situation?

	Initial invest-ment	Production and sales costs in 1st year	Reve-nue in 1st year	Production and sales costs in 2d year	Reve-nue in 2d year	Production and sales costs in 3d year	Reve-nue in 3d year	Production and sales costs in 4th year	Reve-nue in 4th year
A	1,000	800	1,400	700	1,200	0	0	0	0
B	1,500	700	900	500	1,500	100	400	0	0

27. One characteristic of the product search procedures discussed in the chapter was that subjective estimates by experts are a significant part of the analysis. Since we are using subjective judgments anyway, why don't we just ask for overall product judgments in the first place and not bother to go through all of these steps?

28. A model for the determination of *minimum market requirement* for a new product is given below (adapted from the Stillson and Arnoff reference):

Total cost/year = fixed equipment charge (amortized) +

expenses + production cost

Let
  A = fixed equipment charge for year
  P_i = percentage of gross sales accounted for by ith size
  i = subscript running over three sizes ($i = 1, 2, 3$)
  C_i = unit production cost for ith size as a percentage of selling price
  x = gross sales (dollars)
  z = net rate of return applied to gross sales
  E = total expenses as percentage of gross sales (sales, discounts, etc.)

a. Write an expression for total cost/year, using these symbols and allocating expenses to product sizes in proportion to gross sales.
b. Use this expression to write an expression for net profit, using these symbols.
c. Solve b for the minimum market requirement in order to obtain a net rate of return z.
d. Use the data below to calculate minimum market requirement in terms of E.

  Equipment cost = $10,000 (depreciate over 5 years using straight line)
  Net rate of return desired = 20%

Size	Percent of gross sales	Unit production cost	Unit selling price
1	20	0.50	1.00
2	50	0.60	1.20
3	30	0.70	2.10

e. Assume that E has been at a level of 10 percent normally. Determine the sensitivity of your solution (minimum requirement) to unit changes in this level.

REFERENCES

Berg, T. L., and A. Shuchman (eds.): *Product Strategy and Management*, Holt, Rinehart and Winston, Inc., New York, 1963.
Grant, E. L., and W. G. Ireson: *Principles of Engineering Economy*, The Ronald Press Company, New York, 1960.
King, W. R., and T. A. Wilson: "Subjective Time Estimates in Critical Path Planning—A Preliminary Analysis," *Management Science*, vol. 13, no. 5 (January, 1967), pp. 307–320.
Kline, C. H.: "The Strategy of Product Policy," *Harvard Business Review* (July–August, 1955), pp. 91–100.

MacCrimmon, K. R., and C. A. Ryavec: "An Analytic Study of the PERT Assumptions," *Operations Research,* vol. 12, no. 1 (1964), pp. 16–37.

Malcolm, D. G., J. H. Roseboom, C. E. Clark, and W. Frazer: "Application of a Technique for R and D Program Evaluation (PERT)," *Operations Research,* vol. 10, no. 6 (1962), pp. 808–817.

O'Meara, J. T., Jr.: "Selecting Profitable Products," *Harvard Business Review* (January–February, 1961), pp. 83–89.

Pessemier, E. A.: "New Product Search and Preliminary Evaluation," paper presented at 1963 American International Meeting of the Institute of Management Sciences, New York, Sept. 12–13, 1963.

————: *New-Product Decisions: An Analytical Approach,* McGraw-Hill Book Company, New York, 1966.

Richman, B.: "A Rating Scale for Product Innovation," *Business Horizons* (Summer, 1962), pp. 37–42.

Stillson, P., and E. L. Arnoff: "Product Search and Evaluation," *Journal of Marketing,* vol. 22, no. 1 (July, 1957), pp. 33–39.

CHAPTER *6*

Market Analysis and Test Marketing

At the opening of Meredith Willson's famous play *The Music Man*, a group of salesmen purveying such diverse wares as anvils and band instruments discuss the problems of "drummers" and agree that to be successful, "ya gotta know the territory." It is not less true of the marketing manager than of the salesman that information concerning the market forms the basis for most marketing decisions. To have good information is to have potentially good decisions, and to have poor information is to reduce the likelihood of good decisions.

Two general sources of the kind of market information necessary to good decision making may be distinguished. *Internal* sources provide the cost, production, and financial data which are generated within the organization. *External* sources provide information on the market and environment which originates outside the organization. Generally the responsibility for the collection and compilation of internally supplied data is the function of the nonmarketing units of an enterprise. On the other hand, the collection, compilation, and dissemination of external data are largely the task of the marketing function.

Market analysis is that portion of the marketing function which is concerned with all of the external factors which can affect the sale of the organization's products. Generally such factors are uncontrollable by the organization. One view of market analysis, in the general framework of marketing analysis,[1] is, then, that it deals with the uncontrollable variables which influence marketing decisions.

Test marketing is concerned with a body of procedures used to test consumers' acceptance of a new product by marketing the product in a limited number of areas. Its place in the chronology of the marketing process is close to product planning. Yet the emphasis placed on the uncontrollables of the marketplace allies test marketing with market analysis.

Markets and Market Areas

Before proceeding with a discussion of market analysis, we should make clear the meaning of the term "market." *A market is a group of potential consumers who have something in common.* We speak, for instance, of the color television market, the replacement market for automobile tires (in contrast to the original equipment market), the Cleveland market, and the teen-age market. In each case we define a category of potential consumers on some basis—in these examples, desire for the product, the motivation for purchase, geographic location, and age group. Of course, the most common interpretation given to the term "market" by laymen is a geographically defined market such as Pittsburgh, Rochester, or the eastern United States. This concept is of great significance in market analysis because the mechanics of the distribution process require that market decisions be made on a geographic basis. One makes the decision to introduce the latest teen-age fad in the New York area, even though the primary market is teen-agers rather than New York. In addition, much of the data which serve as an input to market decision processes is compiled by the Federal government, trade associations, and individual organizations on a geographic market basis.

To illustrate the significance of geographic market areas, let us consider the problems involved in defining a set of disjoint areas, covering the entire United States, which can be used as a basis for data gathering and market planning. Our likely, and logical, first step would be to investigate what others have done along similar lines. We would find that a great deal of effort has been expended on such a definition of market areas; so much, in fact, that we would be very likely to decide that to attempt to begin

---

[1] Care should be taken to distinguish between *marketing analysis*, which is concerned with all of the functions and decisions involved in the transfer of a product from the producer to the ultimate consumer, and *market analysis*, which is one particular aspect of marketing analysis.

anew would be both foolish and beyond the scope of our limited resources. Therefore, we would probably begin by using preestablished market areas and attempt to alter them in some respects to conform to the distribution and marketing organization of our firm.

For the moment, however, let us consider the problem of defining a set of market areas of our own without drawing on the work of others. Although it would not be practical for us actually to do this, the exercise will facilitate greater understanding of market areas, their virtues, and their vices. Our first problem would be to determine a basic building block—a territorial unit to be used in "building" market areas. Since most political units are not defined in terms of the practicalities of modern life, our initial reaction would be to overlook political boundaries and to search for some other territorial unit defined in terms of the business functions carried on within. Yet we quickly realize that this is just about what we envision a market area to be. Since we can hardly define market areas in terms of themselves, we return to the consideration of political boundaries, for, indeed, much of the information which we will later desire to know concerning the market areas is available through government sources on the basis of political territories. An investigation into government and other data which are available will reveal that the county is the smallest political unit for which a great wealth of market data are already available. As such, we choose the county as our basic building block.

Because we realize that market centers in metropolitan areas are of great significance to the marketing process, we would be likely to attempt next to distinguish the nation's important market centers. If our interest is in consumer goods, we might do this on the basis of the volume of retail sales in each city, for example. In defining these centers, we have led ourselves to a natural method for constructing market areas, for if we simply aggregate the counties surrounding a market center, the resulting area should define a meaningful market area. The only question remaining is how to do this. Should we take all counties which are adjacent to the county in which the center is located? Suppose two centers claim the same county under this rule of thumb. What do we do then?

For the answer we must go back to our objectives in developing market areas and to the decisions in which market areas will play an important part. The two most significant of these decisions involve *sales* and *advertising*. The marketing manager is frequently faced with geographic decisions relevant to sales and advertising: "Where to advertise?" "Where to sell?"

As a result we would be likely to base our determination of

which counties to attach to which marketing centers on the reach of media advertising emanating from each center and the attraction of the city as a trading center. To this end we might determine the limits of coverage of television and radio stations, newspapers, and periodicals which are based in each market center. We would undoubtedly discover that most centers have overlapping coverage areas; e.g., Dayton receives television advertising from both the Columbus market center and the Cincinnati market center. We would find many counties, however, which receive a vast proportion of media advertising from a single market center. Even in the populous megalopolis of the eastern United States, many counties are largely oriented, through media advertising, to a single urban market center. Each of these counties would be included in the market area to be constructed around its related market center.

The only remaining question is what to do with the counties which have divided loyalties in terms of media advertising influences. One way to handle this question is to conduct interviews in the counties, the interviews being designed to determine the retail trade orientation of the people. We might have our survey organization ask questions related to the locations at which people shop for food, clothing, and consumer durables, where they go for entertainment, etc. In this way, we could decide either that a particular county is predominantly oriented toward a particular market center or that it should be broken up into areas oriented to different market centers. At any rate, using these survey results, we should be able to complete our definitions of U.S. market areas.

Of course, all this has already been done in one form or another. Curtis Publishing Company, for example, has defined about 500 areas and publishes data on each in its *Market Areas in the United States*.[2] The United States government has defined similar entities called "Standard Metropolitan Statistical Areas" (SMSAs), each of which is an "integrated economic and social unit with a recognized large population nucleus in which a large volume of daily travel and communication is carried on between the central city and outlying parts of the area."[3] These SMSAs do not cover the area of the United States exhaustively, but the reams of market data available in them make them extremely useful. In some sections of the country, for example, we might take the SMSA to be representative of a larger area. Since over 60 percent of the population of the nation resides within the SMSAs, such an approximation might well be warranted.

[2] At this writing, the last of these "annual" reports is several years old. The market areas, which the report is based on, remain quite useful, however.
[3] *Measuring Metropolitan Markets*, U.S. Department of Commerce.

In any case we shall probably desire to begin with already defined areas and alter them in such a fashion that available data can be used and the peculiarities of our own organization's distribution system are accounted for. Having accomplished this, we would wish to distribute maps, detailed descriptions of the areas, explanations of their advantages from the viewpoint of data availability, etc., throughout the organization to obtain the comments of all potential users of the basic areas. Almost invariably we will have neglected some aspect of the marketing process in our design of the market areas and this will result in the complication of the lives of some organizational elements. Eventually we should be able to arrive at a set of areas which are acceptable enough to the organization to serve as a basis for market planning and analysis.

Market Analysis

The study of the uncontrollable factors which affect the sales results of the organization's products must begin with a definition of the market for the product. Is there a market for the product? If so, where? What is its size? What are its characteristics?

To obtain an answer to the basic question concerning the existence of a market invariably requires ingenuity and patience. To have predicted the existence of a large market for skateboards in the mid sixties, for example, one would have had to determine first that surfing and skiing would increase in the favor of teen-agers and young adults. Then one would have had to know the skateboards represent the same sort of recreational fulfillment to those masses not fortunate enough to live near sea or slope. It is highly unlikely that such a "second-order" determination of the market's existence was made by anyone.

To imply that the process of defining an existing market requires clairvoyance, mysticism, or just plain luck would be misleading, however. In most less explosive situations a thorough search for data on which to base logical, if imprecise, inferences is both feasible and useful.

AN ILLUSTRATION OF MARKET ANALYSIS

Consider, for example, the manufacturer who is contemplating the production and sale of outboard motorboats. His problem is to determine the future existence, size, and characteristics of a currently existing market. The market analyst in this case would undoubtedly first study the past performance and current status of the market as it has been, and is being, promoted and garnered by those with whom he may eventually be in competition. To do this, he would make use of the trade magazines of manufacturers, wholesalers, and retailers and the many publications which are aimed at the boating public. In fact, total circulation figures for such publications might be meaningful as an indication of the size of the potential market, for by a quick survey of retailers he could obtain the information that many enthusiastic subscribers to boating publications are not boat owners. These people, who are presumably kept from boat ownership because of their location or income, are potential consumers of boating products. In addition, although the market analyst could expect little help from his potential competitors, individual manufacturers of recreational equipment closely associated with boating—water skis, life preservers, trailers, engines, etc.—would be likely to supply him with data and to recommend other data sources.

After compiling historical information, the market analyst must make some estimate of the size and characteristics of the future market. He would like to know who it is that buys boats for recreational use in terms of their sex, marital status, income, etc. Having found such data, he wants to project the size of these groups into the future and to obtain information concerning possible changes in the character of the market. Publications of the U.S. Department of Commerce would be a starting point for his basic data. If, for example, he found that the typical boat buyer was a married couple in the 30 to 40 age group with income in the $10,000 to $20,000 per year bracket, he would seek out population projections for that age and income group. These, together with a decreasing age tendency of the typical boat buyer, would enable him to make a rough estimate of the size of the potential market.

Another viewpoint might be gained from forecasts of total industry sales based on historical data. One might determine the size of the potential market in the past and view the trends in the proportion of the market which was actually penetrated. Because of the phenomenal growth experienced in this industry in the 1950s, the analyst would undoubtedly choose to consider the effect of the growing replacement market. To do this, he would require data on the useful life of boats and on how long boat owners keep the same boat before trading up. Some insight into the latter factor might be gained by considering family size and the consequent need for more spacious boats by boating families.

To determine the location of the market, the analyst would require data on the geographical distribution of boat ownership, motor ownership, marinas, boat clubs, and magazine subscriptions. The National Association of Engine and Boating Manufacturers and the ˙Boating Industry Association could supply some of these data. The Census Bureau's *County and City Data Book* would supply information on the geographical distribution of income and per capita income which would help in projecting future growth areas for boating.

The impact of the entry into the boating market by the analyst's own organization would also need to be considered. If the market in question were one characterized by a large latent demand being filled by a few sellers, the additional promotion engaged in by an additional seller might bring greater latent demand to sales fruition and serve to enlarge total industry sales. In another kind of market the entry of a new producer might simply mean that the sales pie is cut into a larger number of smaller pieces. This would likely be the case in the pleasure boat market, since it is characterized by a large number of companies, many of which cater to local or specialized markets.

This point suggests a potentially fruitful approach to entering the boating market. Perhaps the segmentation of the pleasure boating market into smaller markets for special varieties of boats should be considered with a view toward possible entry into a particular segment of the total boating market. The automobile industry has, on the basis of careful analysis, divided the automobile market into compact, sports, regular-size, small-foreign, and luxury segments, for example. Most automobile manufacturers try to increase their total sales by producing more and more models designed to appeal to a particular segment. The success of Ford's Mustang was largely based upon the recognition of a market for a relatively economical sports car. In the boating situation the analyst might find that women are becoming more important in the boating market, in terms of both individual female purchasers and their influence on purchases by men. A small survey might reveal that women desire luxurious spacious quarters while "at sea." More importantly, they desire boats which are solidly built and do not pitch and roll. These data might indicate that the organization should consider the large luxury boat segment of the boating market.

The next step would involve a reanalysis of the potential size and character of this segment of the boat market. Moreover, some preliminary calculations using the break-even evaluative approaches of the previous chapter might indicate whether this segment is large enough to support a producer of only luxury boats.

In doing all this, the analyst would unquestionably arrive at several predictable conclusions—that there is no standard approach to market analysis, no end to the complexities which may be considered, and no sources of information which are always useful. The process almost always requires equally as much skill and cunning as it does logic and deduction. If he is both skillful and logical, however, the market analyst can often produce results which constitute a valuable basis for marketing decisions.

MARKET ANALYSIS AND MARKETING DECISIONS

One of the primary benefits which may be derived from market analysis is an indication of how the uncontrollables of the market-place determine that the market will respond to the controllables of the marketing manager. To obtain such indications, special attention must be paid to the characteristics of the market for the product in question.

Characterizations of the market by age distribution, income levels, occupational characteristics (white-collar, blue-collar, professional, etc.), and educational levels are invaluable in selecting advertising media, for example. For reaching an executive market *Fortune* and the *Wall Street Journal* would be more likely to be effective vehicles than would *Mechanix Illustrated* or the *Reader's Digest*. On the other hand, the hobbyist market would be reached better through *Mechanix Illustrated*. To reach the teen-age market, distribution through downtown department stores might be less satisfactory than selling through suburban outlets, and advertisements on disc jockey radio shows might be better than prime-time television sponsorship.

Some of these conclusions are obvious, of course. However, the significant point is that we need to know the basic character of a market in order to reach the obvious (and many not so obvious) conclusions which are relevant to significant marketing decisions. This is the role which is played by market analysis in the context of marketing decision making.

MARKET DATA SOURCES

The sources of the data which are the bread and butter of market analysis depend on the type of product toward which the analysis is oriented. An organization seeking to market a high-speed digital computer or a milling machine and a manufacturer of pleasure boats require different information—simply because their markets are different.

As one might surmise from the foregoing description of market analysis, there has been little attention paid to the systemization of market information sources. Moreover, because of the rapid obsolescence of market data, a source which is a good one for today's decision problem may not be so good tomorrow. Next year the same source may be totally useless.

The United States government is undoubtedly the most abundant basic source of almost every kind of market data. The Census Bureau's *Statistical Abstract of the United States* is a basic reference guide. One of its supplements, the *County and City Data Book*, provides data from the latest censuses of agriculture, business, government, housing, manufacturers, mineral industries, and population. The scale of this supplement is apparent in its more than 150 statistical items for each county and incorporated city with over 25,000 inhabitants. Additional data are given for regions, states, and SMSAs. The *Survey of Current Business* is the journal of the Commerce Department which includes economic data on income, prices, orders, inventories, employment, and a wide range of industrial data. Its particular usefulness is to industrial marketers who are concerned with business enterprises as customers. The Census Bureau's *Business Cycle Development* is another useful aid in this area. The Commerce Department's Business and Defense Services Administration also publishes a pamphlet *Measuring Markets: A Guide to the Use of Federal and State Statistical Data* and a series—*Facts for Marketers*— published in nine regional editions; the latter series incorporates SMSA data from government sources.[4]

Sales Management's "Survey of Buying Power," carried in the June issue of the magazine, is one of the most significant data sources for most consumer market analysts. It includes data on population, sales (for various categories), income, and indexes of buying power for states, cities, counties, and metropolitan areas. The November issue of this magazine carries projections of various market factors and statistics for industrial market analysis. The January and February issues contain data on newspaper and television markets respectively.

The periodical *Industrial Marketing* fulfills a similar function in the realm of marketing which is oriented to industry. In its annual "Market Data and Directory" issue this journal provides information which serves to define the nature and scope of industries and the media serving each industry. It is organized along the lines of the government's Standard Industrial Classification (SIC) system. The resulting standardization adds to the usefulness of the available data. Another primary data source should also be noted—the *Rand McNally Commercial Atlas and Marketing Guide*, which provides statistics on sales, auto registrations, bank deposits, and other pertinent market data. It is also a source of city and town population figures which are not generally available.

It is not hard to see that the list of market data sources is vir-

---

[4] All of the government publications noted in this paragraph are available through the U.S. Government Printing Office, Washington, D.C. 20402.

tually endless. To conclude here, we may refer the interested reader to an excellent compilation of market data sources by Nathalie D. Frank (see references). This handbook of data sources (published in 1964) is useful both for its specific contents and for its implications for more recent publications from the same sources.

Test Marketing

Test marketing is a method of evaluating consumer reaction to a product through the actual marketing of the product in selected test areas. As noted earlier, it is closely associated with the product planning phase of marketing. Yet, because the problems of test marketing are so closely associated with those of market analysis, it is informative to view the test marketing process in conjunction with market analysis.

The significant decision problems involved in test marketing may be thought of as the design of a product testing program, test market selection, the evaluation of test market results, and the drawing of inferences from the test markets to other potential markets.

THE DESIGN OF A PRODUCT TESTING PROGRAM

In designing a product testing program, the marketing manager must consider the objective which the organization is seeking to attain. If the objective is solely related to the measurement of the product's perceived quality and acceptance by consumers, for example, the best approach might be to test on a more limited basis than test marketing. One inexpensive way to test potential consumer reaction is to allow some of the organization's employees to use the product if this is practical. Responses to the product may be invited through questionnaires—perhaps unsigned, to increase the likelihood of candor. Of course, there is usually little basis for believing that the employees of a company are representative of the consuming public. Yet their responses may indicate positive or negative aspects of the product which should either be emphasized or deemphasized. Through such "concept testing" during the development of a product, significant product improvements can often be made. One particular virtue of a testing approach involving employees lies in the limitation of the risk of conveying valuable information to competitors which is inherent in most other testing procedures.

Another approach to non-test marketing product testing is through the use of *consumer panels*—groups of consumers who have a continuing relationship with a testing organization. These individuals may be called upon to use the product before it is placed on the market and to evaluate it in their own frames of reference. Their reactions may be very valuable in predicting what the market's reaction would be if the product were placed on sale. In many cases the brand name and manufacturer of the product are not revealed to the panel members, so that they may evaluate the product on its own merits. The primary advantage in the use of independent panels rather than employee groups is in the possibility of their being representative of the potential market. Of course, somewhat greater chances of disclosure of vital information to competitors are inherent in panels than in employee groups. We shall say more about the usefulness of such panels as used in conjunction with test marketing later in this chapter.

If it is determined that the objectives of product testing are somewhat more comprehensive than simply eliciting consumer reactions, a test marketing program may be advisable.[5] One possible objective which would not easily be achieved without test marketing is the testing of the market's reaction to the nonproduct controllables which are the province of the marketing manager. If we desire to know the effect of various combinations of prices, promotional appeals, advertising media, etc., in an area, it is usually necessary to test-market the product, for although data on sales of other similar products are valuable, no two products are alike in the minds of consumers.

Moreover, since many products are introduced sequentially into groups of market areas rather than into all of the nation simultaneously, the use of test market results as a basis for choice of future market areas is often of significant importance. Many beers were first introduced in this manner, for instance. In such a situation the first group of areas to be selected for marketing may be treated as the test markets and inferences drawn for other areas which are potential markets for the product.

TEST MARKET SELECTION

The selection of areas in which to test-market is of significance to the results of a test marketing program. Ideally one would wish to choose the test markets randomly from the set of all markets in which the product might potentially be sold. If the markets are chosen in this fashion, each market area has an equal likelihood of becoming a test area. As a result, all of the powerful techniques of probability and statistics, which are based on the assumption of a known likelihood of selecting each possible element, may be

---

[5] Of course, very often employee and panel product testing are combined with test marketing.

brought to bear in analyzing the test market results and drawing inferences to other potential markets.

Of course, the idea of random selection of test market areas often satisfies neither the passions of traditionalists nor the practicalities of marketing. Problems of distribution and promotion make it difficult for many firms to market a product in several widely separated areas which may be the result of a random selection. In such instances these practicalities force the selection of test markets on nonrandom bases. These bases primarily center about the *representativeness* of the test areas in terms of the total market. If, for example, the distribution outlets, advertising media, population, and demography of the Boston area are similar to those which the total market is expected to exhibit, it might be a good choice as a test market. This selection would hinge on other factors also, however. A previous good record as a test market would augur well for a city's selection, as would cooperative advertising media,[6] wholesalers, and retailers.

If a number of areas are to be used as test markets, each may be selected as representative of a larger group. San Francisco might be chosen to represent the Far West, and Denver, Fort Wayne, Pittsburgh, Boston, and Atlanta might be chosen to represent the Mountain states, the Middle West, Appalachia, New England, and the South respectively. Or Cleveland might be chosen to represent a manufacturing center, Sioux City to represent a farm center, New Orleans to represent a trade center, etc. Or a single area may be chosen to be representative of the entire nation. Syracuse and Columbus, Ohio, for example, have frequently been used because of their similarity to national distributions of age, sex, income, ethnic groups, etc.

In any event the basic truism appropriate to this selection process is that *the choice of "good" test markets depends on the product and on the objectives of the test marketing program.* Salt Lake City is not a good place to test market coffee because of its large non-coffee-drinking Mormon population group and St. Louis is not good for another reason—the dominance of the market by a single regional brand. A city with poorly equipped, poorly staffed, or uncooperative television stations might be satisfactory for some test marketing programs, but it would not do for one which had the evaluation of television messages as one of the objectives of the program. The recognition of this dependence on the product and the test marketing objectives leads naturally to the conclusion, contrary to the folklore of marketing, that there is no such thing as a good or

---

[6] Information on past uses of various cities as test markets may be obtained from advertising agencies. *Sales Management* periodically publishes lists of the most frequently used cities.

bad test market. There are only those which are good or bad for a product and for a given purpose.

Some consideration needs to be given here to the logical consequences of the pervasive choice of test markets on some basis having to do with representativeness. When a set of test markets is chosen randomly, the likelihood of selecting each possible market as a test market is known, and as a consequence the theory of probability and all of the statistical techniques based on it are applicable. In nonrandom selection, such as by choosing representative markets, the probability of the selection of any particular market is unknown, and as a result the theory of probability cannot be applied rigorously. Of course, one can always take test markets chosen to be representative and perform the same arithmetic calculations as would be performed for markets chosen on a random basis. The results of these calculations do not have the simple properties relating to bias and confidence, however, as the results of calculations emanating from randomly chosen areas. As a result, although the choice of test markets on a representative basis is usually more practical in terms of the marketing environment than a random choice, the results which can be anticipated from random selection are "better" than those which can be expected from the choice of representative markets.

Of course, selection of test markets on the basis of their representativeness relies on the judgment of individuals. Although we often make use of such judgments in quantitative marketing analysis, the potential fallacy in this method is that it may imply the existence of knowledge about the markets while this is exactly what is being sought in the test marketing program. When an individual chooses test markets on a representative basis, he should probably ask himself whether his idea of representativeness does not subsume the information he is seeking; i.e., he should question the need for doing any test marketing at all.

In any event the practicalities of marketing often require the choice of test markets on a nonrandom basis. Although this practice suffers from some theoretical difficulties, its wide use and success indicate that it may be adopted to give some indication of the response which the product will receive in the total marketplace. Often this indication is far superior to the traditional educated guesses about what will occur when a product is introduced.

EVALUATION OF TEST MARKET RESULTS

Once the set of test markets has been selected and marketing begun in them, the evaluation of sales results becomes of paramount importance, for all of the decisions which are to be made concerning the composition of the total marketing strategy and the selection of future markets are based on an ability to determine the value of the performance displayed by the product in the test markets.

The evaluation of these market results presents two primary problems: first, the question of *measuring* the relevant performance variables in the test markets and, secondly, the problem of *evaluating* performance by some overall measure of utility or performance indicator which one may regard as approximating utility.

The measurement of performance variables is a problem at the practical level. It generally assumes greater significance in test marketing than elsewhere, because the product, and perhaps the organization, is new to the test market area. As a result, the establishment of specific procedures for measuring sales performance may be necessary. If, for example, data on the product's share of the market in the area are deemed necessary to the evaluation of results, the need for measuring the total sales of all competitors in the area is apparent. Some source of these data must be determined before beginning the test marketing programs. If some established sources are available, they should be checked and an agreement reached. If it is necessary to go to government reports, as might be the case with products which are taxed and for which sales must therefore be reported to government agencies, the method of collection and time lags between the reporting and availability of data should be checked in advance for the degree to which they will meet requirements. Since competitors are normally not willing to divulge their own sales data, estimates may sometimes be made on bases other than those of in-store audits, warehouse withdrawal audits, and published information. In one instance the sales of various brands of a product in an area were estimated from counts of the delivery trucks leaving competitive distribution points. The average delivery capacity of a single truck was well known, and this information, coupled with an awareness of union practices governing work loads, led to estimates which were believed to be highly accurate.

All of these comments are appropriate for data which are to be measured in the marketplace itself. These *market data sources* are invariably important since they reflect firsthand information concerning happenings in the total market. In some instances the use of secondhand sources, such as consumer panels operating in the test areas, is an invaluable complement to the basic market sources. The basic point to be made concerning the measurement of performance variables is that an effective procedure should be devised *prior to entry* into the test areas. In some cases the relative availability of data from area to area will be one of the factors governing the selection of the test markets.

In contrast to the practical difficulties involved in measuring performance in the test markets, the problem of evaluating performance is primarily a theoretical one. This problem has already been touched upon in the context of product decisions in Chapter 5. Since we are seeking an overall measure of satisfaction with performance, the theoretical question is how to develop a utility measure for multidimensional outcome descriptors. However, for the particular decision problems involved in test marketing, the development of such a comprehensive measure is usually not necessary. Often single performance measures or unsophisticated combinations of performance measures are sufficient to determine grossly the success or failure of a product in a test marketing program. A basic distinction which must be made between various approaches to performance evaluation in the test markets is on the basis of the source of the data which are used.

Performance Measurement and Evaluation Using Consumer Panels. One of the devices used for collecting data on sales in a test market is the *consumer panel.*[7] Consumer panels used for this purpose are groups of people in the test market who are generally associated with a research organization, rather than a particular manufacturer, and who record their purchases of various product types and periodically report this information to the research organization. It is clearly important that the panel must not be directly associated with the organization which is test-marketing a new product in the area; for example, if Procter & Gamble were to contact a group of people and ask them to record their purchases of toothpaste, there would be no way of ensuring that this contact would not bias the individuals toward (or against) Procter & Gamble's brand or some other brand believed to be produced by them. If the panel is associated only with an independent research firm and if the manufacturer who will use the data which they generate is not identified to them, no such bias will be introduced into their purchases. In such a case the panel may simply be asked to record their toothpaste purchases, without reference to any particular brand. This would presumably avoid any influence which the act of recording data might have on purchase behavior.

The value of using consumer panels is demonstrated by the

---

[7] Note that the previous mention of consumer panels was in the context of pretesting, i.e., supplying products which are not yet being marketed to panel members for their evaluation. In the test marketing situation, the panel members are not given the product, but their purchases in various product categories are recorded and reported. In this way information on the purchasing behavior of consumers is obtained. Usually no specific evaluative comments are elicited from panels in test markets.

recognition that it is virtually *the only method of obtaining information on the time pattern of purchases made by individual consumers.* There are many devices for gathering data on total purchases made in an area or at a particular retail location; but to obtain data concerning individual consumers' purchasing patterns over time, we must necessarily have access to the records of purchases of individual consumers. We might, of course, simply survey customers in a supermarket and ask them when they last bought detergent and what brand they bought. It would be very unlikely that we could obtain accurate information in this way, however, since humans have notoriously faulty memories concerning matters which are trivial to their everyday existence.

The use of consumer panels may at first appear to be an expensive and tedious device for obtaining test market data on the pattern of consumer purchases, but this generalization is not always valid. In many markets which may be used for testing purposes, there are panels established on a permanent basis under the auspices of research organizations. The individuals who comprise these panels may in some cases be chosen on a random basis so that inferences may be drawn from the behavior of the panel to the behavior of the market by using the methods of probability and statistics. In most cases, however, the practicalities of their choice permit the selection of panel members only as a representative basis. Just as test markets are often selected to be representative of other markets, so are panels often selected to be representative of the consumers in their market area. In some situations, for example, the distribution of ethnic backgrounds and income might be of primary concern, and so the panel would be established in a fashion to ensure representation with respect to these two characteristics. It is clear that the manner in which the panel is chosen may depend on the kind of product for which they will record data. Hence, any use of an established panel should be preceded by a thorough investigation of the compatibility of the particular panel and product in question.

After an acceptable panel has been found, one needs only to make an agreement with the sponsor and have the collection of data begun. For many products such as packaged grocery products, the panel may already be recording purchase data, and it will thereby be possible to obtain historical data on purchase patterns for the product under consideration. In any case it will probably be desirable to have the consumer panel begin to record data on the product type in question before beginning test marketing. This will ensure that all of the bugs are out of the data collection system before accurate data are necessary. Panel members who collect or record data improperly can be corrected prior to the beginning of test marketing, for instance. This procedure will also provide a standard for comparison, for the test marketing organization may wish to estimate the pre-test-marketing shares of the market held by each competitor and to compare them with market shares achieved during test marketing.

The market share measure. One of the simplest measures of performance which may be used as a basis for the evaluation of test marketing results is *market share*. We have already discussed how data may operationally be obtained directly to estimate the market share as the ratio of the total sales of the product in question in the market to the total sales of all such products in the market. Three further points need to be made here. First, we should recognize that the word "sales" as used in the ratio just described is not precisely defined. We may think of sales as the number of units sold or as the total sales revenue achieved. If a constant price holds for all brands sold in the market, the market shares calculated on the basis of quantity and revenue will be the same. If unit price is not constant, they will differ. Care should always be taken to define which measure of market share is being used so that illogical and incorrect conclusions will not be drawn.

Secondly, we should note that as an alternative to the estimation of market share using data on total sales in the market, we may calculate the market share relative to the consumer panel and, on the basis of the representativeness of the panel, use this value as an indication of the total share of the market. The Market Research Corporation of America is one research organization which has provided market share data for some products on this basis. On the other hand, we might wish to conduct a sample survey of consumers or an audit of retail stores in order to obtain such data. In this case we would be working with a kind of nonpermanent panel.

The third point which we need to recognize is that a *time interval* over which the measurement is made is implicit in every market share calculation. We should be careful to avoid comparisons of market shares which are based upon sales realized in time periods of varying length. The dangers in such comparison are fairly apparent. Not the least of these is the simple realization that a market share attained in a short time interval may be very unrepresentative of a long-term share. Since most companies' "sales" data are garnered from shipment records, rather than actual consumer sales, fluctuations in the inventories held by wholesalers and/or retailers can severely bias short-term sales data. If, for example, a number of retailers stock up during a given period, the manufacturer's shipment records will provide a higher estimate of market share than data on actual sales to consumers will. Over a longer time period the effect of such a situation would be relatively less.

Before giving consideration to the ways in which the market share measure may be used to *evaluate* performance, let us first view two measures of test marketing performance which are obtainable only through consumer panels.

Measures of penetration and repetition. Two other valuable evaluative measures which can be obtained from consumer panel inputs are those involving market *penetration* and *sales repetition.* *Penetration may be defined as the proportion of households (in the panel) which make an initial purchase of the product which is being test-marketed.* The importance of this measure for the entire marketing program for the product is clear, since total sales depend heavily on the number of triers. Even though some households will become *triers* but not habitual *users* of the product, those triers who do become users represent the source of the most significant portion of continuing total sales revenue.

Of course, the penetration that we are discussing here is that which is achieved in the entire market, not simply that which is attained among the relatively small number of panel members. It is usually not possible for the marketing manager to measure penetration in the market, since he obtains data on total sales and has no idea as to what proportion of the sales is to triers and what proportion to repeat users. As a result he must use the data emanating from the panel as representing data for the entire market. At the same time he may consider adjustments to account for inadequacies in the panel's degree of representation of the total market.

The term "households" as used in this definition of penetration implies that the product in question is one for which the household is the purchasing decision unit. Toothpaste is an example of a product for which this would be the case, since one individual (probably "mom") is usually assigned the task of buying toothpaste for the entire family. On this basis one would not wish to consider each individual in the market area as a potential purchaser, but only each household. In the case of teen-age clothing products, this might not be so, and we would consequently replace the word "households" with "individuals over 12 years of age" in the definition of penetration.

The penetration which is achieved in a test market area (as measured via the consumer panel for that area) may be illustrated as a function of the time since the beginning of sales in the area, as in Figure 6-1. This graph indicates that about 12 percent of households have made at least an initial purchase in the first month, over 8 percent additional in the second month, etc. After some period of time, the total penetration appears to reach a total of

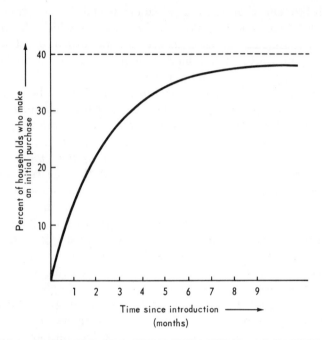

FIGURE 6-1   PENETRATION OF A NEW PRODUCT DURING TEST MARKETING

about 40 percent and further increases thereafter appear to be small. This conforms to what one might expect, since there are some individuals who will never use the product (e.g., no household having only bald male members is likely ever to make a purchase of hair tonic), and the major portion of those households that will ever use the product will undoubtedly do so in a relatively short time after its introduction.

It should be emphasized that the penetration measure is the proportion of households that have purchased the product *at least once*. When a particular household makes its first purchase, it is included in the proportion. Thereafter all subsequent purchases by that household have no influence on this measure.

Penetration on a market is not, however, the sole factor contributing to success. Repetitive purchasing is the lifeblood of most consumer products. As a result, an auxiliary measure of the repetitive sales performance of a product in a test market is necessary. One set of measures which may be used is repeat ratios. *The first repeat ratio for a product is defined as the proportion of initial purchasers who make a second purchase.* Similarly, *the second repeat ratio* is defined as *the proportion of first repeaters who make a second repeat.* Other repeat ratios may be similarly defined as necessary.

Two points should be emphasized concerning these ratios. The first point has to do with terminology. A "first repeater" is a "second purchaser" and a "second repeater" is a "third purchaser"; thus, the *second* repeat ratio might equally well have been defined as *the proportion of second purchasers who make a third purchase.*

Secondly, the simplicity of these ratios in no way implies that they are easy to measure in the test panel. The difficulty is that one is never sure that a first purchaser will not make a second purchase, even if he has failed to do so for a very long time. This, together with the fact of life that we must draw some conclusions from the panel after observing them only for a finite length of time, indicates that some estimate of true repeat ratios for the panel must be made. Of course, the numerical values determined from panel data are then used in making estimates of the repeat ratios for the market area.

Let us consider a numerical example to illustrate the problems which arise in making estimates of repeat ratios for the panel.[8] Suppose that we observe the panel for a year and find that 6,000 households in the panel made at least one purchase during the year. Of these, 2,000 made at least two purchases. On the face, one would therefore be tempted to estimate the first repeat ratio to be 2,000/6,000 or $\frac{1}{3}$. However, we must recognize an omission; some of the 6,000 first purchasers will make a second purchase subsequent to the year for which we have recorded data. They will make their second purchase then for one of two reasons: either they have a usage rate for the product which dictates rather long intervals between successive purchases, or their first purchase was not made until late in the year and they have *not had an opportunity to make a repeat purchase.*

Figure 6-2 illustrates these points for two panel members, purchaser A, who will purchase the product approximately every $1\frac{1}{2}$ years, and purchaser B, who will purchase the product every two months. In this case, the product was introduced on January 1 and purchaser A tried it almost immediately. Purchaser B, on the other hand, did not make his first purchase until December. Both of these individuals will not make their second purchase until after the one-year observation period is complete—purchaser A during the following June and purchaser B during the following February. As a result neither individual will be counted in the 2,000 who were observed to make a second purchase during the year-long observation period, and both will be counted in the number of first purchasers observed during that year. Both of these purchasers

---

[8] Some of the ideas used here are developed in a somewhat different context in the Fourt and Woodlock paper in the references.

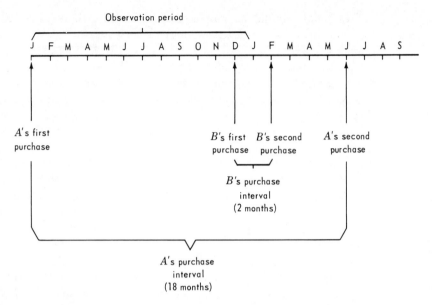

FIGURE 6-2   PURCHASE INTERVAL ILLUSTRATION

will contribute to our making a simple calculation of a repeat ratio that is an underestimate of the true underlying repeat ratio.

In terms of the ratio which may be calculated from the data available from the consumer panel, we may redefine our operational repeat ratio estimates to be

$$\text{First repeat ratio} = \frac{100(\text{number of first repeats in observation period})}{\substack{\text{number of first purchasers who had an} \\ \text{opportunity to repeat in the observation period}}} \tag{6-1}$$

$$\text{Second repeat ratio} = \frac{\substack{100(\text{number of second repeaters in} \\ \text{observation period})}}{\substack{\text{number of first repeaters who had an} \\ \text{opportunity to repeat in observation period}}} \tag{6-2}$$

etc. If we define the "opportunity to repeat" in terms of the purchase interval, we can account for the omission of both categories of individuals which were previously discussed. We may accomplish this by considering the distribution of the interval between purchases for the consumers in the area. Table 6-1 is an illustration of such a distribution. This table illustrates that 10 percent of the consumers have a purchase interval which is less than one month in length, 30 percent have a purchase interval between one

TABLE 6-1   DISTRIBUTION OF INTERVAL BETWEEN PURCHASES

Purchase interval, months	Proportion of consumers, percent
0–1	10
1–2	30
2–3	20
3–4	20
4–5	10
5–6	5
6–12	2
12–18	2
18–24	1

and two months in length, etc. None have purchase intervals greater than two years in length.

The most significant way in which this distribution oversimplifies the real world is that it implies that each individual has a fixed purchase interval, i.e., that each person purchases at a regular interval without variation. This is not truly descriptive of the somewhat haphazard purchase intervals exhibited by most consumers. If we were to think of the distribution as representing the distribution of purchase intervals which we would find if we took a very large number of observations in the market—i.e., if each observation used in developing the distribution were one *instance* of the display of an interval between purchases by *some* individual rather than the invariant interval which an in•ividual always displays—the oversimplification would be obviated and the distribution could be utilized just as easily.

Thinking in these terms, we may observe the distribution and note that a purchase interval of four months or less includes 80 percent of those which were observed. If we thereby arbitrarily declared that *any individual who made his first purchase in the last four months of the observation year had not had an opportunity to repeat within that year,* we would not be grossly incorrect.

Using this as a working definition of the "opportunity to make a first repeat purchase," we may calculate the first repeat ratio by using the 2,000 first repeaters who were observed as the numerator of expression (6-1) and inserting only the number of purchasers who made their first purchase in the first eight months of the observation period in the denominator. In other words, if, 5,000 of the 6,000 first purchasers who were observed in the entire year made their first purchase in the first eight months of the one-year observation period (and consequently 1,000 made their first purchase in the last four months), only the 5,000 would be used in the denominator of expression (6-1), the assumption being that the 1,000 had not had the opportunity to repeat. The first repeat ratio would then be 2,000/5,000, or $\frac{2}{5}$.

To illustrate these points a bit further and to bring the discussion closer to the real world, let us consider the hypothetical data of Table 6-2. These data may be thought of as those collected from a consumer panel during the one-year observation period. In the first two columns of this table, we observe that 6,000 initial purchasers were observed, 2,000 second purchasers, 1,000 third pur-

TABLE 6-2   CONSUMER PANEL PURCHASE DATA

Purchaser type	Number observed during year	Average interval till next purchase, months
First purchaser	6,000	5
First repeater	2,000	4
Second repeater	1,000	3
Third repeater	600	2
Fourth repeater	400	1
Fifth repeater	100	1

chasers, etc. It is important to note that these are the number who made *at least* that number of purchases; i.e., the 2,000 who made a second purchase (first repeat) are also necessarily among the 6,000 who made a first purchase. Similarly the 1,000 who made a third purchase are also among the 2,000 who made a second, etc. The third column illustrates an aspect of the real world which we have not previously considered: *that the average interval between the various sequential purchases may vary widely.* In this case the average interval varies from five months for that between the first and second purchases to one month between the fifth and sixth purchases.

Because of this variation, we might choose to use a different correctional time period at the end of our observation period to determine the number of individuals who had an opportunity to make the repeat purchases required by the various repeat ratios. One way might be simply to use a time period equivalent to the average interval between purchases in every case. To show this, we may expand the previous table into Table 6-3. In this table we can observe that the fourth column repeats the third since we have decided to use one average purchase interval as the period at the end of the year-long observation span in which we will

TABLE 6-3 CONSUMER PANEL PURCHASE DATA AND CALCULATIONS

Purchaser type	Number observed during year	Average interval till next purchase, months	Lack-of-opportunity period at end of year, months	Number observed in period preceding lack-of-opportunity period	Repeat ratio
First purchaser	6,000	5	5	4,500	
First repeater	2,000	4	4	1,800	$\dfrac{2,000}{4,500} = 0.444$
Second repeater	1,000	3	3	900	$\dfrac{1,000}{1,800} = 0.556$
Third repeater	600	2	2	500	$\dfrac{600}{900} = 0.667$
Fourth repeater	400	1	1	300	$\dfrac{400}{500} = 0.800$
Fifth repeater	100	1	1		$\dfrac{100}{300} = 0.333$

consider that no opportunity for repeat was available; i.e., if an individual made his first purchase in the first seven months of the year, he is considered to have had an opportunity to make his second, and if he made his initial purchase in the last five months, he is considered not to have had an opportunity to make a second purchase. Similarly, if he made his second purchase during the first eight months of the year, he is considered to have had the opportunity to make a third purchase, and if he made his second purchase in the last four months, he is considered not to have had an opportunity for his third. The fifth column of Table 6-3 simply indicates the number who were observed to make the respective purchase in the period of the year preceding the lack-of-opportunity period. Of the 6,000 first purchasers in the year, 4,500 made their first purchase in the first seven months of the year and are therefore to be counted as having had the opportunity to make a second purchase. Of the 2,000 first repeaters, 1,800 made their repeat in the first eight months of the year and are therefore to be counted as having had the opportunity to make a second repeat; similar interpretations apply to the other numbers.

The various repeat ratios which may be calculated are given in the last column of Table 6-3. The first repeat ratio is simply the number of first repeaters (2,000) divided by the number of first purchasers who had an opportunity to repeat (4,500), or 0.444.

The second repeat ratio is the number of second repeaters (1,000) divided by the number of first repeaters who had an opportunity to make a second repeat (1,800), or 0.556, etc.

These examples should serve to illustrate the concepts and calculations involved in the market penetration and sales repetition measures. In the next section we shall concern ourselves with using these measures to evaluate test marketing results.

Performance evaluation using consumer panel measures. The important measures which may be developed on the basis of consumer panel data may be used for a number of purposes other than "pure" evaluation of test market results. These purposes—early prediction, control, and drawing inferences to the performance of future markets—will be discussed in subsequent sections. Here we are concerned only with the question: How may we evaluate performance in a test market based on consumer panel data? The most obvious method for accomplishing this is to *compare a single performance measure with a standard.*

In the case of the penetration measure and repeat ratios, we may set goals to be used as a standard for evaluation. It is clear that the initial goal should involve penetration, for it is only through consumers' trying the product that there is any hope for long-term sales success. After most of those who will try the product have already done so, the goal should shift to the repeat sales ratios. Gross evaluations may be made on the basis of numerical values established for the general product class. For example, Fourt and Woodlock (see references) have found that first repeat ratios of less than 0.15 for packaged grocery products "almost always spell failure" and that some very successful products have demonstrated repeat ratios of about 0.50.

In particular we shall want to know whether a sufficient proportion of total sales are accounted for by repeaters rather than triers as time passes. Although penetration is quite important, it is apparent that most products will not achieve success for very long simply by persuading new people to try the product. The supply of triers will soon be exhausted. Long-term success relies on repeat purchases. Hence our particular concern is whether sales in the test markets continue to be dominated by new purchasers or the balance begins to shift toward repeaters. If the latter occurs, the product probably has appeal for consumers which foretells future success.

In the case of the market share measure, we might also simply compare the estimated market share with some goal which has been previously established and base our evaluation on this simple comparison. Undoubtedly we would have previously set our goal as a function of time; i.e., we would have realized that the ultimate market share goal could not be achieved instantaneously and would express our goal in terms of the market shares achieved at various points in time subsequent to introduction.

Alternatively, we might wish to consider explicitly that the various test markets will have different attainable market shares, depending on the product's local appeal, the number of competitors selling there, etc. To do this, we might refer to the historical market share data provided by the consumer panel for the time periods immediately preceding the beginning of test marketing. The share which the leading brand attained in that period might then be chosen as our ultimate goal if we are rather optimistic about the appeal our product will have. Or if there are several competitive brands, we might choose the *median* market share attained. If, for example, nine brands attained market shares of 2, 4, 5, 6, 10, 12, 15, 18, and 28 percent, the median market share—the one having an equal number of greater and lesser shares—is 10 percent. Obviously, using this as an ultimate goal is considerably less ambitious, yet probably more realistic, than using the highest share, 28 percent. However, if the new product is well differentiated from the others in the market—i.e., if the ways in which it differs from the other brands already in the market are apparent (to the consumer) and if it is designed to appeal to large segments of the market—there might be no valid reason for being realistic rather than optimistic.

The basic concepts involved in such comparisons may be incorporated into a single procedure if sufficient past data are available. This might be the case if the consumer panel had been collecting data on competitive brands for some years prior to our interest in the area, for instance. If such data are available, they might be arrayed as in Table 6-4. In this table, data are presented for a two-year period in a market in which four brands have been in competition. At the beginning of 1966 only three brands had been sold; brand *D* was introduced in April of that year. Brand *D* quickly captured about a fourth of the total market and the other brands were accordingly hurt. Brand *A* quickly recovered

TABLE 6-4   COMPETITIVE MARKET SHARE DATA (Percent)

	1966												1967											
	J	*F*	*M*	*A*	*M*	*J*	*J*	*A*	*S*	*O*	*N*	*D*	*J*	*F*	*M*	*A*	*M*	*J*	*J*	*A*	*S*	*O*	*N*	*D*
Brand *A*	30	25	30	25	20	45	50	35	35	30	35	30	25	15	15	15	15	10	10	10	10	15	10	15
Brand *B*	30	40	35	40	30	20	20	25	30	35	25	30	30	30	30	35	35	40	40	40	40	40	45	40
Brand *C*	40	35	35	30	30	10	15	10	10	10	5	10	10	15	10	5	10	15	10	10	5	5	10	5
Brand *D*	0	0	0	5	20	25	15	30	25	25	35	30	35	40	45	45	40	35	40	40	45	40	35	40

from this decline. Brand *B* eventually recovered, although not so dramatically as *A*. Brand *C*, on the other hand, did not recover its lost share during this two-year period. Near the end of the period brand *A*'s performance began to fade, leaving two predominant brands, *B* and *D*.

Using these data, we may calculate the *one-month changes in market share* for all brands and all months during this period and display them as in Table 6-5. This table simply depicts the month-to-month *changes* in market share experienced by the brands in the market during 1966 and 1967; e.g., brand *A*'s share changed from 30 to 25 percent between January and February of 1966, as indicated by the —5 entered at the upper left of Table 6-5.

Similarly, we may calculate the *two-month net changes in market share* as in Table 6-6. For example, brand *C* experienced a net change in market share of —15 percent between May and July of 1966. This may be found as the M–J entry for brand *C* in Table 6-6. It may be verified in Table 6-4 by noting that brand *C*'s market share in May of 1966 was 30 percent and in July had declined to 15 percent—a 15 percent loss.

Referring again to the one-month changes of Table 6-5, we may note that the *largest one-month increase in market share* during this period is 25 percent, realized by brand *A* between May and June of 1966. Similarly the *largest one-month decrease in market share* is the 20 percent loss incurred by *C* in the same period. Correspondingly we can determine from Table 6-6 that the *largest two-month net increase in market share* is 30 percent and the *largest two-month net decrease* is a loss of 20 percent.

If we were to construct corresponding tables for three- and four-month net changes, we would find that the largest increases and decreases are those given in Table 6-7.

Plotting the values of Table 6-7 with the largest increases at the bottom and decreases at the top and with the time increments from right to left (for reasons which will become apparent later), we obtain the curves of Figure 6-3. These curves represent the *outermost limits of changes in market share which have occurred in the observation period for all brands in the market area.*[9] As such, they represent a kind of limit for such changes; i.e., any comparable changes which violated these limits would probably be considered very unusual and of significant importance. It is clear, of course, that in no way do these curves represent an inviolable limit, for, indeed, any brand in any given period might

[9] Elizer Shlifer and Wayne Marshall applied ideas similar to these in a research project at Case Institute of Technology. The interested reader will find that the technique to be described is closely related to statistical techniques referred to as "CUSUM." (See the Barnard and the Johnson and Leone references.)

TABLE 6-5 ONE-MONTH CHANGES IN MARKET SHARE, 1966–1967

	J-F	F-M	M-A	A-M	M-J	J-J	J-A	A-S	S-O	O-N	N-D	D-J	J-F	F-M	M-A	A-M	M-J	J-J	J-A	A-S	S-O	O-N	N-D
Brand A	-5	5	-5	-5	25	5	-5	0	0	5	-5	-5	-10	0	0	0	-5	0	0	0	+5	-5	5
Brand B	10	-5	5	-10	-10	0	0	5	5	-10	5	0	0	0	5	0	5	0	0	0	0	5	-5
Brand C	-5	0	-5	0	-20	5	-5	0	-5	-5	5	0	5	-5	-5	-5	5	-5	0	-5	0	-5	-5
Brand D	···	···	5	15	5	-10	10	-5	0	10	-5	5	5	5	0	-5	-5	5	0	5	-5	-5	5

TABLE 6-6 TWO-MONTH NET CHANGES IN MARKET SHARE, 1966–1967

	J-M	F-A	M-M	A-J	M-J	J-A	J-S	A-O	S-N	O-D	N-J	D-F	J-M	F-A	M-M	A-J	M-J	J-A	J-S	A-O	S-N	O-D
Brand A	0	0	-10	20	30	-10	-5	-5	0	0	-10	-15	-10	0	0	-5	-5	0	0	5	0	0
Brand B	5	0	-5	-20	-10	5	-5	10	-5	-5	5	0	0	5	5	5	5	0	-5	-5	5	0
Brand C	-5	-5	-5	-20	-15	0	-5	0	-5	0	5	5	0	-10	0	10	0	-5	-5	0	5	0
Brand D	···	5	20	20	-5	5	5	-5	10	5	0	10	10	5	-5	-10	0	5	5	0	-10	0

TABLE 6-7   SUMMARY OF LARGEST CHANGES IN MARKET SHARE

	Period			
	1 *month*	2 *months*	3 *months*	4 *months*
Largest increase	25%	30%	25%	25%
Largest decrease	−20%	−20%	−25%	−25%

well exceed them. The use of the term "limit" here is in terms of *expectations,* or likelihoods. One would not expect any particular brand in a particular time interval to exhibit a change in market share which constituted a greater increase than those on the lower curve or a greater decrease than those on the upper curve; i.e., the likelihood of such a change is considered very small.

To make use of these curves, we plot the market share achieved by the product being test-marketed as a function of the number of months since introduction, as is done in Figure 6-4 for a product which has been sold in the market for four months. This figure shows that a 10 percent share of the market was achieved in the first month, a 15 percent share in the second month, and a 20 percent share in the third and fourth months after introduction.

If we superimpose the curves of Figure 6-3 on the graph of market share (Figure 6-4), placing the curves in a horizontal posi-

FIGURE 6-3   CURVES SHOWING LARGEST INCREASES AND DECREASES IN MARKET SHARE OF ANY BRAND IN MARKET

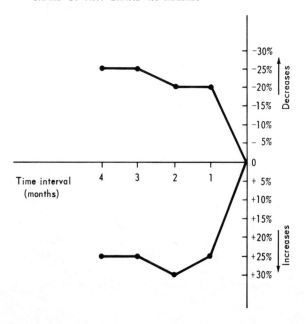

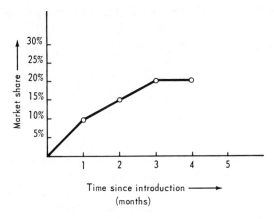

FIGURE 6-4   MARKET SHARE ACHIEVED IN TEST MARKET

tion with the vertex at the last value of market share achieved, we obtain Figure 6-5. In this figure, termed situation A, we note that the actual change in market share which has taken place over the four-month interval, as represented by the difference between the 0 percent market share at the beginning of the period of measurement and the 20 percent point achieved at the end of four months, falls within the limits prescribed by the bounding curves. Moreover *the net change in market share in the most recent three months,* as represented by the difference between the 10 percent

FIGURE 6-5   CURVES OF PAST LARGEST INCREASES AND DECREASES IN MARKET SHARE SUPERIMPOSED ON MARKET SHARE GRAPH—SITUATION *A*

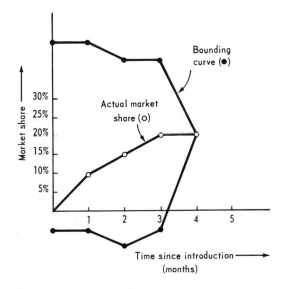

point and the 20 percent point at which the vertex rests, falls
within the bounds, as does the net change in market share which
has occurred in the most recent two- and one-month periods.

The potential significance of this information is apparent. Simultaneously we have compared the last month's change in market
share with *all* one-month changes in market share which were
experienced in the two-year observation period by any brand in
the market, the most recent two-month net change with all past
two-month net changes, the most recent three-month net change
with all past three-month changes, etc. In this case, since the limits
are not violated by any of these, we may infer that our product's
performance is neither exceptionally good nor exceptionally poor.

In another market the set of curves which bound the largest
increases and decreases in market share would be different, because
the past history of any two markets is usually quite different. If
the curves describing another area are superimposed on the market
share graph for the area, the result might be that shown in Figure
6-6. In this instance, situation B, we would be likely to conclude
that the market share actually achieved indicated great success
for the product. This would probably be the conclusion because
three of the four actual changes in market share are greater than
comparable changes which have occurred in the past in the market.
This is indicated by the fact that three of the four changes intersect
the lower bounding curve. The most recent one-month change,
as indicated by the rightmost line segment on the "actual market
share" graph, is the only recent change which does not exceed all
past changes over comparable periods. The most recent net two-
month change, as denoted by the *two* rightmost line segments

FIGURE 6-6   MARKET SHARE AND BOUNDING CURVES—SITUATION *B*

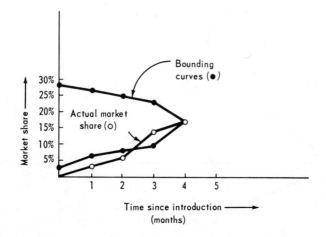

on the "actual market share" graph, cuts through the bounding curve, indicating that this change is greater than any net two-month increase which has occurred in the past. Similarly the net three-month and net four-month changes also intersect the lower bounding curve.

In comparing situation *A* of Figure 6-5 and situation *B* of Figure 6-6, we should note that we assess the actual performance in the latter case as good and the actual performance in the former case as mediocre, *even though the actual market share achieved after four months in situation A is greater than that achieved in the same length of time in situation B*. This vividly illustrates that marketing performance is probably best thought of as a *relative* concept rather than an absolute one; e.g., a market share of 10 percent achieved in four months in an area in which no comparable change has ever occurred is probably more indicative of good performance than a 15 percent share achieved in an area in which changes of such magnitude are common. Of course, no attention is paid here to *profitability*, which is, after all, the most comprehensive single measure which is available.

An intersection of the market share graph with the *upper* bounding curve indicates a change in this performance measure which is significantly *worse* than those which were experienced by any brand in the observation period. Situation *C* in Figure 6-7 illustrates such an occurrence. Note that in this situation the bounding curve is intersected by the most recent one-, two-, and three-month changes; i.e., the plot of recent changes lies above the bounding curve at these three points. This is a strong indication of poor

FIGURE 6-7 MARKET SHARE AND BOUNDING CURVES—SITUATION *C*

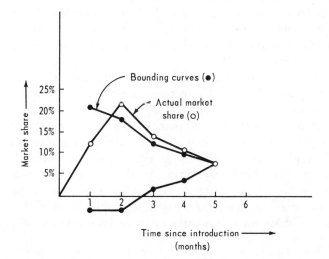

performance, since it indicates that the product is operating outside the boundary of occurrences for other brands during the historical observation period. Further, this illustrates the primary value of this approach and method of presentation for a performance indicator such as market share. Since the most recent changes for varying lengths of time are being viewed together and simultaneously compared with corresponding changes which have occurred in the past, it is possible to get a much clearer picture of the developing situation. Additionally, the significance of a wild performance level in any particular month is reduced in the context of the big picture displayed in a presentation of this kind.

One note of caution should be put forth, however. If we are to use such a procedure for a new product (since, of course, there is nothing which precludes the use of the same procedure with an existing product), we should be careful to determine whether any new product introductions occurred in the market during the observations period. In the case of the data of Table 6-4, a single new product introduction (product D) occurred early in the observation period, for example. If no new product introductions are included in the observed data, we may need to modify over-optimistic performance indications since one would expect that any new product, starting from a base of zero, would make immediate gains which are at least of the same order as the changes which are constantly experienced in a relatively stable market.

One way of amplifying the information conveyed by the bounding curves is to develop similar curves connecting the changes in market share which bound 90 percent, 50 percent, etc., of all changes which occurred in the observation period. In this context, the bounding curves previously used are those which bound 100 percent of past changes in market share, and we now wish to develop those curves which bound other proportions of the total number of increases and decreases which have occurred in the past. Such a system of bounding curves might look like those in Figure 6-8 (situation D).

In this figure the most recent one-month increase which has actually been experienced is compared with curves which bound 50, 90, and 100 percent of past increases and decreases, and it is found to be greater than 90 percent of past one-month increases. This is so because the rightmost line segment on the "actual market share" graph intersects the 90 percent bounding curve, but not the 100 percent bounding curve. Similarly, the most recent net two-month change is greater than 50 percent of past increases because the line segment connecting last month's market share with the market share for two months previous intersects the 50 percent

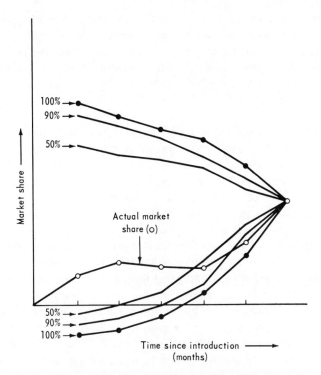

FIGURE 6-8 **MARKET SHARE AND BOUNDING CURVES—SITUATION** D

bounding curve and fails to intersect the 90 percent bounding curve.

The added information and sensitivity provided by the use of several curves which bound different proportions of past changes in market share is apparent. It is simpler to arrive at gross qualitative evaluations of performance when several bounding curves are used than when only the 100 percent curve is used as a basis for comparison.

Of course, we might just as well use any other meaningful performance measure in the same manner for, indeed, all that this procedure entails is the comparison of the levels attained via test marketing with a standard. In the case of the use of bounding curves defined over time, our standard is simply the limits of past performance changes in the area. Penetration, as previously discussed, would be a performance measure which could equally well be used in this fashion. A similar approach might also be taken by using repeat ratios, the distinction between the approach using penetration and that using repeat ratios being dictated by the fact that *the estimation of repeat ratios requires data collected over some substantial time period after entry into the market* whereas *penetration may be estimated very early in a test marketing program.* This in no way implies anything about the usefulness of the two measures, since the initial objective is to get consumers to try the product and the long-term objective is to have them continue to buy it.

Performance Measurement and Evaluation Using Market Data. There is no need to base evaluations of performance in test markets solely on data generated by consumer panels operating in the test areas. In many cases it may be either impractical or undesirable to make any use of panels for this purpose. If no consumer panel data are available, performance evaluations must be based on data generated in the marketplace. We have previously discussed how a performance measure, such as market share, may be determined in a test market even if no direct data source exists. Here we shall be concerned with other performance measures and the evaluations of test marketing results which may be made from them.

Performance measures. One of the basic objectives in choosing the performance measures to be used in evaluating test marketing results is their *comparability between areas.* "Sales revenue" is a valid performance measure, for instance, and yet it would be a poor measure on the basis of its comparability from area to area. A sales revenue of a million dollars in Los Angeles may or may not indicate performance superior to that demonstrated by a revenue of $300,000 in Des Moines, because of the relative sizes of the two markets. Of course, this same objective applies to performance measures obtained from consumer panels, but it is more obvious there because one is always aware that he is dealing with only a small fraction of the total market. The *time interval* which the measure represents is also important. As we have already pointed out with respect to consumer panel measures, comparisons of two numerical values of the same measure, but made over different time periods, are usually not valid.

As a result, use is usually made of *ratios and percentages,* which obviate the problem of comparing areas, and *rates,* which obviate the time interval difficulty. Revenue, for example, may be incorporated into a per capita ratio and expressed as a monthly rate, i.e., "per capita monthly sales revenue rate." Alternatively we might express a performance measure in consumption units, i.e., "per capita monthly usage rate." Sometimes such rates are expressed in annual terms because they are easier to interpret. For instance, one might read in the financial section of the newspaper that "sales for the month were at an annual rate of $250,000." This statement simply means that the month's sales have been projected on a yearly basis, usually with some correction for the effect of seasonality.

Alternatively, as a device for incorporating market share information and the performance of a product relative to other brands in the area, one might choose the "ratio of our product's sales to the sales of the leading brand in the area" (over the same time period) as a performance measure. This measure would be especially valuable for comparing performance in areas in which a different number of competitors operated. If in test area A we achieved a market share of 25 percent and the only other brand in the area had a share of 75 percent, this ratio would be $\frac{1}{3}$.[10] In test area B, in which many competitors operated, a market share of 10 percent, relative to a leading share of 30 percent, would give a comparable ratio.

One of the disadvantages of ratios such as these is that they do not account for the costs of obtaining the performance level which is indicated. Recognition of this fact leads to a problem like that discussed in Chapter 4, having to do with utilities for comprehensively defined outcomes rather than utilities defined solely over "pure" outcome measures. If the cost (disutility) associated with each marketing strategy is different, as it necessarily would be if the strategies involved various expenditures, the preference associated with "pure" outcomes and strategies may be handled by using outcome descriptions which relate to both the performance indicated and the cost associated with achieving that performance. This is especially important in test marketing if one of our objectives is to test the effects of various controllables, e.g., levels of advertising expenditure. In such a case, a measure such as "sales revenue per advertising dollar" satisfies both the criterion of comparability across areas and that of incorporating the performance achieved and the cost of obtaining that performance into a single measure. Of course, both portions of this ratio would be measured for the same (or comparable) time periods. The word "comparable" is inserted here because some organizations prefer to use the advertising expenditure made in some previous time interval in calculating this ratio. The basis for this judgment is that there is some lag involved in the sales effect of advertising; hence, if we believed that the lag was about one month, the calculation might be made by using sales data for the first three months of the year and advertising expenditures made in December, January, and February. The assumption here is that the December–February expenditures affected the sales realized in January through

---

[10] A ratio of market shares is equivalent to a ratio of sales quantities or revenues since the denominator of both shares is the total quantity (or revenue) for all products in the area.

March. In any case, the length of time interval used would always be the same for both portions of the ratio.

Another measure which satisfies the same general criteria is the "ratio of market share and advertising share," the latter quantity being calculated on the basis of advertising expenditures in the market. Of course, the necessary data on total advertising expenditures are not always readily available. However, accurate estimates of competitors' expenditures may often be made through counts of the number of television commercials, newspaper advertisements, etc., coupled with a knowledge of the prevailing rates for such messages. A numerical value greater than unity for this ratio, which indicates that proportionately more is being garnered from the market than is being invested in promoting it, is clearly desirable. A value less than 1 indicates that proportionately less is being gained than expended. If a market share of 25 percent is achieved by a new product in a period in which $500,000 was spent on advertising and $1 million was spent by competitive advertisers, the ratio would be 25 percent divided by $500,000/ $1,500,000 times 100, or

$$\frac{25\%}{100\dfrac{\$500,000}{\$1,500,000}} = 0.75$$

Although this value would not normally indicate good performance, consideration that the product is new, and therefore can be expected to require heavy initial advertising, might not necessarily lead to a negative evaluation. Just as with the previous measure, the time period during which the advertising portion of this ratio is measured may be lagged to account for a lag in the effect of advertising.

An important performance characteristic which is too frequently neglected is the *dynamics,* or time behavior, of market performance. The trend of sales is a good illustration of a measure of the temporal aspect of market performance. Consider two markets, each with the same values of "market share," "sales per advertising dollar," and all the other measures previously noted. If in one of these markets the sales trend is upward and in the other the trend is down, it is probably not illogical to evaluate the performance of the former as superior to that of the latter. This evaluation rests on the idea that the resources expended in an area have had two results: those which have already been realized and those which are yet to be realized. If we did not "count" the upward trend as being a favorable performance factor, we would be neglecting the latter aspect of performance.

The specific trend measure which we shall make use of is simply the "average change in sales rate" over some recent period. If we have experienced sales of $100,000, $150,000, $175,000, and $200,000 in the past four months, the trend measure is simply

$$\frac{\$200,000 - \$100,000}{4} = \$25,000 \text{ per month}$$

If the monthly sales rate had declined from $100,000 to $50,000 over the same period, the trend would be negative, i.e.,

$$\frac{\$50,000 - \$100,000}{4} = -\$12,500 \text{ per month}$$

Of course, we might equally well express a similar trend measure in terms of quantities rather than dollars if these data are more readily obtainable or if this measure would be more easily understood by those who will deal with it.

The reader will note that for the sake of consistency and simplicity, each of the performance measures—whether of the "pure" performance type or the "hybrid" type which incorporates the costs of achieving performance—is so defined that higher numerical values of the measures are more desirable than lower values. We might reverse the numerator and denominator in any of the ratios to develop an equally good measure (since it would convey exactly the same information) which does not possess this simple property. If we were to do so, we would necessarily have some performance measures in which higher numerical values were preferable and some in which lower values were preferable—a rather confusing situation! To avoid this, we shall always define performance measures in such a way that higher numerical values are preferable to lower numerical values.

Performance evaluation using market data. We have already alluded to the two most significant factors of the use of market data to evaluate test market performance. First, the measures should be comparable from area to area. Secondly, there is no single measure which incorporates all aspects of performance (other than a conceptual utility measure, of course).

The methods for the estimation of a utility measure for the multiple outcome descriptors relevant to test marketing are not well developed or practiced. As a result, at the operational level one is left with the problem of evaluating and comparing performance in test markets on the basis of a number of performance measures. Most often, as we shall see, we are reduced to either making qualitative evaluations using these numerical measures or simply ordering (ranking) the performance levels achieved in the various test markets.

It is usually possible to obtain partial rankings of test areas based on the numerical values of the performance measures achieved in the areas. For example, we might be using only two performance measures, "market share" and "sales revenue per advertising dollar." Suppose that the test areas displayed the values of these two performance variables, as shown in Table 6-8. The partial performance

TABLE 6-8   PERFORMANCE ACHIEVED IN TEST AREAS

Test area	Market share	Sales revenue per advertising dollar
1	20%	5.6
2	15%	4.9
3	20%	5.1
4	15%	5.3
5	15%	5.5

rankings which would evolve from these data may be summed up by the simple statements that performance in area 1 is superior to area 3 and that area 5 is superior to area 4, which is in turn better than area 2. In other words, the partial rankings are

(1,3)
(5,4,2)

in which we are simply listing areas in rank order of performance. Note that these groupings are based on the equality of the "market share" measure for areas within groups. Hence, since the value of market share is equal for areas 1 and 3, their relative performance is indicated solely by the value of the other performance measure. Area 1 (with a "sales revenue per advertising dollar" ratio of 5.6) may therefore be declared superior to area 3 (with a corresponding ratio of 5.1).

One implication of this approach has to do with the concept of *Paretian optimality*. This idea, put forth in a different context by the Italian economist Vilfredo Pareto, focuses on one of the several performance measures being utilized. The principle states that *an area is to be deemed preferable to another area on the basis of one of the performance measures if it possesses a higher value of that measure in conjunction with values of all other variables which are no worse than those displayed by the other area.* This, of course, assumes that the performance measure is defined in terms such that larger values are preferable to smaller ones, as is the case with all of those which have previously been noted.

To illustrate this principle, consider the performances indicated for the five test areas (numbered 6 through 10) in Table 6-9. The performance demonstrated by area 6 is preferred to that of area 7 since it is better in terms of the middle measure and equal in terms of the other two. Applying the Paretian principle, we determine that we should prefer the performance of area 6 to that of 7 since it has a higher value of "sales revenue per advertising dollar" and the values of the other variables for 6 are no worse than those for 7. Similarly, area 7 is preferable to area 8 since it is superior in terms of two measures and not inferior in terms

TABLE 6-9   PERFORMANCE LEVELS ACHIEVED IN TEST AREAS

Test area	Market share	Sales revenue per advertising dollar	Sales per capita
6	21%	5.65	0.80
7	21%	5.40	0.80
8	20%	5.25	0.80
9	19%	5.19	0.77
10	19%	5.24	0.76

of the third. Area 8 is superior to area 9 in terms of all measures. The result of these comparisons is that we can develop a preference ranking of the first four areas, which places 6 above 7, 7 above 8, etc., i.e.,

(6,7,8,9)

The performance of area 10 cannot be compared with that of 9, for instance, since it is equal with respect to one measure, better in terms of another, and worse in terms of the third. The Paretian principle offers no help here. However, the performance of area 10 may be compared with that of 6, 7, and 8. In each case, it is clearly inferior, since the numerical values of all performance variables are less than the corresponding values for these areas. Hence, another partial preference ranking may be developed as

(6,7,8,10)

The two rankings taken together indicate that either one of the two complete rankings (6,7,8,9,10) or (6,7,8,10,9) holds, but we do not know which.

To see how we make such determinations in more complex situations let us return to Table 6-8, which we previously analyzed only in terms of those areas which had equal values of a performance variable. Let us introduce a relation called "is preferred to" which is defined over the performance levels in all test areas. We may symbolize this relation with an asterisk. Hence, we would read the statement

1*2

as "the performance in area 1 is preferred to the performance in area 2." If we go to Table 6-8 and make all possible pairwise comparisons of the performance in the test areas by using the Paretian principle, we develop the following preference relationships:

1*2   1*3   1*4   1*5   3*2   4*2   5*2   5*4

To fully comprehend the aggregate meaning of these relationships, we may resort to a sketch in which we allow the left side of the page to represent the direction of increasing preference and in which only order (not distance) is meaningful. The first relationship (1*2) would indicate that 2 should be placed to the right of 1 in the sketch, as in Figure 6-9a. The second relationship (1*3) indicates that 3 should be to the right of 1 on this sketch, but it does not tell us whether it should be to the right or left of 2. The fifth relationship (3*2) determines that it should be

FIGURE 6-9   GEOMETRIC INTERPRETATION OF PREFERENCE RELATIONSHIPS

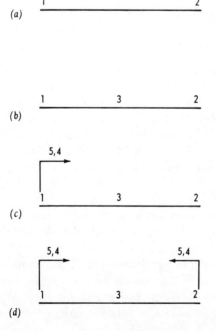

to the left of 2, giving the situation shown in Figure 6-9*b*. The third and fourth relationships tell us that 4 and 5 should both be to the right of 1, but neither their relative order nor their position relative to 2 and 3 is determined. This is indicated by the arrow in Figure 6-9*c*. The last relationship indicates that 5 is preferred to 4, so that the order used in Figure 6-9*c* is valid. The sixth and seventh relationships (4*2 and 5*2) determine that both 4 and 5 are to the left of 2 in such a diagram, as indicated by the arrow introduced into Figure 6-9*d*. This represents the totality of information which may be garnered from Table 6-9 by using the Paretian idea. It is clear from Figure 6-9*d* that any one of the following three complete orderings satisfies all of these relationships:

(1,3,4,5,2)
(1,4,5,3,2)
(1,4,3,5,2)

Hence, we know that one of these orderings must be valid, but we do not, on any objective basis, know which.

There are several possible ways to resolve this dilemma. One would be to develop a comprehensive utility measure defined over the multiple performance measures. Another would be to determine trade-offs for the various performance measures; i.e., how much a unit of one measure is worth in terms of the units of another. If we were able to do this, we could then convert from units of one measure to those of another. This would enable us to determine the relationship between areas such as 3 and 4 in Table 6-8 and 9 and 10 in Table 6-9. Consider the situation of Table 6-8, for example. If we determined that each unit of "market share" was worth 0.5 unit of "sales revenue per advertising dollar," we could then easily calculate that the 5 percent market share superiority of 3 over 4 is equivalent to 2.5 units of the second measure. Since 3 is inferior to 4 by only 0.2 unit of this measure, 3 has an overall degree of superiority to 4. Referring to the three possible preference orderings which are noted in the previous paragraph, we can see that only the ordering (1,3,4,5,2) is compatible with these conclusions. Hence, the only possible logical and consistent preference ordering for the performance displayed by these test markets is this one.

We should keep in mind that this conclusion is based upon our knowledge of the trade-off between the two performance measures, whereas all of the previous partial orderings arrived at on the basis of the Paretian idea do not require any such information. All that is required to develop the partial orderings by using the Paretian approach is a knowledge of the *direction* of increasing preference for each of the performance measures, i.e., whether higher numerical values of the measure are more or less preferred than lower values. We should further recognize that the assumption of knowledge of the trade-offs is not a very realistic one. Usually explicit trade-off relationships between performance variables are unknown. In fact, if we carefully consider the kind of knowledge and understanding which would be required to develop the trade-offs, we realize that the level of knowledge is not very different from that required to develop a utility measure for performance. As a result, it is not unusual to find that in lieu of a formal procedure, such as we have just described, for evaluating market performance, we revert to a subjective evaluation basis. One way of doing this would be first to apply the Paretian idea to all test markets to develop partial orderings. Then we would make note of the further relative performance evaluations which would be necessary to develop a complete ordering. We might obtain these evaluations simply by presenting performance descriptions (in terms of the performance measures) to the relevant executives and ask them to express a preference.

In the situation previously discussed in Table 6-8, it is clear that to develop a unique ranking, we need an evaluation of area 3 relative to area 4 even after having applied the Paretian principle. If we were to query executives concerning their preference for these two performance levels, we might obtain agreement as to which is the more preferred. We would then be able to develop the complete performance ordering which was both logically consistent (as far as that can be evaluated) and consistent with the preference expressions of the executives.[11] Of course, if we can obtain no agreement among the executives, we are left with only the partial orderings which generally result from the application of the Paretian idea. In some cases these may prove to be sufficient.

It should be noted that the process of obtaining subjective relative evaluations from executives is essentially equivalent to asking them to estimate roughly the trade-offs which were previously mentioned. The practical difficulties in making evaluations involving multiple performance measures are great, but they are reduced somewhat by the attendant circumstances that it is never necessary to compare more than two areas at a time and it is usually necessary to make only a very small number of such comparisons.

The use of subjective evaluations in conjunction with the objec-

---

[11] It should be noted that the single comparison of area 3 and area 4 would be sufficient if 3 is preferred to 4 in the example used. If 4 were preferred to 3, another comparison of either 3 or 4 and 5 would be necessary. This may be more easily seen by referring again to the three possible orderings which might exist based on the application of the Paretian principle.

tive evaluations developed with the Paretian principle naturally leads to consideration of the possible usefulness of the approach in making qualitative evaluations of an even lower level. For example, would it be meaningful and useful to take the same approach to categorizing the performance of test areas as "good" or "bad," or "adequate" or "inadequate," rather than the ordering of areas which we have previously discussed? The answer, of course, is that there is no additional conceptual difficulty involved in performing such an evaluation. The practical problems are similar to those encountered previously, for, indeed, one way of viewing this problem is as an extension of the previous one. After we have developed the unique preference ordering on the basis of Paretian ideas and subjective pairwise comparisons, we need only to ask the executives to choose a point in the ordering such that the performance of all areas "above" that point may be termed "good" and the performance at all areas "below" that point may be called "bad."

If we were actually to perform this evaluation procedure, we would be likely to find that three categories would be more suitable, i.e., that the executives would feel better able to classify marketing performance in categories labeled "good," "marginal," and "bad" rather than in only the two extreme categories. In one such trial conducted by the author, the suggestion for three categories emanated from the executives who had been faced with the dichotomous situation. In another trial, the "marginal" category proved useful as a device for resolving disagreements among the executives; the areas which all executives agreed were "good" and "bad" were placed in those categories, and all areas on which there was some disagreement were termed "marginal."

There are some distinct advantages in the introduction of subjective evaluations at a point where objective evaluations have been utilized to their fullest. The most significant of these is that the nonquantifiable factors bearing on performance may be brought into play. The *expectations* of executives are one aspect which undoubtedly enters into their judgment. If, for example, the executive group unanimously expected a particular market to perform poorly and it demonstrated much better performance, it might be desirable to evaluate this fact as favorable to the market's overall performance evaluation. Such comparisons with expectations are probably best done on a subjective basis.

In this section we have discussed the market-based measures of performance and their use in evaluating performance. There is nothing which precludes the use of a method of presentation such as the bounding curves of past increases and decreases (as in Figure 6-6) for market measures, of course. Consequently all of the comments of that section may also be directly applied to a "market share" measure calculated from market data rather than consumer panel data. The same technique might equally well be applied to many of the other measures discussed in this section. Of course, the real value to the marketing process of these measures and methods of presentation lies in the use to which they are put. The important areas in which performance evaluations of test markets are utilized are testing and control, early prediction of success, and drawing inference to the performance in other areas from the test market performance. These uses of the performance measures and evaluations, which we have discussed here, will be treated in the subsequent sections.

TESTING AND CONTROL IN TEST MARKETING

One of the most significant aspects of test marketing is the opportunity which it gives for *testing* marketing actions prior to their use on a national or regional scale. The test marketing organization may desire to try various advertising campaign themes, promotional expenditure levels, prices, or any other of the countless marketing controllables. Of course, the test marketing program itself has as its basic purpose the testing of consumer reaction to the product. Coupled with this may be the second-order objective of testing the effects of other marketing actions.

One way to test marketing controllables may be through an experiment in which various combinations of the controllables in question are used in each market area in a predetermined manner. If we wish to test the effect of various advertising expenditures and prices, for instance, we might use various combinations of levels of each in each area for a predetermined time and attempt to evaluate their effect. We shall consider such experiments in later chapters. Here we are concerned with the less sophisticated, but more practical, tests which are accomplished by changing the level of a single controllable in a test market and noting the result. If such a simple test is repeated in several markets with consistent results, a good, though imprecise, indication of the best level of the controllable in question may be determined.

The *control* aspect of test marketing is very similar to this method of testing. In speaking of control, we recognize that not all actions taken in the test markets are necessarily the result of a predetermined testing procedure. If performance is going badly, we may wish to alter some of the marketing controllables in search of improved performance. The difference between testing and control is therefore a subtle one. In testing we have, as our primary objective, the gaining of knowledge about the effectiveness of various marketing actions; the performance achieved is therefore more or less secondary. In control we seek improved performance, and the knowledge which we might gain, although important, is secondary. The two aspects are not really different. Only the viewpoint taken of them is different.

The techniques which prove useful in testing and control are generally quite similar. Primarily they involve the *detection of changes which are occurring in the marketplace.* In the testing case we wish to take an action and then detect the effect which that action has on market performance. In control we wish to detect performance changes, especially negative ones, and to take actions to correct them. In the one case we view the test market as our laboratory. In the other we recognize that the test market is important both as a laboratory and in the long-term view of the product, for if the decision is made to market the product nationally, the test market instantaneously becomes a full-fledged market. In considering control, we are recognizing that the testing done in the market during the test market phase might permanently impair its performance as a market. As a result, if things are going badly, we wish to take corrective actions which will prevent long-term harm and enable the organization to strive toward a good combination of controllables.

In terms of modern test marketing practice, the testing and control aspects are becoming increasingly important. The primary reason for this is the high cost of test marketing, which has forced companies to show greater concern with test market performance per se. The reduction in emphasis on the "experimental" aspects of the test market is exemplified in the market by market product introductions conducted by some consumer product producers. In such cases a single test market is entered, performance is evaluated, and control actions are taken. Then, a second market is entered only after the test market becomes profitable. The process is then repeated on an area-by-area basis. Clearly, in such a step-wise procedure involving profitability as the primary criterion for taking the next step, testing and control are of vital importance.

One of the techniques which we have previously considered in the context of "pure" evaluation may well be applied to the testing and control aspects of test marketing. This technique concerns the bounding curves which represent the largest increases and decreases in some market performance (such as those shown for "market share" in Figure 6-8). It is principally useful for detecting changes which have occurred in the marketplace but may not be apparent because of the random variations which constantly

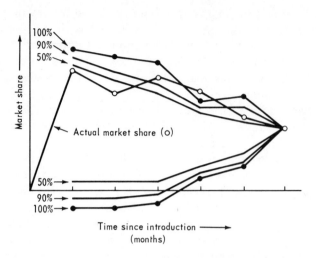

FIGURE 6-10   TESTING AND CONTROL IN TEST MARKETING

occur and because of the complexities which often prevent us from getting a useful perspective of market performance.

Consider, for instance, the situation depicted in Figure 6-10. Here, just as we did previously, we are plotting the values of "market share" attained in each month subsequent to introduction. We have illustrated the situation at the end of six months and have superimposed the curves indicating the levels of past increases and decreases in market share in the market. These curves are developed either from consumer panel data, as described previously, or from market data obtained in the market area over some period in the past. The situation in Figure 6-10 is one in which the last month's decrease in market share is not alarmingly bad— worse than 50 percent of past one-month decreases in the area. However, the last two months' net decrease is worse than 100 percent of past two-month decreases, and the last three month's net change is greater than 90 percent of past three-month decreases. If some new marketing action had been tested beginning three or four months previously, this would probably be sufficient to indicate that it has poor short-term effect and that some control action should be taken so that the market is not permanently affected. This control action might be the simple rescinding of the action which was tested, or it might involve additional promotional expenditures in the area. But if the indicated relative performance were not so bad—say if the one-, two-, and three-month net decreases were no worse than 50 percent of past decreases—no final evaluation or action might be necessary. If, the one-, two-, and three-month net changes were increases, we probably would not consider control action, but rather would try to determine whether the increases were significant enough to validate the action being tested.

Other control actions may be suggested by the penetration measure and repeat ratios which were previously discussed. If a comparison of penetration—the proportion of households that have made an initial purchase—with a standard determines that this aspect is causing difficulty, some action may be taken to increase penetration. Since our underlying objective is to have people *try* the product rather than *buy* it, we might give away samples at appropriate places, e.g., supermarkets for grocery products or football games for hair tonic. This will immediately increase penetration (defined in terms of triers rather than buyers). Or we might offer the kind of "deal" often used with packaged grocery products, e.g., at little or no extra cost we might attach a sample of the new product to a unit of some associated product with a well-known brand name. This would put the new product in the hands of potential customers whom we know, by virtue of their purchase of the associated product, to be interested in the general product class. We would be engaging in such a deal if we were to package a free sample of a new cleanser together with a well-known brand of liquid dishwashing detergent, for example. Alternatively, as a means of increasing penetration, we might simply increase advertising aimed at inducing more people to try the product. If the product truly has appeal for consumers, this kind of action, designed to increase penetration, will allow the product to demonstrate its values and to achieve the desired effect in becoming a long-term success.

If, on the other hand, penetration is deemed to be adequate but repeat ratios are low, the indication is that the product is deficient in its appeal to those who have tried it. In this case, some more drastic step may be called for. We may redesign the product to increase its appeal, for example. If all else fails and low repeat ratios persist, the only reasonable alternative may be to conclude that although the product idea appeared to be good on the basis of prior analysis, its demonstrated performance does not warrant the continuation of sales.

Matthews et al. give a good illustration of a situation in which penetration was adequate but repeat ratios poor.

For example, a few years ago a New England sausage manufacturer learned that a Midwestern manufacturer had successfully launched a new "brown-and-serve" link pork sausage in Chicago. The New Englander decided to do the same thing in his own marketing area. Unlike ordinary link sausages, brown-and-serve sausages were precooked at the factory and required only a few minutes of heating and browning in an ordinary frying pan. Their great virtue was that breakfast sausages could be prepared in 5 minutes during the breakfast rush rather than in 15 to 20 minutes. And there was none of the usual heavy residue of useless and unpleasant fat. In addition, manufacturing costs were lower, so that each ½ pound of brown-and-serve sausages produced 3 cents more profit than ordinary link sausage.

For many years the New England manufacturer had been successfully selling an ordinary link sausage which had wide public appeal. So he made the brown-and-serve sausage to taste the same after only a 5-minute heating and browning.

Its launching was well advertised and was an instant success. First-week sales through food stores exceeded the usual weekly sales of the manufacturer's regular link sausage by 36 percent. But in the second week sales fell to half the ordinary sausage volume, and in the third week they were down to practically zero.

After much investigation the reason was finally discovered. First, New Englanders were much more worried than Midwesterners about contracting trichinosis, a hog disease whose transmission to humans is prevented by thorough cooking. New England hogs at one time had a higher incidence of trichinosis than Midwestern hogs because so much New England pork was fed on refuse and garbage rather than on corn. This made New Englanders cautious, and they tended to cook pork products much more carefully and longer. Although they were attracted by and bought the 5-minute brown-and-serve sausage, they took the precaution of frying it 10 minutes, and this changed the taste and consistency of brown-and-serve to something they did not like. As a result, they did not buy the sausage a second time. And even worse, customers lost confidence in the manufacturer's other meat products, whose sales promptly fell.

When the reason for the failure of the new sausage was finally discovered, the advertising campaign was changed, and the preparation directions on the package were made much more prominent and reassuring. But then the manufacturer had trouble getting retailers to stock and display the brown-and-serve a second time. They did not want to damage the reputations of their stores by offering a product that had disappointed customers. This forced the manufacturer to give special reassurance to retailers and an extra 5-cent profit per package to get them to handle it again. And in order to get housewives to retry the product, a massive and costly promotion program was launched, including free, fresh-browned hot snack samples to all customers in the stores and a special 10 cents off deal. After that the product became a success, and the sales of the manufacturer's other products gradually turned upward.[12]

---

[12] J. B. Matthews, Jr., R. D. Buzzell, T. Levitt, and R. Frank, *Marketing: An Introductory Analysis*, McGraw-Hill Book Company, New York, 1964, pp. 7–8. Quoted with permission.

In this situation, after product redesign had taken place, the problem reverted to one of securing adequate penetration for the "new" product. The resolution of the problem was similar to the method which would be used to increase the penetration of a product being initially introduced.

EARLY PREDICTION OF MARKET PENETRATION[13]

One of the most significant aspects of test marketing is the evaluation of the performance achieved. In addition to evaluating performance which is currently being demonstrated, the marketing manager desires to *predict future performance*. In truth, one of the primary goals of test marketing is the early prediction of eventual success, i.e., the prediction of long-term success, the prediction being made at a point soon after introduction of the product into the test market. Such early predictions permit greater flexibility on the part of the test marketing organization, since they leave open alternatives, such as the early discontinuance of sales in the market, without the great monetary losses inherent in years of poor sales performance, and the use of control actions at a time which is early enough to avoid permanent impairment of the market.

Probably the most significant aspect of eventual success is the repeat ratios which were discussed earlier. Long-term sales revenue is derived primarily from repeat sales, and repeat sales depend on the reaction of consumers to the product. Repeat sales, however, are not subject to a great degree of control by the marketing manager after the commencement of test marketing, since repeat sales are largely determined by the product's inherent characteristics and appeal. Penetration, on the other hand, is subject to manipulation by the marketing manager to a much greater extent, since it depends more heavily on other marketing controllables such as advertising and sales promotion. One important aspect of early prediction is the prediction of the penetration which will eventually be achieved in the market. The penetration model which may be used for this purpose relies on estimates of penetration made from consumer panel data. Hence, the model applies directly only to a test marketing program utilizing a consumer panel.

The Penetration Model. A cumulative penetration model for a test market may be based on a general shape or functional form. Intuition (and observation of many actual penetration curves) tells us that a cumulative penetration curve should have two primary

---

[13] Some of the material in this section is adapted from the Fourt and Woodlock paper in the references.

properties. First, successive increments in the curve should become smaller and smaller, indicating that fewer new triers are attracted to the product in each successive time period. Secondly, the curve should approach some limiting penetration proportion, which is less than 100 percent, indicating that many consumers will never try the product; e.g., bald men will not purchase hair tonic for their own use and vegetarians will never purchase frankfurters.

A useful model having these properties is one which may be verbally described as having the characteristic that *increments in penetration for equal time periods are proportional to the remaining "distance" to the limiting value of penetration.* If the limiting penetration is 40 percent and the proportionality constant is 0.3, the number of new triers in the first period after introduction would be 12 percent (0.3 times 40 percent) according to this model, since the distance to the 40 percent ceiling at the beginning of the first period (when total sales are zero) is 40 percent. At the beginning of the second period the remaining distance to the ceiling is 28 percent (40 percent minus 12 percent). Hence, the proportion of new triers gained during the second period is 8.4 percent (0.3 times the 28 percent distance remaining to the ceiling). Table 6-10 demonstrates these calculations for the general case, where p percent is the limiting proportion and r is the penetration constant, and for the example using 40 percent and 0.3 values for these parameters.

TABLE 6-10   PENETRATION MODEL

Time period	Penetration increment		Cumulative penetration
1	rp	12%	12%
2	$rp(1-r)$	8.4%	20.4%
3	$rp(1-r)^2$	5.9%	26.3%
4	$rp(1-r)^3$	4.1%	30.4%
.	.	.	.
.	.	.	.
.	.	.	.
t	$rp(1-r)^{t-1}$	$0.3(40)(0.7)^{t-1}$	.

The second column of this table illustrates that the first penetration increment will be r times p. The remaining distance to the ceiling at the end of the first period is therefore $p - rp$, or $p(1-r)$. The second period's penetration will be r times this, or $rp(1-r)$, and so on. The expression for the penetration increment achieved in time period t is simply $rp(1-r)^{t-1}$. Analysis of the penetration

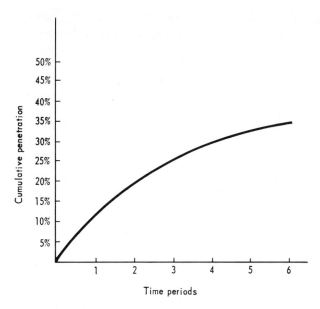

FIGURE 6-11   PENETRATION MODEL

increments in this table indicates that each increment is simply $1 - r$ times the preceding increment. A graph of this situation is shown as Figure 6-11.

To use the model to predict penetration, we must observe the penetration achieved (for a consumer panel) for at least two periods. Consider the situation which exists after exactly two periods. Because each penetration increment is assumed to be $1 - r$ times the preceding increment, a simple estimate of $1 - r$, and hence r, is the ratio of the two successive observed increments in penetration. If the time periods are months and the observed cumulative penetration is 13 percent at the end of the first month and 22 percent at the end of the second, we may estimate $1 - r$ as

$$1 - r = \frac{9}{13}$$

or r is about $\frac{4}{13}$. This is because, as we have already noted, each increment in the penetration model is simply $1 - r$ times the preceding increment, the first increment being 13 percent and the second 9 percent. An estimate of p may be made independently from market research data, or estimates of both p and r may be made as test marketing progresses.

Having estimated these two values, we may then enumerate the numerical penetration increments to be achieved in each succeeding month, using the basic penetration model. With this information we may determine whether the penetration is adequate in two respects. First, is the limiting ceiling adequate? If not, is it imposed by the existing situation (as in the case of bald men using hair tonic) or can it be altered? Secondly, is the time pattern of penetration, as depicted by the graphical expression of the model (as in Figure 6–11), adequate? If not, should the control actions necessary to improve penetration be taken?

DRAWING INFERENCES TO FUTURE MARKETS

Most often the results of test marketing programs are not so clear-cut as one might imagine before entry into the test markets. Of course, there are products which are miserable failures in initial test marketing and do not respond to any sort of control action. On the other hand, a small number of products have tremendous success in test marketing. The subsequent actions to be taken in such cases are obvious: in the first case, vigorously expand sales beyond the test markets as quickly and broadly as financially practicable[14]; in the second case, terminate sale of the product as quickly as possible. In most test marketing situations, the results are much more nebulous. One might, for example, have six test markets, three of which appear to be demonstrating adequate performance, two are marginal performers, and one is below par. The overall question which must be answered is: What action should be taken in the light of these varied performance levels? Indeed, the really significant *decision problem* of test marketing is the post-test market *expansion decision*, i.e., the choice of qualitative alternative from the set "expand sales" or "do not expand sales" and the subsequent choice of areas if the positive expansion alternative is chosen.

If varied performance levels have been displayed in a number of test markets, the choice is not clear. One possible explanation for the varied performance levels might be *inherent differences in the areas* with respect to their reaction to the product in question. Such differences have intuitive appeal; e.g., the product appeals to high-income groups but not to low-income groups, and hence it will sell well in high-income areas and poorly in low-income areas.

---

[14] Of course, this does not imply that other considerations do not enter into such a decision. Often, for example, high introductory levels of advertising and retailer discounts are used in test marketing. This always leads to the suspicion that test market success may not carry over to other markets on an economic basis. Implicit in much of what we shall say in the remainder of this section is the assumption that the same levels of promotion are to be carried on in the test markets as in potential expansion markets. If this is not the case, the necessary qualitative (but not quantitative) adjustments which may be made are apparent.

In analyzing the expansion problem, one quickly realizes that the tactical program of drawing inferences to potential sales areas is of paramount importance because the strategic choice between the alternatives "expand sales" and "do not expand sales" depends on the organization's basic objectives and financial outlook. An organization with a bleak financial outlook might not be tempted to expand sales into new areas if the inferences which could be drawn concerning acceptance in the new areas were not strongly positive. On the other hand, a strong organization might be willing to venture into new areas in which the indications of success were good, but not extraordinary. As a result, the tactical problem of selecting the "best" areas for expansion of sales by using the information gained in the test markets, i.e., of "drawing inferences" from test market performance to potential performance in other markets, must be attacked first. The solution to this problem then becomes the input to the strategic problem of whether or not to expand at all.[15]

If the performance demonstrated in the test markets has varied from area to area and it is believed that the explanation for this might lie in inherent differences in the areas, the problem of selecting areas for expansion might be approached by observing the test areas in which success has been achieved, detecting the pattern of characteristics which these areas have in common, and choosing areas for expansion which have similar characteristics. Additionally, to make the best use of the information available, one might wish to observe similar characteristics in the test areas in which success was not achieved. Of course, in practice, common patterns of characteristics are generally not very easy to detect.

The basic hypothesis in drawing inferences from the test areas to potential market areas revolves about *intrinsic area characteristics which account for differences in performance among areas and which can be used to predict performance in potential market areas.* If one were able to make such predictions for all potential areas, the "best" areas for marketing expansion could be selected. This procedure could then be applied sequentially over time; i.e., choose the best areas for immediate expansion and then make similar predictions at a later date as a guide for future expansion; or if there is no plan for national marketing of the product, the method could be used to select those market areas in which the product is to be placed on sale at some future time.

Generally the area characteristics which are believed to reflect differences in product success among market areas are cultural

---

[15] Note how this approach resembles the approach to the timing of a product introduction in the last section of Chap. 5. See the King (1963) paper in the references.

or economic. Several aspects of such area characteristics explain the difficulty in detecting patterns among them. First, usually more than one area characteristic is involved. The situation in which high per capita income foretells success in an area and low per capita income foretells failure is too simple to be realistic. The resulting problem of detecting patterns among multiple characteristics is similar to that already discussed for multiple performance variables.

Secondly, the effect of such characteristics on success may be *interactive;* in this case no pattern is usually detectable by any simple means. To illustrate the interaction of effects, two rather abstract area characteristics—"income level" and "cultural awareness"—may be considered. It is readily conjectured that the effect of varying income levels on sales might be dependent on the level of cultural awareness of the population rather than independent of it: a high income in a culturally mature area might affect classical record sales much more than a comparable income in a culturally void area, where discretionary income would be spent on other entertainment forms. The degree of such interactions—the effect on performance produced by a change in one characteristic being dependent on the absolute level of another characteristic—is usually difficult to detect intuitively and should therefore be analyzed statistically.

Of course, one cannot make a statistical analysis of characteristics such as "cultural awareness" per se. Numerically valued variables which are measures of these characteristics must be defined and measured for statistical investigation. Let us consider the simple problem of drawing inference to performance in potential areas using a single area characteristic which is measured by a variable x. The numerical value of x is known for all areas; let us assume, for example, that the relevant characteristic is population density as measured by the "population per square mile," x. In the test areas, both the numerical value of x before the beginning of sales and the sales performance are known. The numerical value of x for all potential areas is also known. The unknown quantity is the performance which could be achieved in the potential market areas.

The approach to the problem of drawing performance inferences in such circumstances depends on the *performance measure which is to be used in the test areas and to be predicted in potential areas.* There are two alternatives which must be considered. First, it may be decided that one of the multiple performance measures is sufficiently comprehensive to use a sole performance descriptor. If this is the case, the performance in the test areas will be evaluated by using this measure and this measure should be the one which is predicted for potential areas. Secondly, performance in the test areas may be evaluated in terms of a ranking of areas or in terms of categories developed from the combination objective-subjective approach discussed earlier. In either of these cases a similar evaluation will be the appropriate output from the process of drawing inferences to potential market areas.

Drawing Inferences Using a Single Quantitative Performance Measure. If a single quantitative performance measure is chosen to be indicative of the performance displayed in market areas, the problem of drawing performance inferences from the test areas to potential areas, on the basis of intrinsic characteristics of the areas, may be thought of as a four-stage process: (1) measuring the variables indicative of relevant area characteristics in the test areas prior to introduction, (2) measuring the single performance variable in the test areas as some predetermined time after introduction, (3) measuring similar area characteristics in each potential area, and (4) predicting the level of the performance measure which will be achieved in each potential area if full-scale marketing is embarked upon there.

Consider the case of a single variable describing an area characteristic—say, affluence as measured by "per capita annual income"—and a single performance measure "monthly per capita sales revenue rate." If x denotes "per capita income" and y denotes "per capita revenue," the various combinations of these variables as measured for each of the test areas may be displayed as in Figure 6-12. In this figure, called a *scatter diagram*, each test area is represented by a single point indicating the income level x, as measured prior to the beginning of sales, and the performance level achieved, as measured by the variable y. Observing Figure 6-12, one notices that although there is a good deal of apparently random scatter, the points tend to lie in a generally upward-sloping direction; i.e., test markets having high values of "per capita income" also tend to have high values of "per capita revenue" and test areas having low values of x also tend to have low values of y. Variables having this property are said to have a high degree of positive *correlation* between them.

The term "correlation" is a part of almost everyone's vocabulary. Like many other terms used in scientific analysis, its precise meaning is somewhat different from its more nebulous meaning in everyday usage. "Correlation," as used here, is the degree of *linear dependence* existing between the variables in question, i.e., the degree to which the points on the scatter diagram tend to lie in a straight line.

In this situation our objective is to predict performance y in

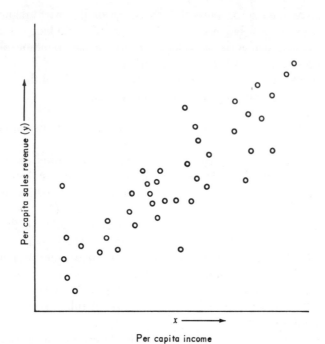

FIGURE 6-12   SCATTER DIAGRAM FOR TEST AREAS

potential markets, given a knowledge of the variable x. Let us
consider, for example, one market area, San Francisco, which was
not one of the original test areas. Supposing the "per capita income"
in San Francisco is $2,000, what level of "per capita sales revenue"
can be expected there? To answer this question, we must develop a *prediction model*. The model which we shall use is based,
like all models, on a set of assumptions about the real world. We
shall later refer to some of the more technical assumptions. The
most important one for us to recognize at this point is that there
are no conditions which might bear on our performance measure
in San Francisco which are drastically different from the conditions
in the test markets, i.e., that there is a stability which exists from
area to area. In developing a prediction model, we shall take advantage of the information which the scatter diagram conveys to
us concerning the degree of correlation existing between x and
y in the test areas.

To make predictions of the value of y which might be achieved
in San Francisco, we shall use the linear prediction model of expression (6-3). In this expression y^* is the predicted value of y for
San Francisco, $\bar{y}$ is the mean value of y for the test areas, b is
a coefficient to be determined, and $x - \bar{x}$ is the difference between

the observed value of x in San Francisco and the mean value of x for the test areas; the mean value of x is indicated by an x with a horizontal bar above it to be read "x-bar."

$$y^* = \bar{y} + b(x - \bar{x}) \qquad (6\text{-}3)$$

In this expression y^* and b are the only unknown quantities, since $\bar{x}$ and $\bar{y}$ are determined from the data on area characteristics and performances in the test areas and x is the known value of the area characteristic variable for the potential area, in this case San Francisco. The determination of y^* is the objective of the prediction model; hence, if b were known, the objective could be achieved.

To determine a numerical value for b, we should consider the nature of the prediction which we wish to make for San Francisco. Ideally we would like a prediction which is as close as possible to the actual value of y which is eventually achieved in San Francisco. The difference between the predicted and actual values of y may be written as

$$y^* - y \qquad (6\text{-}4)$$

and termed the *prediction error*. The loosely stated criterion of making this error as small as possible determines the manner in which we may estimate the parameter b.

Since we shall not know the value of y for San Francisco until after we market the product there (if, indeed, we do so at all), it is not possible for us to choose, before the fact, a value for b which will directly make expression (6-4) as small as possible. We can, however, use various numerical values of b to make hypothetical predictions of performance for the *test areas* (rather than for San Francisco). Having done this, we may determine the test area prediction errors by applying expression (6-4) *to the test areas*. The particular numerical value of b which gives the smallest prediction error for the test areas may then be chosen as the "best" value of this parameter.

There is one slight complication in this procedure, however. When we are speaking of a prediction error for a single area, the prediction error is uniquely defined, by expression (6-4), as simply the algebraic difference between the actual and predicted values. When we are speaking of the errors for the entire group of test areas, the idea of total prediction error is not so well defined. Is it, for instance, simply the sum of the individual errors associated with each test area? A little reflection shows that such a simple sum would mask the effect of overestimates and underestimates; i.e., an overestimate of a given amount in one test area will be canceled by a corresponding underestimate in another test area, resulting in a total numerical error sum which is uninformative. One way of circumventing this problem would be to use the absolute value of each error in the sum, i.e., to count only the magnitude, and not the algebraic sign, of each error and to sum these magnitudes. Absolute values, however, are difficult to manage in algebraic expressions. Therefore, for mathematical simplicity, we shall choose the *sum of the squared values of the prediction errors in the test areas* as a measure of the "goodness" of particular values of b; i.e., we shall choose b to be that value which results in the smallest value of

$$\sum_{i=1}^{n} (y_i^* - y_i)^2 \qquad\qquad (6\text{-}5)$$

where the i subscript runs over the n test areas, y_i^* is the prediction of y for the ith test area, and y_i is the actual value of y observed in the ith test area. Since this sum of squared errors conveys all of the information (as does the sum of absolute values), avoids the algebraic sign problem, and is mathematically tractable, its use as a measure of the "goodness" of each b value has both theoretical justification and intuitive appeal. The idea of choosing a value for b which minimizes this sum of squared errors is appropriately termed a *least-squares* principle.

FIGURE 6-13   LEAST-SQUARES ANALYSIS

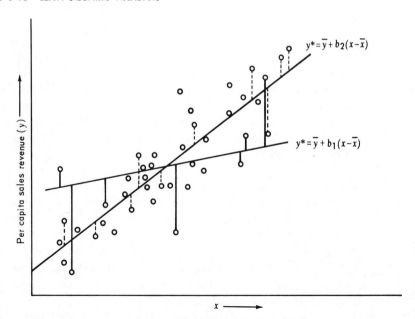

Per capita income

The procedure is graphically illustrated in Figure 6-13. In this figure two possible values of b, as represented by the lines which are the graphs of the two predictive equations

$$y^* = \bar{y} + b_1(x - \bar{x})$$

and

$$y^* = \bar{y} + b_2(x - \bar{x})$$

are superimposed on the scatter diagram. The individual predictive error for each value of b is simply the vertical distance between the point describing each test area and the predictive line. The magnitudes of *some* of these errors are represented by solid lines for the value b_1 and by dotted lines for the value b_2. It is easily seen that b_2 has a much smaller sum of squared errors than b_1. If we were to try all such values for b, we would arrive at that b for which the sum of squares of the errors is a minimum. This is the value which we would use in the prediction model of expression (6-3) for predicting the performance to be achieved in potential areas.

It is not actually necessary to calculate the sum of the squares of the errors for a large number of predictive equations. The problem of selecting the best value of b has been solved mathematically and the following expression obtained.[16] The reader may refer to the Appendix for greater detail.

$$b = \frac{\Sigma x_i y_i - (\Sigma x_i \Sigma y_i)/n}{\Sigma x_i^2 - (\Sigma x_i)^2/n} \tag{6-6}$$

In expression (6-6) all sums are over the values for the n test areas.

To demonstrate this calculation, let us tabulate the individual summations using, for simplicity, only the hypothetical values of x and y for four test areas as given in Table 6-11.

TABLE 6-11   VALUES CHARACTERISTIC AND PERFORMANCE VARIABLES FOR FOUR TEST AREAS

Test area	Characteristic variable, x_i	Performance variable, y_i
1	2,000	10
2	3,000	16
3	1,000	8
4	900	7

[16] Completeness demands that we note at this point that the use of b as determined from expression (6-6) in expression (6-3) may be justified on statistical grounds which are entirely unrelated to the principle of least squares. In this instance both approaches give identical results. See the text by Dixon and Massey in the references.

The appropriate summations which may be used in expression (6-6) are

$\Sigma x_i = 2,000 + 3,000 + 1,000 + 900 = 6,900$

$\Sigma y_i = 10 + 16 + 8 + 7 = 41$

$\bar{x} = \Sigma x_i/n = {}^{6,900}\!/_4 = 1,725$

$\bar{y} = \Sigma y_i/n = {}^{41}\!/_4 = 10.25$

$\Sigma x_i \Sigma y_i = 6,900(41) = 282,900$

$\Sigma x_i y_i = 2,000(10) + 3,000(16) + 1,000(8) + 900(7) = 82,300$

$\Sigma x_i{}^2 = (2,000)^2 + (3,000)^2 + (1,000)^2 + (900)^2 = 14,810,000$

$(\Sigma x_i)^2 = (6,900)^2 = 47,610,000$

Hence,

$$b = \frac{82,300 - {}^{282,900}\!/_4}{14,810,000 - {}^{47,610,000}\!/_4}$$

or about 0.004.

If we were to use this meager data from only four areas to predict the performance to be achieved in San Francisco, or any other potential area, we would use the prediction equation

$$y^* = 10.25 + 0.004(x - 1,725)$$

which may be seen to be exactly the form of expression (6-3) with the appropriate numerical values for this example inserted. Since the value of x for San Francisco has already been determined to be $2,000, the prediction of the performance level to be achieved there would be

$$y^* = 10.25 + 0.004(2,000 - 1,725)$$

or 11.35. We would therefore predict that a rate of $11.35 of "monthly per capita sales revenue rate" would be achieved if marketing were undertaken in San Francisco. If we were to make such predictions for each of the areas considered to be potential market areas for our product, we could judge which would produce the best results on the basis of this single performance measure and hence, we would be better able to decide which areas should be entered.

In using this approach, *we should always ensure that the numerical value of the potential area's predictor variable is within the range of those test area values on which the estimate of b is based.* If this is not the case, we would, in using this approach, be predicting results which are outside our range of experience—at best, a doubtful undertaking.

Drawing inferences by using a single quantitative performance measure and multiple predictor variables. In most instances there will be more than one area characteristic which is considered indicative of marketing performance. In one situation, for example, company executives were quick to hypothesize about thirty area characteristics which they felt might be related to the penetrability of an area. Some of these characteristics were simple enough, e.g., a belief that areas differed in the product "user versus nonuser" proportional composition of their population. Others were more subtle, e.g., a hypothesis that a high proportion of foreign-born citizens in an area was intrinsically negative for any new product since these people were less likely to be reached effectively by advertising messages which were in the English language and reflected American middle-class values. Most of the characteristics dealt with cultural and economic factors, but competition was not neglected. One executive proposed the logical hypothesis that an area in which a large number of competitors were already operating was more difficult to penetrate than one with a smaller number of competitors, simply because of the greater number of minds, bodies, and other resources working against the new product.

In a situation involving more than one area characteristic, the first question to be faced is whether appropriate quantitative variables which reflect these characteristics can be defined and measured. In the situation discussed, for example, per capita consumption of all products of the product class was taken to be a quantitative measure related to the "user versus nonuser" composition of the population. Similarly, the percentage of foreign-born and the market share of the major competitor were taken to be quantitative indicators of the "foreign-born" and "number of competitors" hypotheses. Of course, it is usually the case that some characteristics are either so vaguely or so qualitatively hypothesized that no meaningful quantitative indicator can be determined for them; in this case they must simply be omitted from the objective analysis and borne in mind for later subjective analysis of the results.

If a number of variables reflecting area characteristics believed to be indicative of performance are to be used, a *multivariate linear prediction model* of the form

$$y^* = \bar{y} + b_1(x_1 - \bar{x}_1) + b_2(x_2 - \bar{x}_2) + \cdots + b_p(x_p - \bar{x}_p) \qquad (6\text{-}7)$$

may be employed. In this expression $x_1, x_2, \ldots, x_p$ are the area characteristic variables; $\bar{x}_1, \bar{x}_2, \ldots \bar{x}_p$ are the means of the values of these variables which were observed in the test areas; $\bar{y}$ is the mean of the values of the single performance variable observed in the test areas; and $b_1, b_2, \ldots, b_p$ are parameters to be determined, as was the single parameter b in expression (6-3). Of course, y^* is the prediction of the performance to be achieved in a potential area having predictor variable values $x_1, x_2, \ldots, x_p$.

It is often simpler to write this expression in terms of the variables themselves, rather than using the deviations of the variables from their means. If this is the case, the linear prediction model may be written as

$$y^* = a + b_1x_1 + b_2x_2 + \cdots + b_px_p \qquad (6\text{-}8)$$

where the constant a is simply

$$a = \bar{y} - b_1\bar{x}_1 - b_2\bar{x}_2 - \cdots - b_p\bar{x}_p \qquad (6\text{-}9)$$

We shall give the expressions for determining the numerical values of the b's for the case where p equals 2—i.e., there are only two predictor variables—and refer the reader to the Ezekiel and Fox text for the formulas for the cases involving more than two predictor variables. The parameters b_1 and b_2 may be determined from the simultaneous solution of equations (6-10) and (6-11).

$$b_1\Sigma(x_1 - \bar{x}_1)^2 + b_2\Sigma(x_1 - \bar{x}_1)(x_2 - \bar{x}_2) = \Sigma(y - \bar{y})(x_1 - \bar{x}_1) \qquad (6\text{-}10)$$
$$b_1\Sigma(x_1 - \bar{x}_1)(x_2 - \bar{x}_2) + b_2\Sigma(x_2 - \bar{x}_2)^2 = \Sigma(y - \bar{y})(x_2 - \bar{x}_2) \qquad (6\text{-}11)$$

All of the sums in expressions (6-10) and (6-11) are taken over the test areas. The constant a is then determined from the b's, using expression (6-9). In the case of a number of predictor variables much greater than 2, the calculations become quite extensive. However, computer programs for making these calculations are readily available.[17]

Drawing Inferences by Using a Qualitative Performance Measure.[18] If no single performance measure appears to be sufficiently comprehensive to use in the manner of the previous section, a qualitative performance measure may be used to evaluate test area results. Correspondingly, the qualitative performance measure would then be the one which is predicted for potential areas.

In this section we shall consider two distinct kinds of market areas: those in which the product in question will demonstrate good sales performance and those in which it will perform poorly. For simplicity, let us think of two disjoint groups of areas—those which are "good" and those which are "bad" for the product. We shall label these two groups of areas G and B respectively.

In most cases some of the test areas will be type G and some type B. It is the function of the performance evaluation phase to determine which are which, i.e., to label each test area as having performed well enough to be labeled "good" or poorly enough to be labeled "bad." As we have noted earlier, this evaluation may

---

[17] One should seek out a program labeled "Multiple Regression Analysis."
[18] Some of the material in this section is adapted from the King (1963 and 1965) papers in the references.

be made on the basis of the objective-subjective approach discussed earlier or on a purely subjective basis.

If we were to illustrate graphically the values of an area characteristic measure x (say, "per capita income") in conjunction with the qualitative (G versus B) performance evaluations for each of the test areas, we might discover a situation like that given in Figure 6-14. In this figure, labeled situation I, each test area is represented by a letter, indicating its evaluation as either a type G or a type B area, which is located at a point on the x scale indicating the value of the predictor variable for that area. The lowest observed value of x in the test areas was slightly above $1,000, and the performance achieved in that area was such that it was evaluated as "bad," as indicated by the leftmost B on the x scale. The highest observed value of x was somewhat above $3,000 in an area whose performance was evaluated as "good," as indicated by the rightmost G on the x scale. It is apparent that high values of x are generally associated with "good" areas and low values are associated with "bad" areas. If this very unrealistic situation were the case, it would be relatively simple to predict the performance of a potential area, i.e., to predict whether a potential area was the "good" or the "bad" type. One could simply determine a critical value of x—say x_0—which is between the two groups of test areas and predict that a potential area would be "good" if its value of the predictor variable x were above x_0 or that it would be "bad" if its value of x were below x_0. In this case x_0 would probably lie somewhere in the range $1,800 to $2,400.

We should recognize two important points concerning this situation. First, in the real world we would be very unlikely to encounter a situation in which a single variable had such great predictive power as that illustrated in Figure 6-14. The situation would more likely be similar to that in Figure 6-15 (situation II), in which there is an overlap between the "good" and "bad" test areas; high values of x are still generally associated with "good" areas and low values of x with "bad" areas, but values of x in the middle of the range have no clear "good" or "bad" tendency.

Secondly, even in the overly simple case of Figure 6-14, the exact location of the critical value x_0 would be determined by

FIGURE 6-14 "GOOD" AND "BAD" TEST AREAS AND THEIR ASSOCIATED VALUES OF x—SITUATION I

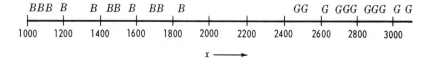

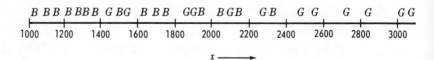

FIGURE 6-15 "GOOD" AND "BAD" TEST AREAS AND THEIR ASSOCIATED VALUES
OF x—SITUATION II

the relative cost of making the two possible kinds of prediction
error. These are: error 1, predicting that an area which is really
type G will be type B; and error 2, the converse, predicting that
an area which is really type B will be type G. If the cost of error
1 is termed C_{GB} and the cost of error 2 is termed C_{BG}, we may
recognize in Figure 6-14 that if C_{GB} were greater than C_{BG}, the
x_0 value would be farther to the left; i.e., the cost of predicting
that a "good" area is "bad" is relatively great, and so we wish
to take little chance of doing this. We may reduce this chance
by predicting areas to be "good" even if they are closer to the
B group of test areas than they are to the G group. For instance,
if an area had an x value of \$2,000, we might wish to predict
that it would be "good," even though it is closer to the B group
than to the G group. We might do this because the cost of making
the only possible error in this case (predicting a "bad" area to
be "good") is relatively small compared to the cost of the converse
error. To do this, we would simply set x_0 to be less than \$2,000.

The outcome array for the decision problem involving the choice
of a prediction to make for a potential area is given as Table
6-12. In this table, the cost associated with a prediction which

TABLE 6-12   OUTCOME TABLE IN TERMS OF ERROR COSTS

Strategy	N_1 (Area is really G)	N_2 (Area is really B)
S_1: predict area to be G	0	C_{BG}
S_2: predict area to be B	C_{GB}	0

turns out to be correct is assumed to be zero.

If the probability distributions[19] of the variable x for areas of
types G and B and the relative frequency of occurrence of G and

[19] The reader who is unfamiliar with probability distributions may think of
the distribution of the variable x, symbolized $f(x)$, as a function under which
areas represent probabilities; i.e., the area under the curve $f(x)$ and above
the x axis between any two points a and b represents the probability that
the variable x assumes a value between a and b.

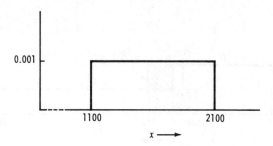

FIGURE 6-16 PROBABILITY DISTRIBUTIONS FOR TYPE *B* AND TYPE *G* AREAS

B areas can be estimated, the determination of a "best" value for x_0 is not difficult. Suppose, for example, that the probability distribution of x for the *B* areas is determined to be that shown in Figure 6-16, i.e., uniform over the interval $1,100 to $2,100. Further suppose that the distribution of x for the *G* areas is of the same type—uniform between $1,500 and $3,500. If this distribution were superimposed onto the previous one, Figure 6-17 would result. Note that the two distributions overlap between $1,500 and $2,100, indicating that it is possible for an area having an x value in this range to be either type *G* or type *B*. The critical value x_0 will lie somewhere in this overlapping range, since if it were greater than $2,100, we would be predicting some areas to be "bad" when we know they are "good"; e.g., any area having an x value greater than $2,100 is known, with certainty, to be of the "good" type because the probability distribution of x for *B* areas does not extend above $2,100. Similarly, if x_0 were below $1,500, we would be predicting some known "bad" areas to be "good."

Suppose x_0 were set to equal $2,000. The probability of predicting a "bad" area to be "good" would be 0.1—the area under the *B* distribution to the right of $2,000. This is the correct probability because any area having an x value greater than $2,000 will be

FIGURE 6-17 PROBABILITY DISTRIBUTION FOR TYPE *B* AND TYPE *G* AREAS

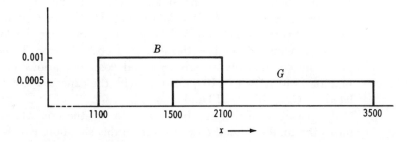

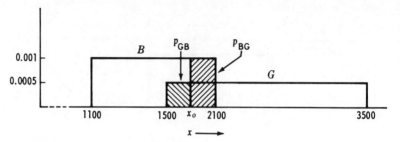

FIGURE 6-18   PREDICTION ERROR PROBABILITIES

predicted to be "good," and if an area is really "bad," the probability of its displaying an x value above \$2,000 is given by the area under the B distribution to the right of \$2,000. Correspondingly, the probability that an area which is really "good" will be predicted to be "bad" if x_0 is \$2,000 is 0.25—the area under the G distribution to the left of \$2,000. These two areas are indicated in Figure 6-18 and labeled p_{BG} and p_{GB} respectively for some undefined value of x_0.

If q_B and q_G are the respective relative frequencies of occurrence of areas of type B and type G, the *expected cost of error* for a predictive model using the critical value x_0 is

$$E \text{ (cost of error)} = q_G C_{GB} p_{GB}(x_0) + q_B C_{BG} p_{BG}(x_0) \qquad (6\text{-}12)$$

Each of the two terms in this expression represents one of the two kinds of predictive error. Each term is simply the product of the cost of making the error, the likelihood that an area is of the type necessary for the error to occur (e.g., to predict that a "good" area is "bad," the area must be of the "good" type, and this occurs with probability q_G), and the probability of making the error using x_0. The latter probabilities are written $p_{BG}(x_0)$ and $p_{GB}(x_0)$ to indicate that their numerical values depend on the location of x_0.

A rational method for choosing a value of x_0 to use in predicting whether potential areas will be of the "good" or "bad" type would be one which results in the smallest value of the expected cost of error in expression (6-12). We might do this by trying values for x_0 between \$1,500 and \$2,100 and approximating the location of the x_0 which minimizes the expected cost of error. Table 6-13 gives the components of this expected cost for various x_0 values ranging from 1,500 to 2,100, together with the expected cost (calculated for $C_{GB} = C_{BG} = 10$ and $q_G = q_B = \frac{1}{2}$).

It is easily seen that the best value for x_0, the one which minimizes the total expected cost of error, in this situation, is $x_0 = 2,100$

TABLE 6-13   COMPONENTS OF EXPECTED COST OF ERROR

x_0	(1) $p_{GB}(x_0)C_{GB}q_G$	(2) $p_{BG}(x_0)C_{BG}q_B$	$E(C) = (1) + (2)$
1,500	0.0	$0.60C_{BG}q_B$	3.00
1,600	$0.05C_{GB}q_G$	$0.50C_{BG}q_B$	2.75
1,700	$0.10C_{GB}q_G$	$0.40C_{BG}q_B$	2.50
1,800	$0.15C_{GB}q_G$	$0.30C_{BG}q_B$	2.25
1,900	$0.20C_{GB}q_G$	$0.20C_{BG}q_B$	2.00
2,000	$0.25C_{GB}q_G$	$0.10C_{BG}q_B$	1.75
2,100	$0.30C_{GB}q_G$	0.0	1.50

with its associated expected cost of 1.50. One can also determine that all potential values for x_0 which are outside the range shown in Table 6-13 will have an expected cost which is greater than 1.50. Hence $x_0 = 2,100$ is the critical value of the variable x which should be used in predicting areas to be of either the "good" or the "bad" type in this case; i.e., if an area has a value of x which is less than 2,100, we shall predict it to be "bad" and if it has an x value greater than 2,100, we shall predict it to be "good." If its value is exactly equal to 2,100, the prediction is arbitrary.

Moreover, in the situation involving these simple probability distributions, a little reflection and/or some review of elementary calculus leads one to the conclusion that the best value of x_0 will be either 1,500 or 2,100, regardless of the numerical values of q_G, q_B, C_{GB}, and C_{BG}. Of course, if the probability distributions are not so simple, the best value of x_0 may well not be one of the boundary values.

Technical Note on the General Case of Drawing Inferences by Using a Qualitative Performance Measure.[20] Of course, one need not actually make all of the calculations involved in Table 6-13 in order to determine the best value of x_0. We may express the probabilities $p_{BG}(x_0)$ and $p_{GB}(x_0)$ as areas under the respective probability distributions $f_G(x)$ and $f_B(x)$, i.e.,

$$p_{GB}(x_0) = \int_{-\infty}^{x_0} f_G(x)\,dx \tag{6-13}$$

$$p_{BG}(x_0) = \int_{x_0}^{\infty} f_B(x)\,dx \tag{6-14}$$

---

[20] This section requires only a knowledge of elementary calculus. It is set off as a technical note because this level of mathematics is beyond that required in all previous sections of this chapter. Coverage of this section is not necessary for a general understanding of the principles involved in drawing inferences by using a qualitative performance measure.

The total expected cost of error may then be written as

$$E(C) = q_G C_{GB} \int_{-\infty}^{x_0} f_G(x)\, dx + q_B C_{BG} \int_{x_0}^{\infty} f_B(x)\, dx \qquad (6\text{-}15)$$

where $f_G(x)$ and $f_B(x)$ are the probability distributions of x for "good" and "bad" areas respectively, and the positive and negative infinities in the integration limits may be interpreted as "the lowest and highest numerical values which the variable x may take on."

To determine the best x_0, we must minimize $E(C)$. We may accomplish this by setting the derivative of the expected cost with respect to x_0 equal to zero,[21] i.e.,

$$\frac{dE(C)}{dx_0} = 0 \qquad (6\text{-}16)$$

We may then solve expression (6-16) for x_0 to determine the value which minimizes the expected cost and which is therefore best to use as the critical value of x for predicting that a potential area is either the G or the B type.

Of course, since the probability distributions in Figure 6-18 are uniform, expression (6-15) is linear in x_0. We should recall from calculus that the basic theorem relating to the determination of the relative maxima and minima of a function, which lies behind expression (6-16), says nothing about the situation in which a minimum point occurs at the end point of the interval of definition of the function. This is the situation in the simple case illustrated by Figure 6-18. Hence, expression (6-16) will not determine the best value of x_0 for us in that case. (One should refer to the treatment of maxima and minima in any good calculus book for a comprehensive explanation of these points.)

Drawing inferences by using a qualitative performance measure and multiple predictor variables. It is often desirable to predict the eventual performance of a potential market area by using a number of predictor variables rather than just one. One reason for this, of course, is that there are very few situations in which a single variable would have sufficient predictive ability to be useful.

If multiple predictor variables are available, the situation is likely to be similar to that shown in Figure 6-19 for two predictor variables x_1 and x_2. In this figure each test area is indicated by a point (x_1, x_2) and its associated performance evaluation is indicated by a G or B at the point. In this case we can see that "good" areas generally tend to have high values of both variables, as indi-

---

[21] The second derivative needs to be checked to ascertain that the x_0 value which is determined actually represents a relative minimum of the expected cost function. See any basic calculus book for a further explanation of this point.

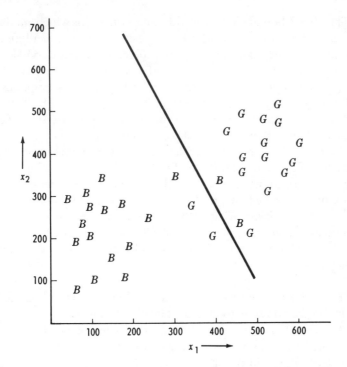

FIGURE 6-19   VALUES OF x_1 AND x_2 FOR TYPE G AND TYPE B TEST AREAS

cated by the preponderance of G's in the upper right, and "bad" areas tend to have lower values of both. There is, as might be expected, some overlap between the "good" and "bad" test areas.

The prediction problem in this situation is to determine some boundary line between the G and B areas which can be used to predict whether a potential area will be type G or type B. Such a line is shown in Figure 6-19. If this were the one to be used for prediction, we would simply observe the values of the predictor variables x_1 and x_2 for a potential area and predict it to be "good" if the point described by these values were to the right of the line or predict it to be "bad" if it were to the left. A potential area having x_1 and x_2 values of 200 and 100 would therefore be predicted to be "bad" and one having values of 600 and 400 would be predicted to be "good."

There is no reason that the line which "discriminates" between G and B areas should be a linear function; the "best" such discriminator might well be curved. We shall limit ourselves to linear boundary lines, however. In doing so, we are making assumptions which are completely analogous to the use of a linear prediction model in the case of a quantitative performance measure.

The "best" boundary line in this case is the one which minimizes the expected cost of error. In the next section we shall develop the mathematical solution to this problem. The reader who does not have a background for the mathematics may apply the result by (1) developing an understanding of the listed assumptions and determining whether his situation is compatible with them and (2) referring to the solution, as indicated by expressions (6-21) and (6-22) and their associated paragraphs.

Technical Note on the Prediction of a Qualitative Performance Measure by Using Multiple Predictor Variables.[22] To develop the best discriminator between the *G* and *B* test areas, we shall think of the prediction problem as taking a single potential market area and classifying it as either type *G* or type *B* on the basis of measurements on a number of predictor variables which describe the area. The assumptions and formal statements of the problem elements are:

1. There are two mutually exclusive and exhaustive subpopulations in the domain of market areas (labeled *G* and *B*).
2. A vector $X_k = (x_{1k}, x_{2k}, \ldots x_{pk})$ consisting of measurements on p predictor variables is known for all test areas and for the potential area in question.
3. It is desired to define two mutually exclusive and exhaustive regions, R_G and R_B, in p-dimensional predictor space so that the "point" described by X_k falling into R_G or R_B corresponds to a classification of the market area as type *G* or type *B* respectively.
4. The regions R_G and R_B are to be defined in a manner which will minimize the expected cost of error in classifying an area.
5. The test areas are considered to be randomly selected from the *G* or *B* subpopulations; each test area is known to be either type *G* or type *B*.
6. The probabilistic properties of the variables are described by joint probability distributions $f_G(X)$ and $f_B(X)$ for the *G* and *B* areas respectively.
7. The probability that a randomly selected area is type *G* or type *B* is known—q_G and q_B, respectively ($q_G + q_B = 1$).
8. The cost of making the two possible kinds of errors is known[23] to be C_{GB} for taking a "good" area and classifying it as "bad," and G_{BG} for taking a "bad" area and classifying it as "good."

---

[22] Full understanding of the mathematics of this section requires a knowledge of multiple integration, vector and matrix notation, and basic statistical principles. It is not a prerequisite to any other material to follow.
[23] We shall see later that this assumption may be relaxed.

The probabilities of making the two kinds of error [corresponding to expressions (6-13) and (6-14) for the single predictor variable case] may be written as $p_{GB}(R)$ and $p_{BG}(R)$ since they depend on the partition of the predictor space into the regions R_G and R_B rather than on a single point as they did in the single variable situation. The expected cost of error for classifying an area is therefore

$$E(C) = q_G C_{GB} p_{GB}(R) + q_B C_{BG} p_{BG}(R) \qquad (6\text{-}17)$$

which is completely analogous to expression (6-12) in the one-variable case. The probabilities $p_{GB}(R)$ and p_{BG} are

$$p_{GB}(R) = \int_{R_B} f_G(X)\, dx \qquad (6\text{-}18)$$

$$p_{BG}(R) = \int_{R_G} f_B(X)\, dx \qquad (6\text{-}19)$$

where the integral signs represent multiple integrals over the regions R_B and R_G. Expression (6-18) has the form shown because we are dealing with areas which are really type G and hence have the probability distribution $f_G(X)$, and we make an error in classification if the point describing the area falls within the region R_B. A similar interpretation may be given to expression (6-19).

The solution to the problem is a choice of the regions R_G and R_B, in a fashion which minimizes $E(C)$. The expected cost may be written as

$$E(C) = q_G C_{GB} \int_{R_B} f_G(X)\, dX + q_B C_{BG} \int_{R_G} f_B(X)\, dX$$

and recognizing that

$$1 - \int_{R_B} f_B(X)\, dX = \int_{R_G} f_B(X)\, dX$$

it may be further simplified to the form

$$E(C) = \int_{R_B} [q_G C_{GB} f_G(X) - q_B C_{BG} f_B(X)]\, dX \qquad (6\text{-}20)$$

in which a constant term has been omitted.

Expression (6-20) is minimized if R_B is chosen to include all points for which the integrand is negative and to exclude all points for which the integrand is positive. This is equivalent to stating the condition for choice of R_B, and hence *the region for classification as B,* as being

$$\frac{f_G(X)}{f_B(X)} < \frac{q_B C_{BG}}{q_G C_{GB}} \qquad (6\text{-}21)$$

The corresponding *condition for a choice of R_G* is

$$\frac{f_G(X)}{f_B(X)} \geq \frac{q_B C_{BG}}{q_G C_{GB}} \tag{6-22}$$

where the equality case has arbitrarily been assigned to R_G.

The use of this condition is extremely simple. We simply calculate the constant on the right side of the previous inequalities and evaluate the left-side ratio for the potential area in question. If the ratio of probabilities is less than the right-side constant, we predict that the area will be type B, according to expression (6-21). If expression (6-22) holds, we predict that the area will be type B.

It should be recognized that these conditions involve only the cost *ratio* C_{BG}/C_{GB}. Hence, assumption 8 may be relaxed in that we need to know only the relative size of the error costs and not absolute dollar quantities; e.g., to apply these conditions one needs only to know that it is twice as costly to call a "bad" area "good" as to call a "good" area bad," making the cost ratio simply 2.

9. If we further assume that $f_G(X)$ and $f_B(X)$ are p-variable normal probability distributions with expectation vectors U_G and U_B and a common variance-covariance matrix D, the condition for a choice of R_B may be written

$$XD^{-1}(U_G - U_B)^* - \tfrac{1}{2}(U_G + U_B)D^{-1}(U_G - U_B)^* \geq \ln \frac{q_B C_{BG}}{q_G C_{GB}} \tag{6-23}$$

where the asterisk indicates vector transposition and D^{-1} is the inverse of the D matrix. The natural logarithm on the right side enters in simplifying the normal probability function.

However, a knowledge of the theoretical parameters U_G, U_B, and D is not generally available. If we use the maximum likelihood estimates $\overline{X}_G$, $\overline{X}_B$, and S, the resulting condition [written in the form of (6-23)] is not known to be optimal. It does appear reasonable to expect it to be "good," however, since the limiting distribution of this quantity can be shown to be the distribution of the left side of expression (6-23). [See King (1963) for more details on this point.]

COMPETITIVE TEST MARKETING

Each of the analyses related to test marketing in the preceding discussion has been of the strategy-state of nature variety. The question which naturally arises is—what about competitive test marketing actions? The best answer to this is that competitive factors are becoming increasingly important in test marketing.

One aspect of competition arises naturally. There are a number of favorite areas for test marketing consumer products—Albany, Syracuse, and Columbus, Ohio, for example. As a result, so many products are tested in these areas simultaneously that it calls into question the value of test marketing. After all, how representative of anything is a market in which many products are being simultaneously tested?

The second important element of competition in test marketing is the purposeful "jamming" or "spoiling" of test market results. Increasingly, companies are introducing products for test marketing into areas only a few days or weeks after the launching of test marketing of rivals' products. Thus, Duncan Hines and Betty Crocker cake frostings were both test-marketed in Phoenix within a few days of one another. The apparent purpose of such test market competition is the "jamming" of results. With two heavily promoted new products in an area, the additional difficulties in evaluating performance and drawing inferences beyond the test markets are clear. One marketing manager of a large household products company is quoted by the *Wall Street Journal* as saying that his competitors ". . . make special price deals on their own products with the stores, they expand their advertising heavily, and they'll even do things like yanking the number one brand off the shelves temporarily just to foul up sales comparisons. . . . One firm even bought up most of the available radio and TV spots when word leaked out we were going into a city for test marketing."[24]

The competitive aspects of test marketing are so troublesome that some companies are placing more emphasis on pretesting. However, the volume of new consumer products and the volume of test marketing continue to increase. The analysis of such competitive aspects of test marketing is not well developed. Some of the competitive analyses to be described in subsequent chapters are applicable, however. When the analytic techniques are developed in other contexts in later chapters, we shall refer back to the problems of competitive test marketing.

Summary

Market analysis is concerned with the delineation of those uncontrollable elements which can affect the outcome of marketing decisions. Although this process is of great importance to all marketing decisions, it is perhaps the least scientific of all aspects of marketing analysis, for although there are some logical and consistent ways of proceeding with the search for important factors and for quanti-

[24] *Wall Street Journal,* "Figures Don't Lie?: Test Market Results Are Clouded by 'Spoiler' Tactics of Competitors," May 24, 1966, p. 1.

tative data on these factors, there is no single approach which is generally applicable.

Test marketing, on the other hand, has a precise theoretical basis which must sometimes be "bent" in practical applications. The process of test marketing involves selling and promoting a new product in a limited fashion in order to make preliminary evaluations of its potential. Both the product and its associated marketing strategy can be evaluated in this way, so that a better overall strategy can be determined for use when the product is marketed on a larger scale.

Test marketing programs may make use of consumer panel information, direct aggregate market data, or both. *Evaluating* test market performance data is usually more difficult than making performance measurements.

The testing and control aspects of test marketing involve the use of test markets for evaluating the overall product strategy. Thus, while one of the basic objectives is to test the product itself, the organization may wish also to evaluate the relative merit of various advertising themes, levels of expenditure, etc. An approximation to the best overall strategy may be approached through a control scheme for a test market, i.e., a procedure of identifying poor performance and of reacting to alter such performance by changing the levels of the various controllables. Thus, if a product is doing badly in a test market, a greater advertising expenditure may be the cure and the level of advertising which will be necessary to make the product successful on a large scale can be predicted accordingly.

The problem of making early predictions of eventual marketing performance is also critical to test marketing. Since it is impractical to wait years until conclusive long-term performance evidence is available in the test markets, some method for predicting long-term performance from introductory performance data is a necessity.

So, too, is a method for drawing inferences from the test markets, for the basic objective of test marketing is a prediction of how well the product will perform if it is distributed on a wider basis. The drawing of inferences may be accomplished on a gross qualitative basis, but since most products are introduced sequentially into various sections of the nation, the significant inferences which need to be drawn involve the answers to the questions: Should we expand from the test markets? If so, where?

EXERCISES

1. What are the advantages and disadvantages of using political entities such as counties as the basic building blocks in constructing market areas?

2. Why are formally defined market areas necessary, or, indeed, are they really necessary?

3. What is the primary factor which determines whether or not a consumer panel is necessary to obtain test data?

4. What test objectives are relevant to the choice between some restricted product testing scheme, such as obtaining employee reactions, versus full-scale test marketing?

5. A set of test markets which is selected on a random basis may be far from representative of all possible market areas. Why, then, is random selection to be desired?

6. Aside from the obvious danger of a complete inability to obtain desired information, what is the primary reason for requiring that all data sources and data collection procedures be formalized prior to the commencement of a test marketing program?

7. What is the difference between the measure of market penetration used in this chapter and the market share measure?

8. A cartoon depicts a father and son busily engaged in dishwashing. The son says, "What's wrong with the soap people, Pop? They aim all their dishwashing commercials at women." Comment on this in the context of market analysis.

9. Data from a consumer panel observed for a year is given in the table below.

	Number who purchase during first 3 months	Number who purchase during second 3 months	Number who purchase during third 3 months	Number who purchase during fourth 3 months	Average purchase interval, in months	Number of average purchase intervals in end lag	Repeat ratio
First purchaser	2,000	1,000	1,500	800	3	1	
First repeat	400	600	400	400	1.5	2	
Second repeat	200	200	200	200	1	2	

The "end lag" referred to in the next-to-last column is the portion of the year in which purchasers are considered as not having had the opportunity to repurchase. Calculate the two *repeat ratios* which are possible, using this data.

10. Product A and product B have penetration curves as given below (x is the maximum percentage which will ever be attained by the product).

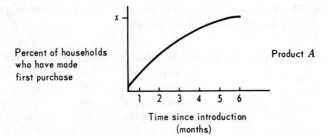

Percent of households who have made first purchase

Product A

Time since introduction (months)

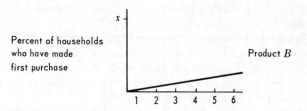

Repeat ratios for the two products have been estimated as:

	Product A	Product B
First repeat ratio	0.17	0.50
Second repeat ratio	0.03	0.25
Third repeat ratio	0.01	0.10
Fourth repeat ratio	0.001	0.03

    a. Diagnose the difficulties (if any) *and possible causes* of these difficulties for the two products.

    b. Suggest how these difficulties might be remedied.

11. Using the data of Table 6-4, determine the 50, 90, and 100 percent bounding curves as illustrated in Figure 6-8.

12. The market shares achieved by three products during an introductory period are shown below. Using the curves developed in the previous exercise, evaluate these early performances.

Products	Market share in various months				
	1	2	3	4	5
A	5%	10%	15%	25%	25%
B	25%	20%	25%	20%	15%
C	10%	10%	15%	30%	30%

13. Illustrate, in terms of the market share measure and the bounding curves of the previous two exercises, how any technique which focused on only one-month changes might lead to fallacious performance evaluations.

14. The following data were observed for a consumer panel for an introductory period of one month.

    Number of people who, during the month, purchase exactly:

once	120
twice	80
three times	60
four times	20

In calculating the first repeat ratio, we wish to exclude from consideration those people who made a first purchase during the last week of the month. Assuming that 25 percent of first purchasers made their initial purchase during this last week:

a. Calculate the first repeat ratio.

b. Explain the need for excluding those who made their first purchase in the last week.

15. Why might the "average purchase interval," as calculated from the data of Table 6-1, not be a good indicator of the period during which the opportunity to make a repeat purchase had been denied to customers?

16. Comment on the validity of a "profit" measure as the sole one for evaluating marketing performance.

17. If one tried to order automobiles in terms of their license plate numbers, one could obtain a sequence of partial rankings. Similarly, military ranks represent a partial ordering of servicemen. Which of these kinds is similar to the partial ranking of five market areas which we previously depicted as (1,3) and (5,4,2)? Why?

18. An individual expresses his preferences for gin, bourbon, scotch, and vodka in specified quantities as $S*B$ (i.e., scotch is preferred to bourbon), $G*V$, and $B*V$. What ranking of his preferences, if any, does this imply?

19. Distinguish between the testing and control aspects of test marketing.

20. An analyst has determined that a particular variable x may be used to predict success or failure in a new market area. He knows that the variable x has the distribution

$$f(x) = 1 \quad 1.5 \leq x \leq 2.5$$

in the areas which will be successful and

$$g(x) = \tfrac{1}{2} \quad 0 \leq x \leq 2$$

in the areas which will be failures. He decides to use a cutoff point of $x = 1.75$ to predict successful versus unsuccessful areas (e.g., if x is less than 1.75, he will call the area a failure and if it is greater than 1.75, he will call it a success). Suppose that 20 percent of all areas are successful and 80 percent are unsuccessful, and that the costs involved are:

Cost of calling a successful area unsuccessful = $100,000
Cost of calling an unsuccessful area successful = $50,000

What is the expected cost of misclassification if his cutoff point is used?

21. Comment on the following statement made by a marketing manager: "Sampling to obtain estimates of purchasing rates is very effective, but to be perfectly accurate, we should observe the purchasing rates of the entire population rather than just a sample from it."

22. Consider the approach to market analysis which you might use if you worked for Wham-O Manufacturing Company, the maker

of novelty products such as the hula hoop, Super-Ball, flying saucer, etc., and the introduction of a similar new product were being considered. In what ways would your market analysis be similar to the one described in the chapter? In what ways would it be different?

REFERENCES

Barnard, G. A.: "Control Charts and Stochastic Processes," *Journal of the Royal Statistical Society* (B), vol. 21 (1959), pp. 239–271.

Dixon, W. J., and F. J. Massey, Jr.: *Introduction to Statistical Analysis,* 2d ed., McGraw-Hill Book Company, New York, 1957.

Ezekiel, M., and K. A. Fox: *Methods of Correlation and Regression Analysis,* John Wiley & Sons, Inc., New York, 1963.

Fourt, L. A., and J. W. Woodlock: "Early Prediction of Market Success for New Grocery Products," *Journal of Marketing,* vol. 25, no. 2 (October, 1960), pp. 31–38.

Frank, N. D.: *Market Analysis,* The Scarecrow Press, Inc., New York, 1964.

Johnson, N. L., and F. C. Leone: *Statistics and Experimental Design in Engineering and the Physical Sciences,* John Wiley & Sons, Inc., New York, 1964.

King, W. R.: "Marketing Expansion—A Statistical Analysis," *Management Science,* vol. 9, no. 4 (July, 1963), pp. 563–573.

———: "Performance Evaluation in Marketing Systems," *Management Science,* vol. 10, no. 4 (July, 1964), pp. 659–666.

———: "Toward a Methodology of Market Analysis," *Journal of Marketing Research,* August, 1965, pp. 236–243.

Lorie, J. H., and H. V. Roberts: *Basic Methods of Marketing Research,* McGraw-Hill Book Company, New York, 1951.

Matthews, J. B., Jr., R. D. Buzzell, T. Levitt, and R. Frank: *Marketing: An Introductory Analysis,* McGraw-Hill Book Company, New York, 1964.

Sales Management: "Now on Your Grocer's Shelves: Test Market Turmoil," August 1, 1966, pp. 43–45.

Tintner, G. A.: *Econometrics,* John Wiley & Sons, Inc., New York, 1952.

7

Purchasing Decisions

It is not unusual to find that a textbook introduces each new main topic or area in the way that a specialist in that particular area might view his domain. To the specialist all other problems of the business enterprise revolve about his problems. There is no question for him that his problems are the most critical and that all other problems, compared with those he faces, pale in significance. While this is undoubtedly a proper viewpoint for the specialist, there is some question as to whether the adoption of this attitude by a textbook author actually contributes to better communication. In some cases the author appears to be crying wolf again and again in insisting that each new topic is of such great significance.

In this chapter we shall discuss the purchasing decisions which are made by business organizations. We shall not attempt to argue that purchasing decisions are *the* most significant problems of marketing. Indeed, many would argue that in some sense they are relatively insignificant. The potential savings to be made through the use of "best" purchasing strategies is usually not of the magnitude of the savings which may result from "best" product choices or "best" advertising strategies, for example.

Nonetheless, purchasing decisions permeate the marketing process. From the purchase of raw materials by the manufacturer of a product, through the purchases of the wholesaler and retailer, to the transactions between the retailer and ultimate consumer, the marketing process may be described by a series of purchasing decisions. Because of the pervasive nature of such decisions, they are of great importance in marketing, the degree of significance depending on the particular product and organizational element being dealt with.

Because purchasing decisions pervade the marketing system, the marketing manager must thoroughly understand their structure and implications. This is so because, even though the marketing manager may not himself be responsible for purchasing decisions, he must deal with those who are. If he is to operate in the marketing environment effectively, he must be able to take the viewpoint of others in the system and to understand their actions in terms of their motivations. If he does this, he can exercise one of the basic functions of management—the motivation of others to seek the goals of his organization.

This aspect of the manager's role is particularly important in the marketing environment because a majority of marketing managers operate in distribution systems which they do not directly control. The typical consumer product is distributed through non-owned (by the manufacturers) channels made up of independent profit-seeking enterprises. Thus, to effectively motivate the wholesalers, distributors, and retailers with whom he deals, the manager must take a subtle approach. To be able to do so, he must understand these organizations and their motivations. He can do this in no better way than by understanding the structure of the important decision problems which these enterprises face—purchasing decisions.

Thus, whereas the importance of purchasing decisions to marketing is based on the large number which take place each business day, the importance of the *analysis* of purchasing decisions rests on the dual base of their pedagogic importance to the marketing manager and the amazing success which has been achieved in developing and using "best" purchasing strategies. Many business enterprises make use of models for the analysis of large-scale repetitive purchasing situations. Westinghouse Electric Corporation is reported[1] to have cut inventory levels in half through the use of models of this type, resulting in an estimated savings of over $3 million annually. In many up-to-date applications *decision rules* which are developed from models are applied in mechanical repetitive fashion by clerks or computers because purchasing decisions are too numerous to analyze individually and because the "best" strategies have proved to be so useful.

Although the purchasing decisions made by the consumer are the very basis of all of marketing, we shall not be concerned with

[1] "Companies Spur Drives to Fill Orders Faster, Cut Distribution Costs," *Wall Street Journal,* vol. SLVI, no. 43 (Dec. 14, 1965), p. 1.

them here. The primary reason for this is that we wish to focus our attention in this chapter on analyses of purchasing decision problems done by (or on behalf of) a rational decision maker. It is a basic premise of much of marketing and advertising that the consumer does not, could not, and perhaps, should not base his purchasing actions on any rational analysis. As a result, we shall not endeavor to solve the problems of the neglected consumer. Rather, we shall devote ourselves to large-scale organizational purchasing decision problems.

It should be noted, however, that an understanding of consumer purchasing problems and behavior needs to be acquired by those organizations wishing to sell their products to consumers, if only because of the basic military tenet "know your enemy." We shall treat such analyses in a later chapter which deals with the promotional decision situations in which they have been most useful.

In this chapter we shall be concerned with organizational purchasing decision situations. The models which we shall use are closely associated with those of inventory theory, for indeed, one cannot deal with purchasing decisions without inventory considerations. The similarity and interaction of purchasing and inventory considerations are demonstrated in Figure 7-1. If the stock depicted there is a supply of goods on the shelves of the retailer, the input to stock is the quantity of finished goods which the retailer purchases and the output represents purchases by consumers. Since the primary organizational decision to be made here concerns purchasing (by the retailer), this would be called a purchasing decision situation. It includes inventory aspects since the items purchased become the retailer's inventory while they are on his shelves. Consequently, he has decisions relevant to inventory management which he must consider. [For example, he must decide the order in which the items are to be sold to consumers; in this case his alternatives might be random, first in first out (FIFO), or last in first out (LIFO).] If, on the other hand, the stock in Figure 7-1 represents a manufacturer's warehouse stock of finished goods, the input results from his production process, the output is the demand placed upon him by wholesalers, and no purchasing decisions on the part of the manufacturer enter directly into the situation. In this case, primary emphasis is on production and inventory management.

FIGURE 7-1   INPUT-OUTPUT STOCK SITUATION

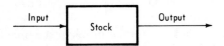

The interdependence of purchasing and inventory is an inherent part of purchasing decisions. However, when variable prices are significant in a purchasing situation, the potential price savings to be gained from intelligent purchasing actions often overshadow the inventory management aspects. In some such situations we shall even find it fruitful to neglect these facets as incidental to the overall decision problem.

In any case, the basic decision problem of organizational purchasing, whether the manufacturer's purchase of raw materials or the wholesaler's and retailer's purchase of finished goods, may be formally stated in a simple way. If $a(t)$ is the input (purchase) rate, $b(t)$ the output (demand) rate, and $I(t)$ the stock level at time t, the basic equation governing the system's operation during some time interval $(0 - T)$ is

$$I(T) = I(0) + \sum_{t=0}^{T} [a(t) - b(t)] \qquad\qquad (7\text{-}1)$$

This expression simply says that the stock level at T is the stock level at time zero plus the sum of the net rate of addition to stock $[a(t) - b(t)]$ over the time periods in the interval T. Since this is basically an expression of a rule of *conservation of inventory*, an obvious requirement is that the dimensions of the components be the same.

The meaning of expression (7-1) is best illustrated by a simple numerical example. If a retailer who has an initial stock level of 20 units buys 200 units per month and sells 190 units per month, his stock level after the elapse of 12 months will be

$$I(12) = 20 + \sum_{t=0}^{12} (200 - 190)$$

or 140 units. In this case the input rate $a(t)$ is 200 units per month, the output rate $b(t)$ is 190 units per month, and the stock level at time zero $I(0)$ is 20 units. Thus the stock level after 12 months is 140 units—the 20 initial units plus the 10 excess units which have been taken into stock each month for 12 months.

The basic purchasing decision problem is to select $a(t)$, given a "knowledge" of $b(t)$, in a fashion which maximizes the utility (or expected utility) of the organization, i.e., to choose a "best" rate of input to stock, given the rate of output from stock. The kind of knowledge which may be available concerning the output rate $b(t)$ will determine whether the decision problem involves certainty, risk, or uncertainty.

It should be recognized that this formal problem statement represents a very narrow view of the interplay of marketing and the purchasing function, for the only controllable recognized is $a(t)$. One of the primary facets of modern marketing involves attempts to influence (if not control) the output (demand) rate $b(t)$, e.g., through sales promotion, advertising, etc. The analysis of these broader marketing decision situations is left for later consideration. Here we begin by adopting a somewhat restricted viewpoint, but in several later chapters we shall attempt to build on this narrow outlook to gain a more valid systems viewpoint. In this chapter the primary attempt to apply a systems approach will revolve about the integration of the purchasing function of marketing and the inventory management function of production.

The controllable treated here, the purchase (input) rate, may be influenced through either the *quantity* or the *timing* of purchases. In some situations one or the other of these aspects may be predetermined. Consider, for example, the milk purchasing decision faced by the housewife. If she purchases milk through a regularly scheduled deliveryman, the timing aspect of her decision is fixed and she must determine only quantity. If, on the other hand, she purchases through a supermarket, she must decide both *when* to shop and *how much* to buy.

Purchasing with Fixed Prices

Purchasing situations in which the unit price of the items to be purchased is fixed are the simplest to analyze. Such situations are not uncommon in the modern economy in which many industries take the form of oligopolies—markets characterized by a small number of large producers—which display little apparent short-run price competition. Models which assume fixed prices are also often useful approximations to real-world purchasing situations in which prices vary so little as to be of negligible importance.

PURCHASING DECISIONS UNDER CERTAINTY

In purchasing decision situations which may be modeled as decision making under certainty, the output (demand) is known. This would be the situation in the case of a contract calling for delivery of D units during a fixed period T. It would also be a close approximation for an organization with a stable demand which could be well approximated as a fixed quantity over some specified time period.

For simplicity we shall consider the organization as a processor who buys units, transforms them, and then sells the same units. In doing so, we avoid the conceptually simple, yet burdensome, transformations which are involved when the dimensions of the input and output quantities are different. No essential differences exist between this treatment and the one which would be appropriate to a steel mill, which must purchase ore and ship ingots, bars, sheets, and tubes, other than the obvious dimensional transformations which must be applied. In speaking of a processing operation, we shall use the terms "raw goods" and "finished goods" because they are most descriptive. The models are equally applicable to other situations, of course, in which case "raw goods" are to be interpreted as inputs to the organization and "finished goods" as outputs. In the case of a wholesaler, for example, the input and output units might be physically identical.

A Basic Model under Certainty. Four basic assumptions will permit us to develop a simple, yet comprehensive, model of the processors' purchasing decision problem.

1. D units must be purchased, processed, and delivered during the time period T.
2. Delivery of the raw units which are ordered is virtually instantaneous.
3. Processing time is negligible.
4. The output rate of finished goods is to be constant between purchases, but not necessarily constant during the entire period T. A fixed quantity of raw goods q will be ordered each time a purchase is made.

The first assumption is equivalent to saying that the cost of not meeting the requirement for all D units is infinite. We shall later relax this assumption to demonstrate interesting circumstances which may arise in the case of such simple purchasing situations.

The second assumption says that we shall abstract out any raw material delivery delays which may occur. If such delays actually exist and are of known duration, we shall show later that it is simple to incorporate them into the analysis.

The third assumption, that processing time is negligible, would be descriptive of the wholesaler's or retailer's purchasing situation in which "processing" simply involves receiving the goods and placing them into stock. In the case of a "process" in which the goods are physically changed, the processing time may not be negligible. A later model will demonstrate how to handle this situation.

The last assumption says that our output of finished goods will be taken to be at a constant rate between purchase orders. If the situation involves a contractual agreement, the determination of this rate may be our prerogative. In many cases, a constant shipping rate over some period is a natural consequence of the physical

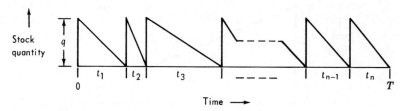

FIGURE 7-2 STOCK QUANTITY–TIME GRAPH

limitations involved. If our output represents consumer demand, a constant demand rate over relatively short periods may be an adequate description of reality. Since we do not require that the output rate always be constant, but only constant *between* purchases, we might consider a high constant rate during one season, a lower constant rate during another, etc., as descriptive of many marketing situations. Further, our choice of a fixed purchase quantity q is a natural one since responsibility for operational employment of such strategies is easily delegated to clerks or computers.

A stock quantity–time graph, such as Figure 7-2, better illustrates this purchasing situation. At time zero a purchase of q units is made, the units are processed, and stock is depleted at a nominal constant rate, as indicated by the line sloping from a stock quantity q to the base line stock quantity of zero after period t_1. Then another purchase of q units is made, bringing the stock level back up to q. During time interval t_2 this stock is reduced at a faster rate. After the third purchase stock is reduced at a slower rate (during t_3). The process continues until all D units have been purchased, processed, and delivered by the end of time T.

The basic decision variable here is q, the purchase quantity. The timing of purchases may also be controllable, depending on whether the output rates between purchases are imposed upon the organization by demand or determined by it. In any event, we shall focus attention on the order (purchase) quantity q.

But why is q important? Why don't we just purchase all D units and ship them at a constant rate as shown in Figure 7-3? The answer, of course, has to do with *the costs involved in implementing the various purchasing strategies*. Two significant costs relate to this situation. The first, the *inventory holding (or carrying) cost*, is the cost associated with having units of a product in one's possession and under one's ownership. An organization's carrying cost may be composed of any, or all, of the investment cost (i.e., the opportunity return on the money which is tied up in inventory), the costs associated with storage (warehouse rental, handling, insurance, taxes, etc.), and expected costs of damage

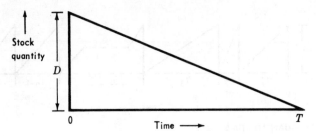

FIGURE 7-3   STOCK QUANTITY-TIME GRAPH

or obsolescence (e.g., goods which we own may be decreased in worth by damage or obsolescence, whereas such losses on goods not owned by us are not incurred by us). The significance of carrying costs becomes apparent when one realizes that they usually represent between 10 and 25 percent of the value of an item on an annual basis. One major producer of electrical equipment estimates, for example, that it costs 20 cents to hold a dollar's worth of finished goods in inventory for a year—a 20 percent annual rate.

It is easy to see how the total of these carrying costs would vary according to the size of q. If we let the order quantity equal D, as in Figure 7-3, we tie up a large amount of money through purchase of the large quantity, we incur large storage costs, since we must provide space for storage of all D units at once, and we take a high risk of loss due to damage or obsolescence, since we immediately own D units of the product. This would seem to indicate that D ought to be very small rather than large. Perhaps we should order only one unit at a time, for instance, since this would result in low investment, storage, and risk-associated costs. Indeed, if carrying costs were all that were involved, this would be our best strategy.

However, we recognize that there is usually a fixed cost associated with each order—*an order cost*—in such situations. This cost involves all of the time and resources involved in purchase order preparation, the phone calls and time used in expediting orders (to assure that our mythical instantaneous delivery will be achieved), etc. The dollar amount of order costs varies widely with products and procedures. In one enterprise the ordering system had become so burdensome and planning so poor over the years that virtually every purchase order was given "special" treatment and order costs were estimated to be over $100 per order.

Order costs are fixed dollar costs which are incurred each time a purchase is made. Consequently total order costs over the period T would be greatest if we were to purchase in single unit quantities. Since we must purchase a total of D units, this policy would require a number of purchases equal to D. On the other hand, if we were to follow the one-order procedure of Figure 7-3, we would incur an order cost only once. Hence, total order costs would be greatest for the strategy requiring D purchases, each of a single unit, and least for the strategy requiring one purchase of the entire D units. However, just the opposite situation holds true for total carrying costs; i.e., when total carrying costs are minimum, total order costs are maximum, and when total carrying costs are maximum, total order costs are minimum.

The outcome array for this decision problem under certainty is shown as Table 7-1. Only one state of nature, represented by the known total demand D, exists. Each strategy has quantity and time components represented by the purchase quantity q and the intervals between purchases $t_1, t_2, \ldots, t_n$. There are a large number of strategies, involving all possible values of q between 1 and D and all possible values for the t's (as long as their sum is T). The outcomes descriptors are the total order costs TOC and total carrying costs TCC which evolve from each strategy; i.e., (TOC_1, TCC_1) represents the costs resulting from S_1; (TOC_2, TCC_2) represents the costs resulting from S_2, etc.

TABLE 7-1   OUTCOME ARRAY FOR BASIC PURCHASING SITUATION

Strategy	State of Nature (D)
$S_1\ (q, t_1, t_2, \ldots t_n)$	(TOC_1, TCC_1)
S_2	(TOC_2, TCC_2)
S_3	
.	.
.	.

The maximization of utility criterion will be approximated here by the *minimization of total relevant costs,* the sum of TOC and TCC. The word "relevant" is used here to denote that, although other costs are involved in this purchasing situation (such as the cost of the units themselves), only order costs and carrying costs are relevant in the sense that they may be affected by the controllable variable q. All other costs are irrelevant because they are incurred in fixed dollar amounts regardless of the value of the order quantity.

The situation in which one cost component is minimum and the other maximum at one extreme of the decision variable (order quantity), while the reverse situation exists at the other extreme, suggests that overall total relevant costs may be minimized at some value of q between 1 and D. We may write total relevant costs as

Total relevant cost = total carrying costs + total order costs   (7-2)

The composition of the "total order costs" term in this expression is simple. There is a fixed cost associated with the placing of an order, say C_0. Since purchases are to be made in quantities of q until a total of D units is purchased, the number of orders is simply D/q and total order costs TOC are simply

$$TOC = C_0 \frac{D}{q} \qquad\qquad (7\text{-}3)$$

The "total carrying costs" term in expression (7-2) is also fairly simple. If we consider that carrying costs are dependent on both the *quantity* held and the *duration* of holding, total carrying costs are both quantity and time dependent. Thus, we shall treat holding costs as behaving as though it costs twice as much to hold two units for a fixed time as it does to hold one unit for the same time and as though it costs twice as much to hold one unit for two months as it does to hold one unit for one month. Both these characteristics are consistent with the nature of the charges which make up holding costs (warehouse rental, taxes, etc.). As a result, we may consider the cost of holding one unit in inventory for one time period to be C_H. The interpretation given to C_H, if the relevant time units are months, would be the dollar cost of holding a unit for a month.

Now consider the holding cost associated with the first purchase cycle in Figure 7-2. The average quantity held during the cycle from zero to t_1 is simply the quantity held at the beginning of the cycle plus that held at the end, divided by 2, or

$$\frac{q+0}{2}$$

This is another way of recognizing that, because the output rate is constant, the situation in which the stock level goes from q to zero in a linear fashion is equivalent to having held the quantity equal to the average stock level over the entire duration t_1.

If we think in terms of having held an average stock of $q/2$ over the duration t_1, the holding cost associated with the first cycle is

$$C_H \frac{q}{2} t_1$$

which has dimensions

$$\frac{\text{dollars}}{\text{unit} \cdot \text{time}} \cdot \text{units} \cdot \text{time}$$

or simply dollars. A similar holding cost is associated with each subsequent cycle. Hence, total holding costs over the entire period from 0 to T are

$$C_H \frac{q}{2} t_1 + C_H \frac{q}{2} t_2 + \cdots + C_H \frac{q}{2} t_n \tag{7-4}$$

which may be written as

$$C_H \frac{q}{2} (t_1 + t_2 + \cdots + t_n)$$

But since the sum of the intervals $t_1, t_2, \ldots, t_n$ is simply T, an even simpler form for total carrying costs TCC is

$$TCC = C_H \frac{q}{2} T \tag{7-5}$$

It should be noted that each term in the summation expression for total carrying costs (7-4) is simply the *area* of the relevant portion of the stock quantity–time graph of Figure 7-2 times the unit carrying cost C_H. Similarly the simpler expression for total carrying costs (7-5) is just the total area under the stock level diagram times C_H. This is no coincidence. We can see that since the carrying cost is dependent on both quantity and time, to compute the total carrying cost, we wish to multiply the cost of carrying one unit for one time period by the total number of unit-time periods which were utilized. This latter quantity is simply the area under the stock level diagram. Since this will hold true as long as carrying costs are dependent on both the quantity and duration of holding, we have a simple device for obtaining holding costs: *we simply calculate the area under the stock level diagram and multiply by C_H to obtain total carrying costs.*

Using the total carrying costs and total order costs obtained from (7-3) and (7-5) in the basic expression (7-2), we obtain total relevant costs TRC as

$$TRC = C_H \left(\frac{q}{2}\right) T + C_0 \frac{D}{q} \tag{7-6}$$

It is our objective to choose a value for q which minimizes total relevant costs, as given by expression (7-6). Before doing this, it will be instructive to view the relationship between the cost components and the decision variable q as shown in Figure 7-4. In this figure, the upward-sloping straight line represents carrying costs. As q increases, greater average quantities are held in inventory and holding costs increase at the rate $C_H T/2$, as seen in expression (7-5). Conversely, as q increases, order costs decrease

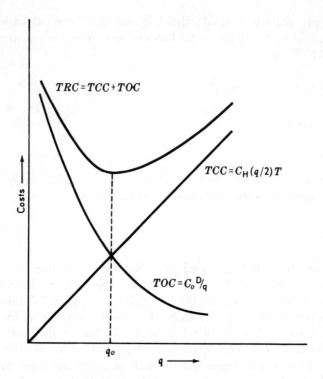

FIGURE 7-4 COST–PURCHASE QUANTITY GRAPH

in a fashion which is inversely related to q, as shown by expression (7-3). The downward-sloping curve in Figure 7-4 represents this relationship. Total costs are simply the sum of these two cost components for any value of q. The upper total cost curve represents the total relevant cost as given by expression (7-6).

Figure 7-4 shows that total relevant costs are relatively high for low values of q, reflecting primarily the high order costs for the many orders required. Total relevant costs are also relatively high for high values of q, reflecting primarily the high carrying costs incurred when high inventories are held. At the intermediate value q_0, total costs are a minimum. It is this value of q which is best (optimal). Hence this is the value which we wish to determine.

We may use the calculus to find q_0 by setting the first derivative of the total relevant cost function equal to zero and solving for q; i.e., q_0 is the solution to the equation

$$\frac{d(TRC)}{dq} = 0 \tag{7-7}$$

Expression (7-7) is

$$\frac{C_H T}{2} - \frac{C_0 D}{q^2} = 0$$

or, solving for q, we obtain the "best" $q - q_0$, as[2]

$$q_0 = \sqrt{\frac{2C_0 D}{C_H T}} \qquad\qquad (7\text{-}8)$$

Hence, q_0 is the best purchase quantity we can use; i.e., we should adopt a policy of purchasing q_0 units, processing and delivering these at some constant rate until our supply is depleted, ordering another q_0 units, processing and delivering these at a constant rate (not necessarily the same as the first) until the inventory is depleted, and so forth, for a total of D/q_0 times, at which time a total of D units will have been purchased, processed, and delivered.

A numerical example will help to illustrate this. Suppose that our organization has a contract for delivery of 180,000 units in a year and that we decide to purchase the 180,000 raw material units in a constant order quantity q. The selling price of the finished goods is $2 per unit and it is estimated that it costs $20 to place an order for raw units. The company's inventory holding costs for finished goods are 10 percent of the item's value per year.

Since processing time is negligible, all inventories which are held are finished goods. Hence, the value of C_H is 20 cents for holding each unit a year, i.e., $10\% \times \$2$. Expression (7-8) may be found to be

$$q_0 = \sqrt{\frac{2 \cdot \$20 \cdot 180{,}000}{\$0.20 \cdot 1}} = 6{,}000$$

which says that we should order in quantities of 6,000. It will be necessary to do so $180{,}000/6{,}000$, or 30, times during the year.

It should be noted that the numerical values plugged into expression (7-8) have dimensional consistency, i.e., years and dollars. If the required output had been expressed as a rate such as 15,000 per month, the carrying cost would also need to be expressed in monthly terms—i.e., $0.20/12$ per month, which results in the same numerical value for q_0. If this were the case, the output rate would undoubtedly be defined by the contract to be constant throughout T rather than simply constant within each purchase-delivery cycle.

[2] Of course, we should investigate the sign of the second derivative to ensure that we have found a relative minimum point on the total relevant cost curve. This is left to the reader. (See any calculus text for a treatment of relative minima and maxima of functions.)

The total costs to be incurred by using this optimal value q_0 may be determined by substituting expression (7-8) into the total cost expression (7-6) to obtain, after some simplification, the optimal (minimum) total relevant cost TRC_0:

$$TRC_0 = \sqrt{2C_0C_HDT} \tag{7-9}$$

For the numerical example this quantity is

$$TRC_0 = \sqrt{2 \cdot \$20 \cdot \$0.20 \cdot 180,000 \cdot 1}$$

or \$1,200. We can see that this represents 30 orders at \$20 each, or \$600 in order costs and the holding of an average inventory of 3,000 items, at a carrying cost of 20 cents per item for the year, or \$600 in carrying costs, for a total cost of \$1,200.

This calculation of costs illustrates an interesting feature of the solution which is revealed by viewing Figure 7-4. There *the optimal order quantity q_0 can be seen to lie at the point where total order costs and total carrying costs are equal,* i.e., at the point of intersection of the TOC and TCC curves. This realization permits us a simple alternative solution technique, for we can determine q_0 simply by equating the TOC and TCC expressions (7-3) and (7-5). If we solve that equation for q, we obtain the optimal order quantity as given by expression (7-8).

One of the most interesting and enlightening features of the solution given by expression (7-8) is its counterintuitive nature. The reader is asked to conduct the following experiment, which has been performed by the author many times, always with similar results. Select a classmate or purchasing decision maker and pose to him the hypothetical situation involving the choice of a purchase quantity q, when faced with the need to purchase 10,000 units within a year's time, and ask him what he believes would be the best value of q. Once he has answered, ask him how he would alter this value if the contract called for 20,000 units in the year rather than 10,000. Almost invariably his intuition will tell him to double the order quantity which he previously selected because the total quantity which he must purchase has doubled; e.g., if he thought that he ought to purchase 10,000 units in 1,000 unit lots, he will purchase 20,000 units in 2,000 unit lots. However, reference to expression (7-7) shows that however good or bad his initial choice of q was, whenever D is doubled, the optimal q will increase by a factor of the square root of 2, or 1.414, and not simply double. This square-root relationship is counter to the intuition of most. Many purchasing agents and executives charged with purchasing responsibility have responded in the same fashion as those not in the field when presented with this situation by the author. Indeed, this realization is a bit frightening to some, for since we have posed it in the form of a change in the total quantity from D to two times D, the degree of confidence which we have in our estimates of the costs C_0 and C_H is of no concern. The fact remains that *changes in the best order quantity have a square-root relationship with changes in demand.* Many companies have followed rule-of-thumb policies in which order quantities are directly related to sales by a fixed percentage. This explains the great savings which can be achieved through the application of simple models in these decision situations. In any case, to anyone whose "business world" is built around foundations of decisions which are made on the basis of intuition and experience, the simple solution to a simple problem represented by expression (7-7) poses significant questions about the wisdom of reliance on intuition, for if intuition cannot be relied upon in simple situations, how well can it be relied upon in complex ones?

Nonnegligible delivery times in the basic model. If a known fixed time is required between the placing of a purchase order and the delivery of the goods to the processor, a simple extension of the basic model is in order. If t_L time units were required between order and delivery, the decision maker would no longer wait until he was out of stock before ordering, but would order when his stock level had reached such a point that it would be just used up when the order was delivered. For example, consider the thinking of the decision maker during period t_1 in the situation of Figure 7-4. He knows the rate at which the finished goods made from the first purchase quantity are being shipped out from his organization. Since this rate is constant, he knows that the quantity of finished goods which will be shipped out during the delivery lag t_L is just this rate times t_L. Hence, he will place his next order for q units whenever the stock level falls to this quantity. By doing so, he will ensure that there is no waiting for raw units to be delivered.

In the numerical example we noted the possibility of an overall constant output rate of 15,000 per month, or 500 per day. If the order–delivery lag for raw units were five days, we would simply order 6,000 units each time the stock level dropped to 5 times 500, or 2,500 units. The 6,000 units would arrive five days later, when the stock level had been reduced to zero. Since no additional cost is usually associated with a known fixed order–delivery delay, no additional changes need to be made in the basic model.

Nonnegligible processing time in the basic model. If processing time is not negligible, some significant changes need to be made in the basic model. Consider, for example, that processing takes place at a rate of k units per day and that we have committed ourselves to supplying R units of finished goods per day. Of course,

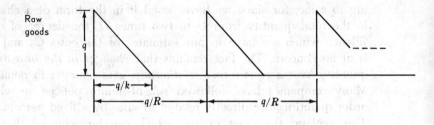

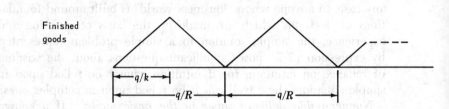

FIGURE 7-5 RAW AND FINISHED GOODS STOCK DIAGRAMS

k must be at least as great as R or we cannot fulfill that commitment.

We shall need to consider two different carrying costs: that applying to finished goods and that applying to raw units. Let us symbolize these unit costs as C_{HF} and C_{HR} dollars per day respectively. Order cost, which almost certainly would involve some sort of production setup components in this case, is still symbolized as C_0 dollars per order. The stock-time diagrams for raw units and finished goods are given as Figure 7-5, where we have supposed that the situation goes on indefinitely rather than only over some fixed time. We have also assumed delivery of raw goods to be instantaneous.

Figure 7-5 may be interpreted by first viewing the stock diagram for raw goods. An order quantity q is received, bringing the raw goods stock level to q. This is then depleted by the processing operation at the rate of k units per day. Hence, in q/k days the raw goods stock will be reduced to zero. It will continue to be zero until the next order of raw goods is received.

Viewing the finished goods diagram in the same way, we see that finished goods will be converted from raw goods and added to stock at the rate k per day and *simultaneously* the output of finished goods at the rate R per day will occur, resulting in a *net* rate of addition to finished goods stock of k minus R units per day. This will continue until all raw goods have been transformed into finished goods, which will require q/k days. Then the finished goods stock is diminished at the rate R per day until the stock is depleted. This will occur in q/R days from the beginning since the total q has been diminished at the constant rate R for the entire period. When the finished goods stock is depleted, another order for raw goods is placed and received and the cycle begins anew.

Since we consider that such cycles continue indefinitely, our objective here should be to *minimize total relevant cost per unit time*. To do this, we first calculate the total cost for a single cycle as the sum of the order cost, the carrying cost for raw goods, and the carrying cost for finished goods. The first of these costs is obviously C_0; there is only one order per cycle. The area under the raw goods stock level diagram for one cycle is $q^2/2k$. Hence, total raw goods carrying costs for a cycle are

$$\frac{C_{HR}q^2}{2k}$$

The area under the finished goods diagram for each cycle may be calculated as the sum of the areas of the two triangles involved if the maximum stock level—the high point of the stock level which occurs at q/k days from the beginning of the cycle—is known. This is easily found to be

$$q - Rq/k$$

since it is the level reached when a total of q units have been produced and after R per day have been shipped out for q/k days. The total area under the finished goods stock level diagram may then be found, after some simplification, to be

$$\frac{q^2}{2R}\left(1 - \frac{R}{k}\right)$$

and the total finished goods carrying cost for a cycle is simply C_{HF} times this quantity.

The total relevant cost per cycle is the sum of these three components, i.e.,

$$\text{Total relevant cost per cycle} = C_0 + \frac{C_{HR}q^2}{2k} + \frac{C_{HF}q^2}{2R}\left(1 - \frac{R}{k}\right) \qquad (7\text{-}10)$$

The total relevant cost per unit time $TRCPUT$ is the total cost per cycle divided by the length of a cycle q/R or

$$TRCPUT = \frac{C_0R}{q} + \frac{C_{HR}Rq}{2k} + \frac{C_{HF}q}{2}\left(1 - \frac{R}{k}\right) \qquad (7\text{-}11)$$

Since we wish to minimize total relevant cost per unit time, we differentiate $TRCPUT$ with respect to the decision variable q and set the result equal to zero, i.e.,

$$\frac{d(TRCPUT)}{dq} = 0 \qquad (7\text{-}12)$$

The optimal purchase quantity is therefore[3]

$$q_0 = \sqrt{\frac{2C_0 R k}{C_{HR} R + C_{HF}(k - R)}} \qquad (7\text{-}13)$$

Suppose that

$$R = 100 \text{ units per day}$$
$$k = 200 \text{ units per day}$$
$$C_0 = \$50$$
$$C_{HR} = \$0.0005 \text{ per day}$$
$$C_{HF} = \$0.004 \text{ per day}$$

According to expression (7-13) the optimal purchase quantity would be

$$q_0 = \sqrt{\frac{2 \cdot 50 \cdot 100 \cdot 200}{0.0005(100) + 0.004(200 - 100)}}$$

which is approximately 2,100. The optimal policy is to purchase 2,100 raw units and to process them for $10\frac{1}{2}$ days. The interval between purchases of raw units will be about 21 days.

The basic model with shortages. The concept of a *shortage* can also be introduced into the basic model to make it more comprehensive. It may be possible, for example, for the processor to fail to meet the delivery terms of a contract. If it is to his advantage to do so, the purchasing model should be flexible enough to allow this possibility.

The most realistic way of treating a shortage is for the processor to incur a charge for each shortage which occurs. Moreover, such a charge may also be time dependent. Some contracts between prime contractors to the government and their subcontractors specify a charge of a given dollar amount for every unit short for a day, for example. In this case the problem of determining the total shortage cost is simple; it is simply this charge times the number of unit-shortage days. In circumstances not involving contractual agreements, the determination of shortage costs may not be so easy. For example, if an auto dealer does not have a model which has been ordered by a customer, the dealer may lose the sale to another who has such a car. Even if no sale is lost, the resultant loss of goodwill from a delayed delivery is hard to evaluate quantitatively.

In considering shortages in the basic model, let us return to the original idea of a contract for delivery of D units of finished

---

[3] Again we must check the sign of the second derivative to ensure that we have found a relative minimum point of the *TRCPUT* function.

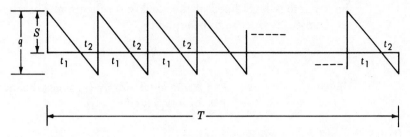

FIGURE 7-6   STOCK LEVEL DIAGRAM WITH SHORTAGES ALLOWED

product within a time period of length T. Let us further suppose that *deliveries are to be made at a constant rate throughout the entire period T*. If delivery and processing time are negligible, the stock level diagram is similar to Figure 7-2 except that we now allow the possibility of a shortage; i.e., the stock level may be negative. Figure 7-6 shows such a situation. In a typical cycle in this diagram, a quantity q is purchased and processed. Shipments of finished goods are *immediately* made in an amount sufficient to satisfy all outstanding shortages, the rest being shipped at a constant rate until the stock is depleted. At that point, shortages begin to build up, as indicated by the horizontal distance of the stock level diagram below zero stock level, until another order is received. The cycle then begins anew. The figure shows that t_1 is the duration of the shipping portion of *each* cycle, t_2 is the duration of the shortage portion, q is the purchase quantity, and S is the stock level achieved after all outstanding shortages have been accounted for. For simplicity we shall neglect the difficulties brought about because this diagram does not exactly reflect the situation at the beginning of the first cycle and the end of the last cycle. This will not be a very great distortion of the real situation if T is fairly long and large quantities of goods are involved.

The area under the stock level diagram for the portion of each cycle having a positive stock level is

$$\frac{St_1}{2}$$

Hence, the carrying cost for each cycle is simply C_H times this quantity. The order cost per cycle is, of course, C_0. The only other cost component which needs to be determined is the shortage cost. If we take it to be dependent on the quantity and duration of shortage, as we did with carrying cost, the shortage cost for a cycle is simply the cost of being short one unit for a specified time C_A times the area above the shortage portion (and below the zero stock level line) of the stock level diagram. This triangular area can easily be seen to be equal to

$$\frac{(q - S)t_2}{2}$$

Hence, the shortage cost for a cycle is simply C_A times this quantity. The total relevant costs for a typical cycle are

$$\text{Total relevant costs per cycle} = C_0 + \frac{C_H S t_1}{2} + \frac{C_A(q - S)t_2}{2}$$

Since there are (approximately) D/q cycles during T, the total relevant cost TRC for T is

$$TRC = \frac{D}{q}\left[C_0 + \frac{C_H S t_1}{2} + \frac{C_A(q - S)t_2}{2} \right] \tag{7-14}$$

The attentive reader will note that *there are two decision variables involved in this extension of the basic model.* If both S and q are known, the process is completely specified. With only one of these, no unique process exists. We must therefore determine a strategy which encompasses two controllable variables. The variables S and q are the simplest, although we could just as well use either t_1 or t_2. This is further exemplified by noting that

$$t_1 + t_2 = Tq/D \tag{7-15}$$

i.e., the length of a cycle is just the total period T divided by the number of cycles D/q, and that

$$t_1 = \frac{S}{q}\frac{Tq}{D} = \frac{ST}{D} \tag{7-16}$$

which is obtained from Figure 7-6 by using similar triangles in the relationship

$$\frac{S}{t_1} = \frac{q}{t_1 + t_2}$$

in conjunction with expression (7-15). Similarly

$$t_2 = \frac{T(q - S)}{D} \tag{7-17}$$

In expressions (7-15) to (7-17) we note that there are only two unknowns; i.e., specification of numerical values for any two of the quantities q, S, t_1, and t_2 uniquely determines all of the rest. There is no way to reduce these expressions so that the determination of a single quantity is sufficient to determine all. Hence, we have two decision variables.

This problem in two decision variables is the first nontrivial example of the determination of the input *rate* as discussed in the chapter introduction. There we stated that the rate had a *quantity and time* portion which must be determined. We might explicitly do that here if we chose q and one of the t's as our decision variables. (For mathematical simplicity, we have instead chosen q and S.) In the basic model and all previous extensions, the determination of one variable, say quantity, uniquely specified the time interval. As such, the previous models were special cases of the more general two-decision variable model.

The total cost for period T may be rewritten from expression (7-14), using (7-16) and (7-17) to obtain

$$TRC = \frac{DC_c}{q} + \frac{C_H S^2 T}{2q} + \frac{C_A T (q - S)^2}{2q} \qquad (7\text{-}18)$$

The optimal values of q and S may be derived, using the calculus, by setting the partial derivatives of the TRC function with respect to q and S equal to zero.[4] The mathematics involved is given in the technical note at the end of this chapter. The solutions developed there are

$$q_0 = \sqrt{\frac{2DC_0(C_H + C_A)}{TC_H C_A}} \qquad (7\text{-}19)$$

$$S_0 = \sqrt{\frac{2DC_0 C_A}{TC_H (C_H + C_A)}} \qquad (7\text{-}20)$$

If we consider the same numerical example used for the basic model with a unit shortage cost C_A of 50 cents per unit year, these expressions become

$$q_0 = \sqrt{\frac{2(180,000) \cdot \$20 \cdot (\$0.20 + \$0.50)}{1(\$0.20)(\$0.50)}}$$

$$S_0 = \sqrt{\frac{2(180,000) \cdot \$20 \cdot (\$0.20)}{1(\$0.20)(\$0.20 + \$0.50)}}$$

or q_0 is about 7,100 and S_0 is about 3,200. This says that we should purchase 7,100 units each time and allow a shortage of 3,900 units to occur before placing the next order.

The implications of this result are interesting. The solution says that a *planned shortage* should be allowed to occur. Although this is contrary to one's intuition, the explanation is that it depends on the relative sizes of the costs involved. In the particular numerical example used here, the relative costs are such that a shortage should be planned. Of course, this assumes that *all* of the relevant

---

[4] Setting these partial derivatives equal to zero is a necessary, but not sufficient, condition that a minimum point has been determined. See the Thomas book in the references.

factors are incorporated into the model. If some other considerations bear on the problem, such as unquantifiable effects of shortages, the decision maker would need to supplement the optimal solution with his estimate of their effect.

The important point is that the model permits him to see, in dollar and cents terms, the cost of not planning for a shortage. Using the same numerical example, for instance, the decision maker can calculate, from expression (7-18), the total relevant cost associated with the optimal solution. He can also calculate the total relevant cost associated with some solution involving no planned shortages. *The difference in costs is the dollar value which the decision maker would be placing on the effect of unquantifiable factors if he were to alter the derived optimal solution on these grounds.*

It is easily demonstrated that the basic model is a special case of the more general model by recognizing that our disallowance of a shortage in the basic model was equivalent to imposing an infinite shortage cost on the model developed in this section. If C_A is allowed to become infinitely large in expression (7-19), the resulting limit is identically expression (7-8), the optimal purchase quantity derived for the basic model.

PURCHASING DECISIONS UNDER RISK

In purchasing decision situations in which there is no contractual agreement or in which demand does not exhibit a sufficiently stable pattern to be approximated as a known fixed quantity, it is often reasonable to consider the quantity demanded as behaving as a random variable. In such a case we are assuming that the demand quantity for each time period of interest is simply the result of a random selection of a numerical value for a variable described by a known probability distribution.

A Model under Risk. We shall introduce the idea of a purchasing decision under risk in the context of a rather special situation—that of an organization dealing with a product which has a *seasonal demand pattern,* such as women's clothes or Christmas decorations. Although illustrated in this context, the general method of analysis is equally applicable to a wide range of purchasing situations.

Consider that orders must be placed for the seasonal article in advance of the season, that delivery occurs prior to the beginning of the season, and that there is no possibility of placing additional orders and receiving the goods before the end of the season. In other words, the organization has a one-shot decision, with no option to change its mind and order additional goods after the season has begun. In such an instance any of three outcomes may result:

1. The quantity ordered may be exactly equal to the quantity demanded.
2. The quantity stocked may be greater than that demanded.
3. Demand may exceed the quantity stocked.

It is highly unlikely that the first outcome would occur unless demand were so stable that the situation would best be treated as decision making under certainty. The second outcome, in which stock remains at the end of the season, is not very desirable because excess stocks of highly seasonal products must either be held over until the recurrence of the season or be disposed of at a reduced price (often less than cost). For most products the holding of excess stock is not good management practice because of high carrying costs. Not only must storage resources be utilized, for instance, but the very nature of seasonal products means that expected obsolescence costs are high. Who, for example, would be willing to take the chance that this year's excess stock of summer dresses could be held over and sold next summer, or that plastic Christmas wreaths will not be superseded by wreaths made of a new material next year? One has only to witness post-Christmas sales of cards and decorations and bargain basement sales of clothing to recognize the attitude taken by retailers on holding such merchandise from season to season. In the third outcome, in which insufficient stocks are available to meet demand, possible profits are forgone because of timid ordering practice.

In such a purchasing situation the costs incurred because of deficient or excess stocks are obviously of great importance. The two kinds of costs considered previously—order costs and carrying costs—are of much less importance. Only one order is to be placed each season, and so *total* order costs are fixed. This condition should be distinguished from the purchasing situations treated previously in which the order cost *per order* was fixed but total order costs depended on the value of one of the decision variables, the number of orders. Carrying costs, on the other hand, do vary with the purchase quantity. As the purchase quantity increases, a larger average inventory is held and total carrying costs are greater. If excess stock is carried over from season to season, the carrying cost components related to obsolescence and deterioration come into play.

If the season is relatively short, the simplest method of handling carrying cost makes use of a unit carrying cost for the "average" unit which is added to the cost price of each item for accounting purposes. If, as is usually the case, excess stock is disposed of at the end of the season, additional consideration of carrying costs is unnecessary. In many cases the season will be so short, and the other monetary amounts involved so large relative to carrying costs, that carrying costs may be entirely ignored without seriously biasing the solution.

The basic purchasing decision problem may then be stated as choosing a season's order quantity q, given knowledge of the probability distribution of the season's demand quantity, in a fashion which maximizes *expected* utility.

Let us consider a numerical example of such a situation. Suppose that the cost price of an item is $48, the unit carrying cost for the average unit is $2, and the in-season selling price is $100 per unit. Suppose further that any item of excess stock at the end of the season can be disposed of for $35 and that all items which are left over will be sold at this postseason price.

If we symbolize the quantity demanded during the season as Z, we may write the form of the expression for profit, if q units are stocked and Z demanded, for the case in which demand exceeds the stock quantity,

i.e.,

$$Z > q$$

as

$$\text{Profit} = 100q - (48 + 2)q = 50q \qquad (7\text{-}21)$$

Expression (7-21) simply states that if in-season demand exceeds the stock quantity, all of the q units which are stocked will be sold at the in-season price of $100, and since total costs of $50 per unit (purchase price plus carrying costs) have been incurred, the total profit is simply $50 times the number of units stocked, q.

If demand is less than the quantity stocked, i.e.,

$$0 \leq Z \leq q$$

the profit expression is

$$\text{Profit} = 100Z + 35(q - Z) - 50q \qquad (7\text{-}22)$$

or simply $65Z - 15q$. Expression (7-22) says that if demand is less than the quantity stocked, the number of units demanded will be sold at $100 (the first term), the excess stock $q - Z$ will be sold after the season at $35 per unit, resulting in a total revenue, as given by the sum of the first two terms. The last term indicates that the total quantity q has associated with it a total cost of 50 times q dollars.

If maximization of expected profit is used as a proxy for maximization of expected utility, we may proceed to write the *expected profit* expression as

$$E(P) = \sum_{Z=0}^{q} (65Z - 15q) f(Z) + \sum_{Z=q+1}^{\infty} 50q\, f(Z) \qquad (7\text{-}23)$$

where $f(Z)$ is the (discrete) probability distribution of the in-season demand quantity Z. For purposes of this illustration this probability distribution will be taken to be that shown in Table 7-2.

The *outcome table* for this decision problem is given as Table 7-3. Since no more than seven units can be demanded during the

TABLE 7-2 PROBABILITY DISTRIBUTION OF SEASONAL DEMAND

Quantity, Z	0	1	2	3	4	5	6	7	8 or more
Probability, $f(Z)$	0.10	0.20	0.20	0.10	0.10	0.10	0.10	0.10	0

TABLE 7-3 OUTCOME TABLE

Strategy	State of nature							
	N_1 $Z = 0$	N_2 $Z = 1$	N_3 $Z = 2$	N_4 $Z = 3$	N_5 $Z = 4$	N_6 $Z = 5$	N_7 $Z = 6$	N_8 $Z = 7$
$S_1: q = 0$	....	1 lost sale	2 lost sales	3 lost sales	4 lost sales	5 lost sales	6 lost sales	7 lost sales
$S_2: q = 1$	1 excess	1 sale	1 sale 1 lost sale	1 sale 2 lost sales	1 sale 3 lost sales	1 sale 4 lost sales	1 sale 5 lost sales	1 sale 6 lost sales
$S_3: q = 2$	2 excess	1 sale 1 excess	2 sales	2 sales 1 lost sale	2 sales 2 lost sales	2 sales 3 lost sales	2 sales 4 lost sales	2 sales 5 lost sales
$S_4: q = 3$	3 excess	1 sale 2 excess	2 sales 1 excess	3 sales	3 sales 1 lost sale	3 sales 2 lost sales	3 sales 3 lost sales	3 sales 4 lost sales
$S_5: q = 4$	4 excess	1 sale 3 excess	2 sales 2 excess	3 sales 1 excess	4 sales	4 sales 1 lost sale	4 sales 2 lost sales	4 sales 3 lost sales
$S_6: q = 5$	5 excess	1 sale 4 excess	2 sales 3 excess	3 sales 2 excess	4 sales 1 excess	5 sales	5 sales 1 lost sale	5 sales 2 lost sales
$S_7: q = 6$	6 excess	1 sale 5 excess	2 sales 4 excess	3 sales 3 excess	4 sales 2 excess	5 sales 1 excess	6 sales	6 sales 1 lost sale
$S_8: q = 7$	7 excess	1 sale 6 excess	2 sales 5 excess	3 sales 4 excess	4 sales 3 excess	5 sales 2 excess	6 sales 1 excess	7 sales

season, no strategy involving the stocking of more than seven units need be considered, because excess units are sold at a loss after the season. The strategies involve order quantities of 0 through 7. The states of nature are represented by the various demand levels which may occur. The outcomes are described in terms of "sales" (the number of units sold), "lost sales," and "excess" (the number of units left in stock at the end of the season).

A mathematical trick which simplifies expression (7-23) may be employed to aid us in dealing with it. To do this, we must note that

$$\sum_{Z=0}^{\infty} f(Z) = 1$$

i.e., the sum of all of the terms of the probability distribution must be unity. The trick involves the simultaneous addition and subtraction of the quantity

$$\sum_{Z=0}^{q} 50q \, f(Z)$$

to expression (7-23). This induces no net change in the value of (7-23), but it changes the form of its component terms. If we add this quantity to the last term, we have

$$\sum_{Z=q+1}^{\infty} 50q \, f(Z) + \sum_{Z=0}^{q} 50q \, f(Z)$$

from which we may factor $50q$ (which does not depend on Z) to get

$$50q \sum_{Z=0}^{\infty} f(Z)$$

or simply $50q$. If we subtract the same quantity from the first term of expression (7-23), we obtain

$$E(P) = \sum_{Z=0}^{q} (65Z - 65q)f(Z) + 50q$$

or more simply

$$E(P) = 65 \sum_{Z=0}^{q} (Z - q)f(Z) + 50q \qquad (7\text{-}24)$$

Having simplified the expected profit expression, we need now to find the value of q for which the expression takes on its maximum value. We may do this in any of several ways. The brute-force approach would simply be to calculate $E(P)$ for various values of q and plot the results; i.e., we would set q equal to zero and note that expression (7-24) has the value zero. If $q = 1$, expression (7-24) has the value

$$65 \sum_{Z=0}^{1} (Z - 1)f(Z) + 50 = 65(-1)(0.10) + 65(0)(0.20) + 50$$

or 43.5. If $q = 2$, expression (7-24) is

$$65 \sum_{Z=0}^{2} (Z - 2)f(Z) + 100$$
$$= 65(-2)(0.10) + 65(-1)(0.20) + 65(0)(0.20) + 100$$

or 74, etc. The plot of $E(P)$ for various values of q is shown as Figure 7-7. Note that $E(P)$ is calculated only for q up to 7 since we know, from the probability distribution of Table 7-2, that we cannot sell more than seven units in a season. Because post-season stocks are sold at a loss, we would be foolish to stock any quantity of units when we know that it will not even be possible to sell all of them.

Examination of Figure 7-7 reveals that total expected profit attains its maximum value when the order quantity is five units, the maximum expected profit being 107. Every other possible value of q results in a lower expected profit. This procedure for determining the value of q for maximum expected profit is identically equivalent to expressing the outcomes of Table 7-3 in terms of profit and then taking the expected value of profit for each of the strategies (rows) across the states of nature (columns).

Solution using marginal analysis. Of course, the brute-force approach would be too laborious if the number of strategies and states of nature were much larger than in this simple example. Consequently we shall present a generally applicable approach to the solution of this model which is based on the principles of *marginal analysis*. The detailed explanation of this approach

FIGURE 7-7 EXPECTED PROFIT FOR VARIOUS ORDER QUANTITIES

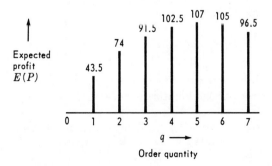

is itself rather tedious. However, it is enlightening and, as we shall show, lends itself well to a shortcut application.

The basic idea of marginal analysis as a method for determining the best level of any activity—in this case the best level of the order quantity q—is the concept of *marginal return*. The marginal return is the *change in return which is generated (caused) by a change in the controllable variable*. In this situation a change in q will influence (since we are dealing with a risk situation) a change in revenue. *The expected marginal return* is simply the mathematical expectation of the marginal return.

The criterion based on marginal return which tells us when we have found the best value of the controllable variable is a simple one. *If the expected marginal net yield (returns minus costs) is positive, an increase in the controllable level should be made, whenever possible.*[5] *If the expected marginal net yield is not positive, no increase in the level of the controllable should be undertaken.* The everyday interpretation of this is obvious. If one can take an action which will return him more than the action will cost (i.e., positive net yield), one should take the action, but if the action costs more than it will return, one should not. In decision problems under risk, analogous statements hold for the *expected* net yield, *expected* cost, and *expected* return. To do anything else is to forgo profit (expected profit), which is clearly not an optimal strategy.

To apply these ideas to the seasonal purchasing situation, consider that we have already calculated $E(P)$ for some particular order quantity q. We may symbolize this expected profit as $E(P_q)$, which would be identically equivalent to expression (7-24). Now, if we were to increase our controllable order quantity (activity level) by one unit to $q + 1$, the expected profit from this new order quantity would be $E(P_{q+1})$ where

$$E(P_{q+1}) = 65 \sum_{Z=0}^{q+1} (Z - q - 1)f(Z) + 50(q + 1) \qquad (7\text{-}25)$$

which is obtained from (7-24) simply by replacing every q with $q + 1$. This is equivalent to

$$E(P_{q+1}) = \sum_{Z=0}^{q} (Z - q - 1)f(Z) + 50(q + 1)$$

since $Z - q - 1$ is zero for $Z = q + 1$; i.e., the last term in the sum of expression (7-25) is identically zero.

---

[5] The phrase "whenever possible" is used because financial constraints may sometimes prevent the increase in activity level. We shall see examples of this in later chapters.

The change in expected profit resulting from a change in order quantity from q to $q + 1$ is simply

$$\Delta E(P_q) = E(P_{q+1}) - E(P_q) \tag{7-26}$$

or

$$\Delta E(P_q) = 50 - 65 \sum_{Z=0}^{q} f(Z) \tag{7-27}$$

In marginal language the expected marginal net yield (profit) resulting in a change in the level of the controllable from q to $q + 1$ is given by expression (7-27). For any q we have shown that it is better to increase the order quantity to $q + 1$ than to leave it at q if the resulting marginal expected profit is positive; i.e., if we gain something in expected profit in altering our order quantity from q to $q + 1$, we should do so. *Conversely, if we do not gain expected profit by increasing our order quantity, we should not do so.* These two statements may be summarized as

$$\begin{aligned} \Delta E(P_q) > 0 \qquad &\textit{Increase order quantity at least to } q + 1. \\ \Delta E(P_q) < 0 \qquad &\textit{Do not increase order quantity beyond } q. \end{aligned} \tag{7-28}$$

Of course, if $\Delta E(P_q)$ is identically zero, we are technically indifferent and would probably choose the strategy requiring the least investment, i.e., q rather than $(q + 1)$.

Using these criteria, we may approach the purchasing decision problem as a series of problems, each of which asks the question: "Should q be increased by one unit?" If we begin with q equal to zero and ask the same question for successive values of q, the criterion of expression (7-28) will enable us to answer each question. Of course, to do this would also be tedious. If we rewrite the criterion (7-28) in terms of (7-27), we obtain the result

$$\begin{aligned} \sum_{Z=0}^{q} f(Z) < {}^{50}\!/_{65} \qquad &\textit{Increase order quantity at least to } q + 1. \\ \sum_{Z=0}^{q} f(Z) > {}^{50}\!/_{65} \qquad &\textit{Do not increase order quantity beyond } q. \end{aligned} \tag{7-29}$$

The application of the criterion written in the simple form of (7-29) involves only a matching of the summed terms of the probability distribution with a constant, ${}^{50}\!/_{65}$. To apply it, we return to the probability distribution given in Table 7-2 and try various values of q, beginning with zero. If $q = 0$, the sum is 0.10. Since 0.10 is less than ${}^{50}\!/_{65}$, criterion (7-29) tells us to increase the order quantity at least to 1. With $q = 1$, the sum is 0.30, which is also less than ${}^{50}\!/_{65}$. At $q = 3$ the sum is 0.60 and at $q = 4$ it is 0.70, both less than ${}^{50}\!/_{65}$. At $q = 5$ the sum is 0.80, which for the first time is greater than ${}^{50}\!/_{65}$. Hence, the criterion (7-29) has told us to go on increasing q from 0, to 1, to 2, etc., up to 5, and it now tells us not to increase beyond 5. Therefore, the best value for q is $q = 5$. This is the same result we obtained by plotting the expected profit as a function of q.

This method, however, is much simpler. As can easily be seen, it involves only the successive summation of the probabilities in Table 7-2 until for the first time the sum exceeds the critical value $^5\!\%_{65}$. The best value for q is that value for which this sum exceeds the critical value for the first time.

Interpretation of the marginal approach in terms of the conceptual framework. In using the marginal approach, we are decomposing the single large decision problem represented by the outcome array of Table 7-3 into a number of smaller decision problems. The smaller decision problems form a series which begins by tentatively setting the order quantity q equal to zero and then deciding whether or not to increase q by a single unit.

The outcome table, in terms of *marginal* profits for the first of these decisions, is shown as Table 7-4. Beginning at $q = 0$, the two alternatives available are "do not increase q by one unit" (do not stock the first unit) and "increase q by one unit" (stock the first unit). There are two relevant states of nature: either the first unit is demanded or it is not.

TABLE 7-4   OUTCOME TABLE FOR FIRST UNIT

Strategy	N_1 (first unit not demanded)	N_2 (first unit demanded)
S_1: do not stock first unit	0	0
S_2: stock first unit	−15	50

The outcomes (in terms of *marginal profit*—the profit contributed by the first unit) are zero if the first unit is not stocked, a loss of $15 if the first unit is stocked and not demanded, and a profit of $50 if the first unit is stocked and demanded.

Since maximization of expected profit is being used as a proxy for maximization of expected utility, we may use this outcome table to calculate the expected marginal profit for the two alternatives and choose that alternative which has the higher expected marginal profit as "best."

This is a simple extension of the basic idea of marginal analysis which can be used to compare two alternatives. In general we choose to increase any activity level which has positive expected marginal net yield. *In comparing two different activities, we should choose to expend resources in that activity which has the highest expected marginal net yield.* To do otherwise would be to forgo (expected) yield.

To apply this idea here, we need to know the probabilities associated with the two states of nature. Looking back to the probability distribution of Table 7-2, we see that the first unit will not be demanded if Z is identically zero. This occurs with probability 0.10. Hence, the probability associated with N_1 is 0.10. Correspondingly, *the first unit will be demanded if total demand is one, two, three, or some higher number.* This occurs with probability $1 - 0.10$, or 0.90, so the probability associated with N_2 is 0.90.

The expected marginal profit for S_1 is therefore

$$E(S_1) = 0(0.10) + 0(0.90) = 0$$

and for S_2 it is

$$E(S_2) = -15(0.10) + 50(0.90) = 43.5$$

Therefore, S_2 is a better choice than S_1 and we should stock the first unit.

Now, *considering that we have already made the decision to stock the first unit,* we pose the next decision problem: "Should we stock the second unit?" The outcome array for this decision is shown as Table 7-5. It should be noted that the numerical out-

TABLE 7-5   OUTCOME TABLE FOR SECOND UNIT

Strategy	N_3 (*second unit not demanded*)	N_4 (*second unit demanded*)
S_3: do not stock second unit	0	0
S_4: stock second unit	-15	50

come descriptors (marginal profits) here are identical to those of the previous table. This is because *these are marginal profits—* the profits made on the *second* unit. *They are not total profit figures for an order quantity of two units.*

State of nature N_3 occurs if total seasonal demand is either zero or one unit, which, from Table 7-2, occurs with probability 0.30. Consequently, N_4 occurs with probability 0.70. The expected marginal profits are

$$E(S_3) = 0(0.30) + 0(0.70) = 0$$

$$E(S_4) = -15(0.30) + 50(0.70) = 30.5$$

Hence, S_4 is better than S_3 and we decide to stock the second unit.

The connection between total expected profit and expected marginal profit can be made clearer at this point by referring again to Figure 7-7. There, *total* expected profit was being plotted. For $q = 1$ the total expected profit was 43.5. This is also the expected marginal profit calculated here for the strategy "stock the first unit." The total expected profit for $q = 2$ from that figure is 74, which is the *sum* of the expected marginal profit for stocking the first unit (43.5) and the expected marginal profit for stocking the second unit (30.5). As we continue, we shall note that *total expected profit for a given value of q is simply the sum of all of the marginal profits associated with stocking units from the first up to and including q.*

Continuing, we form Table 7-6 for the third unit

TABLE 7-6   OUTCOME TABLE FOR THIRD UNIT

Strategy	N_5 (third unit not demanded)	N_6 (third unit demanded)
S_5: do not stock third unit	0	0
S_6: stock third unit	−15	50

and obtain

$$E(S_5) = 0(0.50) + 0(0.50) = 0$$

and

$$E(S_6) = -15(0.50) + 50(0.50) = 17.5$$

Therefore, we decide to stock the third unit. (Checking back to Figure 7-7, we see that total expected profit for $q = 3$ is 91.5, or $43.5 + 30.5 + 17.5$—the sum of the marginal expected profits for the first, second, and third units.)

A similar outcome table holds for the fourth unit, giving us for S_7 (do not stock fourth unit) and S_8 (stock fourth unit),

$$E(S_7) = 0(0.60) + 0(0.40) = 0$$
$$E(S_8) = -15(0.60) + 50(0.40) = 11$$

since the state "fourth unit not demanded" has probability 0.60. We thereby choose to stock the fourth unit.

Similarly, the strategy of stocking the fifth unit has an expected marginal profit of 4.5, and the strategy of not stocking the fifth unit has an expected marginal profit of zero, as may be verified by the reader. Hence, we choose to stock the fifth unit.

The outcome array for the sixth unit is given as Table 7-7. The sixth unit is not demanded if total seasonal demand is for five units or less. The probability of N_{11} is 0.80. The sixth unit is demanded if total seasonal demand is six or more, and so the probability of N_{12} is 0.20. The expected marginal profits are

$$E(S_{11}) = 0(0.80) + 0(0.20) = 0$$
$$E(S_{12}) = -15(0.80) + 50(0.20) = -2$$

Therefore, the strategy of not stocking the sixth unit, with its expected marginal profit of zero, is preferable to the strategy of stocking the sixth unit, with its expected marginal profit of -2.

TABLE 7-7   OUTCOME TABLE FOR SIXTH UNIT

Strategy	N_{11} (sixth unit not demanded)	N_{12} (sixth unit demanded)
S_{11}: do not stock sixth unit	0	0
S_{12}: stock sixth unit	-15	50

Since we have already made the decision to stock the fifth unit and we conclude here that we should not stock the sixth unit, the best overall strategy is to stock five units.[6] The total expected profit for this strategy is just the sum of the marginal expected profits on the first through the fifth units, or

$$43.5 + 30.5 + 17.5 + 11 + 4.5 = 107$$

This is the value shown for $q = 5$ in Figure 7-7.

The total expected profit is lower for $q = 6$ because the marginal expected profit on the sixth unit is -2. Hence, the total expected profit is $107 + (-2)$, or 105. It is easily seen, by looking at the form of the expected marginal profit expressions for the second strategy in each of the decision problems for the first through the sixth units, that once the expected marginal profit for stocking a unit becomes negative, it will never become positive for a subsequent unit. This is how we know that the best value for q is 5 without going on to calculate subsequent expected marginal profits.

---

[6] The reader should be certain that he recognizes the difference between the overall strategy "stock *five* units" which is one of those in Table 7-3 and the strategy "stock the fifth unit" which is one of those in Table 7-6.

A shortcut approach to marginal analysis. The marginal analysis approach, as presented in terms of the conceptual framework, and the mathematical development which results in the criterion (7-29) may be tied together and displayed as one entity if we develop a general expression for expected marginal profit. To do this, we shall define

k = positive marginal profit resulting from stocking the $(j + 1)$st unit if it is sold

l = marginal loss resulting from failure to sell the $(j + 1)$st unit if it is stocked

It should be noted that these are not new quantities, but simply restatements of those which we used to make up the outcome arrays of the previous section. In the numerical value of the example, k is \$50 and l is \$15 (without a minus sign since it is defined as a *loss* rather than a profit).

The expected marginal profit resulting from stocking the jth unit is

$$\Delta E(P_j) = kP(Z > j) - lP(Z \leq j)$$

i.e., the positive marginal profit from selling the $(j+1)$st unit times the probability that the $(j+1)$st unit will be sold minus the marginal loss from failing to sell the $(j+1)$st unit times the probability that it will not be sold.

This is equivalent to

$$\Delta E(P_j) = k - (k + l)P(Z \leq j)$$

because the probability that Z is greater than j is simply 1 minus the probability that Z is less than or equal to j.

If we are to stock the $(j+1)$st unit, this marginal profit must be positive, i.e.,

$$\Delta E(P_j) > 0$$

which is equivalent to

$$P(Z \leq j) < \frac{k}{k + l} \tag{7-30}$$

We should, therefore, stock the $(j+1)$st unit if (7-30) *holds and not stock the $(j+1)$st unit if it does not.*

In terms of the numerical example, $k/(k+l)$ is $^{50}\!/_{65}$. If we return to the criterion of (7-29), we see that this one is identical; i.e., expressions (7-29) and (7-30) give identical criteria for choice of a best order quantity.

In the numerical example it will be seen that

$$P(Z \le 4) = 0.7$$

which satisfies (7-30) and

$$P(Z \le 5) = 0.80$$

which does not satisfy (7-30). Hence, the highest value of j satisfying (7-30) is 4, meaning that we should stock the $(j+1)$st or *fifth* unit.

In terms of (7-29) we have

$$\sum_{Z=0}^{4} f(Z) = 0.7 < {}^{50}/_{65}$$

which tells us to go on at least to $q = 5$ and

$$\sum_{Z=0}^{5} f(Z) = 0.80 > {}^{50}/_{65}$$

which tells us not to increase q to 6. The best order quantity is therefore five.

Expression (7-30) represents a shortcut approach to marginal analysis which is identical in application to (7-29). Either requires us only to calculate a constant $(k/k+l)$ and to add successive probabilities from the probability distribution of demand until this critical point is reached. The procedure which was used to develop the criteria is not operationally necessary; i.e., we do not actually need to begin with $q = 0$ and analyze the first unit, then the second unit, etc. We can simply directly apply either (7-29) or (7-30) (which are really the same thing written in two different ways).

The only cautionary note which must be inserted is that *the validity of the criteria (7-29) and (7-30) depends on the fact that the prices and costs are constant for all units.* If the price or cost for the one unit is not the same as for another, the criteria are inapplicable. In such a case, we can always resort to the direct marginal analysis approach (but not to the shortcut method), since the ideas of marginal analysis are always applicable.

Purchasing with Variable Prices

If the unit price at which goods may be purchased is not fixed, it is of obvious importance to the purchasing decision maker to purchase at as low a price as possible. Consideration of savings in price must be combined with cost considerations in arriving at a "best" purchasing strategy.

There are two common purchasing environments in which variable prices are of great significance. The first, the situation involving *quantity discounts,* may be treated as a decision problem under certainty. The second, involving the purchase of material on a fluctuating market, is usually best treated in terms of risk.

PURCHASING WITH QUANTITY DISCOUNTS

Situations in which vendors set up price schedules giving unit prices as a function of the total purchase quantity may be found in many industrial purchasing contexts. We may term such a purchasing environment as one involving *quantity discounts—unit price discounts which are determined by the total quantity purchased.* This sort of pricing situation is also often referred to as one with *price breaks.*

The "price break" terminology is descriptive of the typical price structure involved in quantity discounts. Figure 7-8 represents a situation in which orders in quantities from zero to q_1 units have an applicable unit price of c_1, whereas orders in quantities from q_1 to q_2 have a unit price of c_2, which is less than c_1. Similarly the successively smaller unit prices $c_3, c_4, \ldots, c_n$ apply to the other quantity ranges shown. The quantities $q_1, q_2, \ldots$ are known as the *price break quantities.* The dots at the left end of the horizontal price lines in Figure 7-8 denote that the unit price for an order of

FIGURE 7-8   QUANTITY DISCOUNT PRICE SCHEDULE

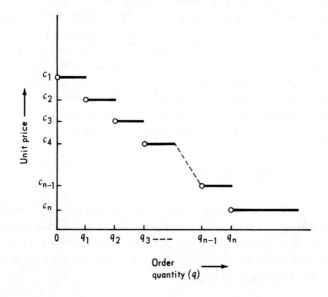

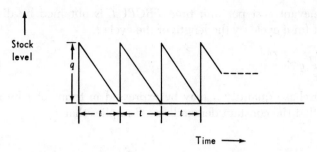

Stock level

q

$|\leftarrow t \rightarrow|\leftarrow t \rightarrow|\leftarrow t \rightarrow|$

Time $\longrightarrow$

FIGURE 7-9   STOCK LEVEL DIAGRAM

the exact quantity q_1, q_2, . . . , q_{n-1} is filled at the lower unit price. Orders for any quantities above q_n are filled at the minimum unit price c_n.

We shall use a slight variation of the basic model under certainty, which was presented previously, to analyze purchasing problems under certainty involving quantity discounts. The reader will recall that the basic model involved purchasing, processing, and delivery of a fixed number of units in a fixed time period. The criterion used was the minimization of total costs over the fixed period. If we assume that the process is a very long one, we may replace the total quantity by a rate, R units per unit time, and make the same basic assumptions of negligible delivery and processing times and a constant delivery rate. In the basic model the cost of the goods to be purchased was fixed and we therefore could omit it as a *nonrelevant* cost when we spoke of total relevant cost. Here, the unit cost varies with the order quantity, and so we must consider, at least, that order costs, carrying costs, and the purchase cost of the goods are relevant cost items.

The basic situation is described by the unending series of stock level "triangles" in Figure 7-9. The total relevant cost per cycle is

$$\frac{C_H}{2} qt + C_0 + qc$$

where C_H is the unit carrying cost, C_0 is the order cost, c is the unit price paid, q is the order quantity, and t is the interval between orders. Because we shall always purchase in quantity q and we have assumed a constant delivery rate, the interval between orders t will be the same for all cycles.

Since the process continues indefinitely, we must *minimize total relevant costs per unit time*, just as we did in one of the earlier variations of the basic model. It is important to note that it is meaningful either to minimize a total relevant cost over a fixed period or to minimize a cost *rate*, i.e., cost per unit time. The total relevant cost per unit time $TRCPUT$ is obtained by dividing the cost for a cycle by the length of the cycle t.

$$TRCPUT = \frac{C_H}{2}q + \frac{C_0}{t} + \frac{qc}{t} \tag{7-31}$$

The purchase quantity q may be expressed in terms of t by recognizing that the constant demand rate R implies that

$$q = Rt \tag{7-32}$$

i.e., we use up the quantity q during time t at a rate R. We may therefore write expression (7-31) in terms of q only, as

$$TRCPUT = \frac{C_H}{2}q + \frac{C_0 R}{q} + cR \tag{7-33}$$

As we have noted earlier, it is common to consider the carrying cost C_H to be some fixed percentage of the cost of the item. If we do so, we may write TRC as

$$TRCPUT = \frac{kc}{2}q + \frac{C_0 R}{q} + cR \tag{7-34}$$

where k is the percentage expressed as a decimal.

Remembering that *the cost c in expression (7-34) depends on q*, we may choose a particular value of c, say c_i, and solve for the best q by taking the derivative and setting it equal to zero. The solution thus obtained is

$$q_{i0} = \sqrt{\frac{2RC_0}{kc_i}} \tag{7-35}$$

However, *the best order quantity determined using unit price c_i—q_{i0}—may not even lie in the range of order quantities for which c_i is applicable.* If it does not, the minimum value of TRC *for that range* is that calculated by using either the lowest or highest order quantity in the range, depending on whether q_{i0} is too small or large. In simple terms, we have found the order quantity which minimizes total relevant costs per unit time, assuming a unit cost c_i. If that optimal order quantity is smaller than the lowest quantity in the range to which c_i applies, the smallest value of total relevant costs per unit time for the range for which c_i is applicable must be that found using the *lowest* order quantity in the range. If the optimal order quantity is larger than the highest order quantity in the range to which c_i applies, the minimum total relevant cost per unit time *for the range for which c_i does apply* must be that found by using the *highest* order quantity in the range. By "using" these order quantities, we mean using them in expression (7-34) to determine total relevant cost per unit time.

This may be summarized for the range of order quantities defined by

$$q_{i-1} \leq q \leq q_i$$

to which the unit cost c_i is applicable; the minimum relevant cost per unit time *for that range* is

$$
\begin{aligned}
TRCPUT(q_{i-1}) & \quad \text{if } q_{i0} \leq q_{i-1} \\
TRCPUT(q_{i0}) & \quad \text{if } q_{i-1} \leq q_{i0} \leq q_i \quad\quad (7\text{-}36) \\
TRCPUT(q_i) & \quad \text{if } q_i \leq q_{i0}
\end{aligned}
$$

These expressions may be interpreted as saying that if the optimal order quantity from expression (7-35) using unit cost c_i is less than or equal to q_{i-1}, the minimum $TRCPUT$ for that range is the one which is calculated by using q_{i-1} in the $TRCPUT$ expression (7-34). Correspondingly, if q_{i0} is within the range, it is the best order quantity for the range, and if it is above q_i, the best order quantity for the range is q_i, with its associated cost $TRCPUT(q_i)$.

The overall minimum total relevant cost per unit time may be found by calculating the minimum for each range of order quantities and selecting the lowest of these. The best order quantity is the one used to obtain the overall minimum cost per unit time.

A numerical example should serve to illustrate this method well. Suppose the demand rate R is 200 units per month, the order cost is $350, the carrying cost is 2 percent per month of the cost of each unit, and the price schedule is given by Table 7-8.

TABLE 7-8   PRICE SCHEDULE

Range of order quantities	Unit price
0–499	$10.00
500–749	9.25
750 or above	8.75

Using expression (7-35), we may obtain the quantities q_{10}, q_{20}, and q_{30} for the three ranges as

$$q_{10} = \sqrt{\frac{2 \cdot 200 \cdot \$350}{0.02(\$10)}} = 837$$

$$q_{20} = \sqrt{\frac{2 \cdot 200 \cdot \$350}{0.02(\$9.25)}} = 870$$

$$q_{30} = \sqrt{\frac{2 \cdot 200 \cdot \$350}{0.02(\$8.75)}} = 894$$

Now, q_{10} is *larger than* the highest value in the range to which it is applicable. Hence, the minimum total relevant cost per unit time *for that range* is the one calculated by using the highest quantity in the range as indicated by $TRCPUT(q_i)$ in (7-36). This quantity is $TRCPUT(499)$ or, from expression (7-34),

$$TRCPUT(499) = \frac{0.02(\$10)}{2}(499) + \frac{\$350 \cdot 200}{499} + \$10 \cdot 200$$

which is about $2,190.

Neither does q_{20} fall within the appropriate range, and so the minimum total relevant cost for the range from 500 to 749 is

$$TRCPUT(749) = \frac{0.02(\$9.25)}{2}(749) + \frac{\$350 \cdot 200}{749} + \$10 \cdot 200$$

or about $2,163.

The quantity q_{30} does fall within the appropriate range and we may calculate the minimum total relevant cost per unit time for that range to be

$$TRCPUT(894) = \frac{0.02(\$8.75)}{2}(894) + \frac{\$350 \cdot 200}{894} + \$10 \cdot 200$$

or about $2,157.

The overall minimum achievable total relevant cost per unit time is the least of $TRCPUT(499)$, $TRCPUT(749)$, and $TRCPUT(894)$, which is $TRCPUT(894)$, or $2,157. The best order quantity is therefore 894 units, which will achieve a total relevant cost per unit time (order, carrying, and purchase costs) of $2,157.

PURCHASING WITH FLUCTUATING PRICES

If purchases are being made in a market environment which fluctuates over time, as would be the case with the markets for commodities, common stock, scrap iron, etc., the purchasing decision problem is quite different from the quantity discount situation in which unit prices are known with certainty. This complex decision problem may be simplified somewhat by recognizing that the order and carrying costs appropriate to most such situations are not of the same order of magnitude as the potential savings which may be gained from buying at a lower price. Because of this, we often completely neglect order and carrying costs and concentrate ex-

---

[7] A thorough understanding of all details of this section requires a somewhat deeper understanding of probability than that given in an earlier chapter.

clusively on the savings which can be achieved by way of the purchase price.

A perfect purchasing policy in a fluctuating market would require that we predict price changes accurately and act to take advantage of them, i.e., to buy at the lowest possible price. Of course, it is not likely that we could ever actually do so. Usually situations of this nature involve a deadline, i.e., we must purchase Q units of a commodity within N purchase opportunities. If we are purchasing raw materials on the commodity market, we might be required to purchase the Q units before 30 days (purchase opportunities) had elapsed, for instance.

There are several simple strategies which we might use in this situation. We might buy Q/N units per day for each of the N days, or we might spend a fixed number of dollars each day, the quantity purchased being whatever quantity that number of dollars will buy at that day's price. This latter strategy, called *dollar averaging* on Wall Street, is usually superior to the former since the variable daily purchase quantity varies inversely with price, thus giving lower total costs; i.e., instead of buying a fixed quantity Q/N each day, by dollar averaging, we buy more on a day when the price is low and less when the price is high.

Here, we will consider a simple kind of policy involving the purchase of all Q units at once. Since the total purchase price will be the unit price times Q, we might just as well speak of making one purchase and use the *total* price as an indication of . when to buy. The purchasing model will involve a single total price quotation being made to us on each of the N days. We must purchase on one of those days at the quoted price. On each day except the last, we have two alternatives—purchase or wait. If we wait, we are waiting for a better price quotation. If we have not purchased by the last day, we must purchase then at whatever price is quoted.

This is not too unrealistic a description of purchasing situations which actually exist in the marketing environment. If purchases of raw materials are being made to supply a production process, for example, there is usually a deadline which represents the scheduled date for start of production (plus any delivery lead times, etc.). The single purchase of the total quantity may be dictated by a policy of avoiding high order costs for small quantities, as is the case with brokerage commissions for purchases in the stock market.

We shall assume that the price quoted for each day is determined by market factors, rather than by some vendor who knows of our requirement and might therefore quote abnormally high prices if we have not purchased until the last day, and that we have some ability to forecast what future prices may be. Of course, we cannot forecast perfectly. It will be sufficient for us to have an estimate of the probability distribution of price for each day. To simplify this example, we shall choose the same probability distribution of total price p for each day. This distribution will be symbolized $f(p)$. In no way is that which we shall do dependent on this distribution being the same for each day, and it is not difficult— only a little more cumbersome—to solve the model in that case.

A reasonable way to decide whether to buy or wait on each day might be to set a critical price for each day. If the price quoted on a particular day is higher than the critical price for that day, the "wait" alternative is chosen. If the quoted price is less than the day's critical price, the "buy" alternative is chosen. In determining this form of decision, we are transforming a decision problem with N strategies (buy on first day, buy on second day, etc.) into, at most, N decision problems, one for each day, each having two strategies (buy or wait). We use the phrase "at most" because once one chooses the "buy" alternative, the remaining decision problems have no meaning.

Let us symbolize the critical price for each of the days as p_1, $p_2, \ldots, p_n$. The determination of these critical prices for each day is equivalent to a solution to the decision problem under the conditions stated, since these critical prices prescribe a well-defined strategy which a clerk or computer could be directed to implement. On the ith day the computer would be directed to take the quoted price p and compare it with that day's critical price p_i. It would then buy or wait accordingly as

$$p \geq p_i: \text{wait} \qquad\qquad (7\text{-}37)$$
$$p < p_i: \text{buy}$$

Consider the critical price for the *last* day, p_n. Since we must purchase at whatever price is quoted on that day, p_n must be infinite, i.e.,

$$p \geq \infty : \text{wait}$$
$$p < \infty : \text{buy}$$

and we shall never choose the alternative "wait" on the Nth day.

On the next-to-last day the critical price p_{n-1} will depend on the likelihood that a better price can be obtained on the one remaining day. One rational way of incorporating this into the critical price is to choose p_n to be equal to the *expected price* on the last day, $E(p)$, i.e., the expected value of the random variable p with probability distribution $f(p)$. This is identically equivalent to our use of the expectation criterion in previous decision problems under risk. We are, in effect, saying that we shall buy on the next-to-last day if the price quoted is lower than the expected price which we are to be quoted tomorrow, and wait if it is higher. The rule is

$p > E(p)$: wait
$p \leq E(p)$: buy

on the next-to-last day.

On the second-to-last day the determination of the critical price is more complex because two purchase opportunities remain if we choose to wait. The critical price should therefore be the expected price for the remaining two-day *process*. If we wait with two days to go and the price quoted is less than the critical price on day $n - 1$, we shall buy at the quoted price p. This portion of the expectation for the remaining two-day process is

$$\int_0^{p_{n-1}} p\, f(p)\, dp$$

i.e., the price on day $n - 1$ is between zero and that day's critical price and we buy at the quoted price p. If, on the other hand, the price quoted on day $n - 1$ is above p_{n-1}, we wait to get an expected price of $E(p)$ on the last day. This occurs with probability

$$\int_{p_{n-1}}^{\infty} f(p)\, dp$$

and *the expected price in the remaining two-day process*, which we equate to p_{n-2}, is

$$p_{n-2} = \int_{p=0}^{p_{n-1}} p\, f(p)\, dp + p_{n-1} \int_{p=p_{n-1}}^{\infty} f(p)\, dp$$

or, since we have already determined that p_{n-1} is $E(p)$,

$$p_{n-2} = \int_{p=0}^{E(p)} p\, f(p)\, dp + E(p) \int_{p=E(p)}^{\infty} f(p)\, dp$$

Similarly, on the kth day, with $n - k$ purchase opportunities remaining, the critical price p_k should be set equal to the expected price from the remaining $(n - k)$-day process, i.e.,

$$p_k = \int_{p=0}^{p_{k+1}} p\, f(p)\, dp + p_{k+1} \int_{p=p_{k+1}}^{\infty} f(p)\, dp \qquad (7\text{-}38)$$

which holds for k between 1 and $n - 2$. The critical prices p_{n-1} and p_n have already been determined to be $E(p)$ and infinity respectively.

To determine the various critical prices using expression (7-38), one must begin with p_{n-2} and *work backwards*, successively determining p_{n-2}, p_{n-3}, etc., until p_1 is determined.

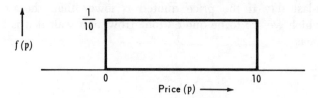

FIGURE 7-10   PROBABILITY DISTRIBUTION OF PRICE (p)

To demonstrate this, let us consider the price distribution $f(p)$ to be the uniform probability distribution (shown in Figure 7-10) in which prices on each day may vary from 0 to 10 with equal probability being accorded to equal intervals between 0 and 10. If we have five purchase opportunities ($n = 5$), we know that

$$p_5 = \infty$$

or 10, since that is the highest possible price in this case, and

$$p_4 = E(p) = \int_0^{10} p\, f(p)\, dp$$

which is 5. Hence, p_4 is 5.[8]

The critical price on the third day may be determined from expression (7-38) by recognizing that we wish to set p_3 equal to the expected price in the remaining two-day process. This expectation will take the form

$$p_3 = \int_0^5 p\, f(p)\, dp + 5 \int_5^{10} f(p)\, dp$$

which is 3.75.[9]

For the second day p_2 will be

$$p_2 = \int_0^{3.75} p\, f(p)\, dp + 3.75 \int_{3.75}^{10} f(p)\, dp$$

which is about 3.04.

[8] Because

$$\int_0^{10} p\, f(p)\, dp = \int_0^{10} p\, \tfrac{1}{10}\, dp = p^2/2(\tfrac{1}{10}) \Big]_0^{10}$$

which when evaluated is $10^2/20 - 0$, or 5.

[9] $\int_0^5 p\, f(p)\, dp$ is found to be equal to 1.25 in the same fashion as shown in the preceding footnote, and

$$\int_5^{10} f(p)\, dp = \int_5^{10} \tfrac{1}{10}\, dp = p/10 \Big]_5^{10} = \tfrac{10}{10} - \tfrac{5}{10}, \text{ or } \tfrac{1}{2}.$$

All other integrations in this sequence involve only changes in the limits of integration.

On the first day the critical price is

$$p_1 = \int_0^{3.04} p\, f(p)\, dp + 3.04 \int_{3.04}^{10} f(p)\, dp$$

which is about 2.58.

Hence, we buy on the first day only if the quoted price is below 2.58. On the second day we buy if it is below 3.04. On the next two days we buy if the quoted price is below 3.75 and 5 respectively, and on the last day we buy regardless of the price quoted.

Note that this increasing value of the critical price is what we would intuitively expect, since on the first day, with many purchase opportunities remaining, we should buy only if the quoted price is quite low (2.58 or below). On the second day, with one less opportunity to go, we become somewhat less stringent as to the highest price which we will pay on that day (3.04). As opportunities pass and the number of remaining opportunities grows smaller, we relax our requirements as to the highest price which we shall pay. If we postpone purchase until the last day, we must buy at whatever price is quoted.

Of course, there are many other models of purchasing decisions with variable prices which may describe particular situations more accurately. Most of these require higher mathematical levels, however, and so they are not included here. The basic objective here is to present the problem and a basic approach to its analysis which may be built upon in order to apply more complex models to real purchasing decision situations.

Technical Note

ON THE MATHEMATICS OF TWO-DECISION-VARIABLE PURCHASING MODELS [10]

The total relevant cost in the two-decision-variable purchasing problem is given by expression (7-18) as

$$TRC = \frac{DC_0}{q} + \frac{C_H S^2 T}{2q} + \frac{C_A T (q - S)^2}{2q}$$

The partial derivatives with respect to the two decision variables are

$$\frac{\partial TRC}{\partial S} = \frac{C_H S T}{q} - \frac{C_A T (q - S)}{q}$$

$$\frac{\partial TRC}{\partial q} = - \frac{DC_0}{q^2} - \frac{C_H S^2 T}{2q^2} + C_A T \frac{4q(q - S) - 2(q - S)^2}{4q^2}$$

---

[10] This technical note requires an understanding of the determination of maxima and minima using partial differentiation.

Setting these partial derivatives equal to zero is a necessary (but not sufficient) condition for the determination of the minimum point of the *TRC* function. The resulting equations are

$$S = \frac{C_A q}{C_A + C_H}$$

$$q^2 C_A - (C_A + C_H) S^2 = \frac{2 C_0 D}{T}$$

Solving this system of two equations for S and q, we obtain expressions (7-19) and (7-20).

Summary

Purchasing decisions pervade the marketing process and the marketing system. Although many marketing managers do not deal directly with purchasing decisions, most deal with people for whom purchasing decisions are of great importance. Since the distribution system for most products is a complex of independent profit-seeking organizations, the understanding of purchasing decisions and the motivations of those who must face them is a little-recognized but significant aspect of marketing management analysis.

Purchasing decision situations may be broadly viewed in terms of the influence of price. The structure of the purchasing problem and the most efficient method of analysis vary accordingly as the unit price of the product in question is fixed or varies in relation to demand. Because of this extreme interdependence of purchasing and pricing, we shall take up the basic elements of pricing decisions in the subsequent chapter. At the conclusion of the pricing discussion, a section treating both pricing and purchasing will serve to define the relationship better.

EXERCISES

1. Suppose that the output of finished goods in the fixed price "basic model under certainty" must be at an overall constant rate during T, rather than simply a constant rate between purchases. How would this change the stock quantity–time graph of Figure 7-2? Solve the model under this additional restriction and compare the solution with that obtained in the text. What are the significant differences?

2. Explain the use of "total *relevant* cost" as an approximation to utility (as opposed to the use of "total cost"). Do so in terms which you might use to make this distinction to your (nontechnical) boss.

3. Note the slope of the "total relevant cost" curve in Figure 7-4 near the optimal point q_0. Does the difference in slope to the right

and left of q_0 suggest anything to you concerning how one should act to minimize the effect of errors which might occur in one's cost estimates?

4. A product must be processed at the approximately constant rate of 5,000 per year. The cost is $2 each, it costs $20 to place an order, and the cost of carrying inventory is 10 percent of the product's value per year. How frequently and in what quantity should orders be placed? If the cost of placing an order has been underestimated (actual cost is $50), how much is this error costing the organization each year?

5. A purchasing agent is faced with purchasing a total quantity of material in 1968 which is four times that which he purchased in 1967. He states: "Assuming my order size for 1967 is optimal (that which minimizes the sum of order and holding costs), I should obviously quadruple each order size in 1968." What should you respond to this?

6. Consider the following purchase situation:

Unit price = $5
Warehouse rental = $0.05 per cubic foot per year
Warehouse space = 50,000 cubic feet
Insurance cost = $1,000 per year
Ordering costs $2 for a secretary's time, $3 for time of transaction approvers, and $1 for other miscellaneous costs (such as phone calls, etc.).

It has been determined that the cost of holding a unit of inventory for a month is 5 percent of the value (price) of the item.

A bright young managerial trainee has proposed that the 12,000 units which must be purchased over the next year should be purchased in quantities of 1,000 on the first of each month. (Assume that usage is 1,000 per month at a constant rate within each month.) An old fogy says that 6,000 should be purchased now and 6,000 six months from now. Whose idea is better? Why?

7. What is the optimal total cost for the strategy of ordering 2,100 units as calculated from expression (7-13)?

8. (Adapted from the Sasieni et al. reference, p. 80.) A processor has a fixed demand with weekly cycles as shown below.

	Number of units
Monday	9
Tuesday	17
Wednesday	2
Thursday	0
Friday	19
Saturday	9
Sunday	14

The processor must take in the same quantity each day to meet this irregular demand. The quantity which he receives on one day is ready for shipment to meet the next day's demand. If shortages cost four times as much per day as a surplus of the same amount, how many items should he have available at the start of business on Monday?

9. Show that the optimal total cost under the model of expression (7-18) is less than that under the model of expression (7-8) (assuming that the cost C_A is not infinite).

10. Determine the optimal total cost for the model of expression (7-18), using the same data as in the numerical example.

11. What is the form of the outcome array associated with the decision problem for which expression (7-18) is a model?

12. A large expensive item may be stocked by a firm. The firm knows that demand during the next year will be for either 0, 1, 2, 3, or 4 items with probabilities as given below:

Number demanded	0	1	2	3	4
Probability	0.5	0.15	0.45	0.25	0.10

The firm knows that the item will be obsolete after one year and that any excess stock will then have to be disposed of at $1,000 unit loss. If sold during the year, each item provides a profit of $3,000. How many items should be stocked?

13. Comment on the validity of the "shortcut approach to marginal analysis" if the seasonal purchasing problem to which it is applied were changed to include quantity discounts.

14. Suppose that the price schedule of Table 7-8 were changed to $10 per unit for quantities below 500 and $9 per unit for quantities 500 or above. Solve the purchasing problem, using these prices.

15. A purchasing agent is faced with three opportunities to purchase a widget on the open market. He believes the probability distribution of price for *any* day is

$$f(p) = 1/2 \qquad 0 \le p \le 2$$

He decides to compare each day's quoted price with a critical value and to buy if the quoted price is less than this value or postpone buying if the quoted price is greater than this value. (If he has postponed purchase till the last opportunity, he must purchase at any price.)

a. What would be a reasonable way for him to choose critical values p_3, p_2, and p_1 (p_3 for the first day, p_2 for the second, etc.)?

b. Determine these critical values by using the above probability distribution and your answer to *a*.

16. Determine the optimal order quantity in the basic model by utilizing the cost equality at q_0 which is depicted by Figure 7-4.

17. Suppose that the decision maker using the model of expression (7-18) decides that the unquantifiable effects of a shortage are so great that he will not permit any to occur, regardless of the optimal solution as given in expressions (7-19) and (7-20). What are the cost implications of this decision?
18. In the purchasing decision situation involving the holding of both raw and finished goods, two holding costs, C_{HR} and C_{HF}, are used in the model. What would be the likely relative size of these two unit costs? Why?

REFERENCES

Baumbol, William J.: *Economic Theory and Operations Analysis,* 2d ed., Prentice-Hall, Inc. Englewood Cliffs, N.J., 1965.

Churchman, C. W., R. L. Ackoff, and E. L. Arnoff: *Introduction to Operations Research,* John Wiley & Sons, Inc., New York, 1957.

Fabian, T., J. L. Fisher, M. W. Sasieni, and A. Yardeni: "Purchasing Raw Materials on a Fluctuating Market," *Operations Research,* vol. 7, no. 1 (January–February, 1959), pp. 107–122.

Hanssmann, Fred: *Operations Research in Production and Inventory Control,* John Wiley & Sons, Inc., New York, 1962.

Morris, W. T.: "Some Analysis of Purchasing Policy," *Management Science,* vol. 5, no. 4 (July, 1959), pp. 443–452.

Sasieni, M. W., A. Yaspan, and L. Friedman: *Operations Research: Methods and Problems,* John Wiley & Sons, Inc., New York, 1959.

Thomas, G. B., Jr.: *Calculus and Analytic Geometry,* 3d ed., Addison-Wesley Publishing Company, Inc., Reading, Mass., 1960.

8

Pricing Decisions

In the introductory remarks to the previous chapter some comments were made concerning the proper elements of a good chapter introduction. In a book about marketing decisions each introduction should include a general statement which reveals the relative importance of the decision area being discussed. No such general statement concerning pricing decisions is practicable.

The price which an organization charges for its products is, like the product itself, of basic importance to the future of the organization, for the relationship between price, revenue, and profit is a direct and inviolable one. The same statement cannot be made, however, about the general importance of *pricing decisions.* Prices and price decisions *are* of central importance in economic *theory,* but the attention paid to pricing decisions by marketing managers has relegated them to a lesser role. There are many possible explanations for this lack of operational significance.

Basic economic theory tells us that in a perfectly competitive market environment with *many* organizations offering the *same* product to *many* customers, the producer cannot have a price decision. He, and all others, will sell at the prevailing market price or not at all. He cannot charge more because buyers can get all they want from his competitors at the prevailing market price, and he would be foolish to charge less, since he can sell all that he can produce at the established price. Moreover, no single organization is large enough to have any influence on the determination of the market price. Is it any wonder, then, that marketing managers have paid scant attention to price decisions?

The answer, of course, is that it is not so obvious at all, since no market environment is truly perfectly competitive; perhaps none are even close approximations. In an advanced technological economy such as ours, there are few industries with perfectly undifferentiated products, for example, and there are few products which are produced by a great many companies.

If a monopoly market environment exists, or if there is collusion among competitors, some price decision aspects exist, but the nature of the decision situation is restricted. Once a price is set in these cases, there is little need for concern with further price decisions. Of course, true monopolies are also rare; the closest resemblance to a monopoly is found in the regulated monopolies, such as telephone or utility companies in a given market. In these cases society has realized that it is sensible to have only one supplier in each community and has avoided the evils of monopoly by placing strict regulations on their policies, activities, and prices.

The pragmatic view of the lack of any general significance being placed on pricing decisions by marketing management might be that pricing decisions are affected by such a wide range of different market factors from product to product and area to area that any general statement concerning their importance is impractical. Government regulation of prices, government pressure on prices, price traditions, and other such factors have great importance in particular situations, but no overall significance. Therefore, pricing decisions may be of crucial importance in one industry and of little significance in another.

Of course, pricing decisions do not simply involve the determination of a price to charge. Just as product strategies involve a myriad of controllable factors, one of which is related to the pricing decision, so do pricing decisions involve many dimensions. The basic price, discounts, price relations within the line, the reaction to be made to competitive price actions, and the relationship of price and promotional activities are only a few of the aspects. In this chapter, however, we shall focus our attention on the basic determination of price. In later chapters we shall take up some of the other considerations.

A Basic Price Model

The basic concepts related to the price which an organization charges for its product are the subject matter of economics. There, the theoretical construct of a perfectly competitive marketing environment is used as a model of real-world markets.

In a perfectly competitive market the market price depends on

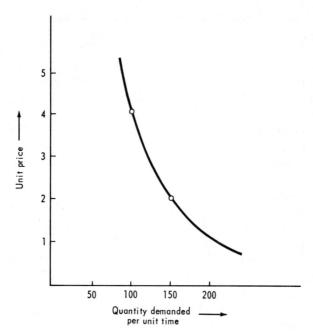

FIGURE 8-1   DEMAND CURVE

the "laws" of supply and demand, i.e., the relationships between
the quantity of a product which is supplied and the quantity which
is demanded at any time. The basic demand concept is that as
the price becomes less, a greater quantity will be purchased (demanded) by consumers, as long as all other factors which affect
the consumer's purchasing behavior are invariant. A *demand curve*
which describes this concept is shown in Figure 8-1. With regard
to the product to which this curve applies, at a unit price of $4
100 units will be sold, and 150 will be sold at a price of $2 (during
a prescribed time period). Such a curve represents the behavior
of the total demand of all consumers at a given point in time.
It may change over time, and it does not necessarily apply to
an individual consumer's behavior, although an individual consumer may often behave in this general fashion. Consider, for example, so commonplace a product as water. If it is very expensive,
I will buy only enough to satisfy my family's thirst. If its price
decreases, I will begin to use it to wash, to water my sun-burned
lawn, etc. If it is very cheap, I will consume large quantities to
fill my pool and hose down my driveway. Yet, regardless of its
price, I can consume only so much. As the price continues to decline, however, *other consumers* will begin to use it for other purposes; e.g., an industrial plant may decide to reduce dust by wash-

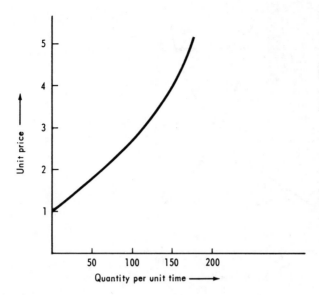

FIGURE 8-2   SUPPLY CURVE

ing its coal fuel before use. These two reactions, extra purchases by present consumers and new purchases by new consumers (as price declines), are the behavioral aspects which the demand curve describes.

The basic supply concepts are equally simple. A *supply curve* such as that shown in Figure 8-2 is based on the idea that as the price at which a product can be sold rises, a greater quantity will be made available (supplied) by sellers, all other things being equal. The rationale for the upward-sloping curve is that additional high-cost production resources will be used as prices rise and it becomes profitable to use them; e.g., U.S. Steel would start up its old inefficient mills if steel prices rose enough to cover the additional costs of using them.

The supply and demand curves jointly determine the market price which will prevail at a given time in a perfectly competitive market. Figure 8-3, which superimposes the supply and demand curves, shows that no price can be stable except at the point of intersection of the two curves—the *equilibrium price*. To demonstrate this, consider that the unit price is higher than the equilibrium price, say $5. In this situation producers will supply 180 units and consumers will demand only 90 units. As stocks pile up in the warehouses of suppliers, the producers will cut the price (because they have already incurred the cost of producing the excess units) and more will be demanded. But they will not continue to do so indefinitely. Consider a price of $1 at which demand

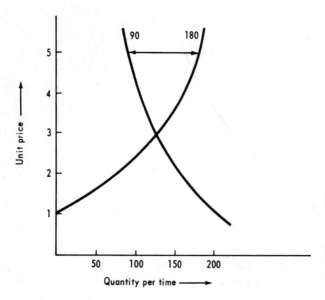

FIGURE 8-3   SUPPLY AND DEMAND MODEL

will exceed supply, for instance. Consumers will tend to bid prices up to get delivery of goods in that situation. In both cases prices will tend to move toward, but not beyond, the equilibrium price, represented by the price at the intersection of the two curves. Here the quantity supplied and demanded are equal. Prices tend to move toward equilibrium in this way precisely because of *competition*—the competition of suppliers to sell in the former case and the competition of consumers to buy in the latter.

These simple ideas constitute the "laws" of supply and demand which form the basis of economic analysis of price decisions. Of course, in any real-world situation the highly idealized conditions required for this model are not met or, perhaps, even approximated. For one thing, the model assumes fixed demand and supply curves. In reality, the existing curves are probably constantly shifting over time, because of changes in the *other factors* which affect the system. The most common demonstration of such a shift used by economic theorists deals with consumer income. A rise in consumer income might cause a change in aggregate consumer behavior which is equivalent to an upward *shift* in the demand curve from *AA'* to *BB'* in Figure 8-4. This means that at any given price, consumers, who now have more available resources, will demand more than they did before the shift. A change in any one of the many variables which affect demand can induce a similar shift in the demand curve; indeed, this is the purpose toward which advertising expenditures, quality change, and the like are directed.

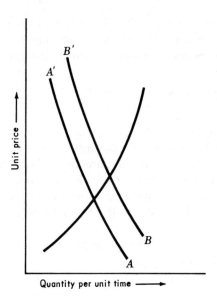

FIGURE 8-4   SHIFTING DEMAND CURVE

The relationship which describes the interdependence of all of these variables with demand is called the *demand function*, as distinguished from the *demand curve* (which relates demand and the single variable, price). We shall have more to say about demand functions later.

DEMAND ELASTICITY

One of the most useful items of information about the demand function is an indication of the *change in demand* which is induced by a change in one of the other variables. In the case of the demand curve, this means the change in the demand quantity D caused by a given change in unit price p, i.e., the marginal demand contribution resulting from a price change. This quantity is symbolized as $\Delta D/\Delta p$ or the corresponding derivative dD/dp. This measure is the reciprocal of the slope of the demand curve so that the flatter the demand curve, the greater will be the marginal demand contribution of a price change.[1]

---

[1] This peculiarity results from the historical convention of drawing demand curves with the controllable variable, price, on the vertical axis. The convention in other cases is the reverse. To get a better idea of this, rotate the book 90° counterclockwise and view the demand curve in Figure 8-1. In doing so, recall that this ridiculous maneuver is made necessary by that peculiar brand of human called an economist.

Because this marginal measure has the dimensions of the product and monetary unit in question, it is desirable to convert it to a percentage basis. The percentage measure is called the *price elasticity of demand*, i.e.,

$$\text{Price elasticity of demand} = \left| \frac{\text{percentage change in } D}{\text{percentage change in } p} \right|$$

The vertical bars indicate that price elasticity is an absolute value; i.e., it is always positive. We adopt this convention for convenience since we shall always assume the same directional behavior: as price falls, demand rises, and as price rises, demand falls. If the price elasticity is less than unity, demand is referred to as being *relatively price inelastic*. If the elasticity is greater than 1, demand is termed *relatively price elastic*.

Computation of the price elasticity is simply accomplished by writing the elasticity as the absolute value

$$\left| \frac{100(\Delta D/D)}{100(\Delta p/p)} \right|$$

which is equivalent to

$$\left| \frac{\Delta D \cdot p}{\Delta p \cdot D} \right|$$

The computational convention for this quantity is to interpret it as the *average* responsiveness to price change over some finite segment of the demand curve. Hence, the average of the end values is used in calculating both D and p. In other words, if we are to consider the segment from (D_1, p_1) to (D_2, p_2) on the demand curve, the elasticity will be the absolute value

$$\left| \frac{(D_2 - D_1)(p_2 + p_1)}{(p_2 - p_1)(D_2 + D_1)} \right| \tag{8-1}$$

The divisor 2, which is used in calculating the average value of p and D, appears in both the numerator and denominator of this expression and is therefore omitted. The corresponding elasticity concept for each point on the demand curve is the limit of this quantity as the segment of the curve extends infinitesimally beyond (D_1, p_1), i.e., the absolute value of

$$\left| \frac{dD}{dp} \frac{p}{D} \right|$$

ELASTICITY IMPLICATIONS

It is clear that the price elasticity of demand has direct bearing on that most significant performance variable—*revenue*. If price goes down, the quantity sold will rise, and yet the total revenue may fall—if the additional volume does not make up the decreased price obtained for each unit sold. Correspondingly, a decreased price can produce an increased revenue.

Whether total revenue will fall or rise depends on the price elasticity. *If demand is relatively price inelastic, an increase in price will increase total revenue. If demand is price elastic, a decrease in price will increase total revenue.* The reader may verify these statements simply by sketching inelastic and elastic demand curves and comparing the rectangular area representing revenue (p times D) for two points on each curve.[2]

An industry faced with an inelastic demand curve would therefore be expected to reduce prices reluctantly, since the result would be an increase in demand (and in total costs) with a decrease in total revenue. An industry with an elastic demand curve, on the other hand, will cheerfully reduce prices (assuming that its cost structure involves significant economics of scale in mass production), since doing so increases revenues while costs increase only modestly.

Such general statements, like all else in the idealized model in the foregoing sections of this chapter, make assumptions that "all else is equal" but, of course, all else is usually *not* equal. The basic ideas developed here are useful, both to the analyst and to the manager, however, *if only because they delineate the kinds of things which should be given attention in pricing decisions.*

PRICE DETERMINATION USING THE BASIC MODEL

Some important aspects of the real world must be considered before any of these theoretical ideas may be put to use in pricing decisions by a single firm. First, we should recognize that we have previously discussed demand curves *for industries or for products.* The individual firm, however, is primarily concerned with the demand curve for its brand. If the demand for a product depends on the brand, for example, the various demand curves for firms within an industry may vary rather widely. Clearly the individual firm is concerned with the price-sales quantity relationship which is applicable to *its* brand rather than some industry "average" demand curve, which may be quite different.

This is equivalent to saying, as we realize from our personal observations, that the real world does not generally exhibit the properties of perfect competition, but rather those of *imperfect*

---

[2] See Samuelson, p. 365, for such a demonstration. Any other basic economics books will contain a similar explanation.

competition, where each supplier is faced with his own demand curve and therefore has some measure of control over the price he will charge.

The range of market situations which may be considered to be imperfectly competitive is wide. The case of many suppliers of a similar, yet differentiated, product, such as toothpaste, is an example of one kind of imperfect competition. Oligopolies, in which a small number of producers sell either differentiated or undifferentiated products, represent the other end of the imperfect competition spectrum. In all such cases the important fact from the viewpoint of pricing decisions is that *the individual firm does exercise some control over price.*[3]

Under conditions of imperfect competition, economic theory assumes that the firm's objective is *profit maximization.* A simple example can be used to demonstrate the methodology for determination of a "best" price in this instance. Suppose that the *firm's* demand curve is given by the equation

$$D = 10,000 - 50p \tag{8-2}$$

where D is in dimensions of units per month and p is dollars per unit.[4] This is equivalent to the rather unrealistic assumption that a maximum of 10,000 units could be given away (set p equal to zero and $D = 10,000$) and that no one will buy if the price is above \$200 (set D equal to zero and $p = 200$).

If the firm has fixed costs of \$50,000 and a unit variable cost of \$70, the total cost expression is

$$TC = 50,000 + 70D \tag{8-3}$$

The sales revenue R from selling D units at a unit price p is simply p times D or, since the sales quantity (demand) is given in terms of p by the demand curve,

$$R = pD = p(10,000 - 50p) \tag{8-4}$$

The reader will note that here we have quickly slipped over a critical point; we have equated the sales quantity and the demand quantity by using the symbol D (demand) in the expression for revenue (which rightly involves the sales quantity). We do this because we recognize that there is no reason for the firm to produce and sell any quantity other than that which is demanded. Because

---

[3] We have previously noted that under conditions of pure competition, the individual firm has no control over price. In the case of monopoly the firm has considerable choice in setting price but is little concerned with price decisions on a continuing basis.

[4] A linear demand curve is used only for simplicity in this example. There is nothing in the analysis which restricts us to this unrealistic kind.

it knows the demand curve (8-2), the firm "controls" demand via the price it sets. To set a price, and hence, a demand quantity, which it could not fulfill would be foolish, since it could always choose a higher price at a demand level which it could fulfill. Conversely, if the firm were to produce more than was demanded, it would incur excessive costs to no avail. In the real world, of course, it may sometimes be desirable to do this, for instance, to overproduce this month and hold stock for next month's demand. Since we know that nothing will change next month in this idealized model, however, we would be foolish to incur carrying costs by overproducing. If total costs *TC* are expressed in terms of the unit price *p* by use of the demand curve, we have

$$TC = 50,000 + 70D = 50,000 + 70(10,000 - 50p)$$

or

$$TC = 750,000 - 3,500p \tag{8-5}$$

This expression may be interpreted by viewing the behavior of the system if price is zero. If price is zero, the demand curve tells us that 10,000 units will be sold (each with a variable cost of $70) for a total variable cost of $700,000. This, added to the $50,000 fixed cost, is the first term in the total cost expression. Each dollar increase will produce 50 fewer units sold—hence, a saving of $3,500 in costs for each dollar increase in price. This is shown as the multiplier of *p* in expression (8-5).

Since we are seeking the profit-maximizing price, we may write profit as revenue minus total costs or, in terms of *p*,

$$\text{Profit} = R - TC = p(10,000 - 50p) - (750,000 - 3,500p)$$

Simplified, this becomes

$$\text{Profit} = 13,500p - 50p^2 - 750,000 \tag{8-6}$$

Since we wish to maximize profit, we take the first derivative of the profit function with respect to the decision variable *p* and equate to zero, i.e.,

$$\frac{d \, (\text{profit})}{dp} = 13,500 - 100p = 0$$

or, the best price p_0 is the solution to this equation

$$p_0 = 135$$

The sales quantity corresponding to this price can be determined (from the demand curve) to be 93,500 units.

An alternative way of viewing this determination of the "best" price would make use of the marginal argument that *the optimal* price is the one at which marginal revenue equals marginal cost; i.e., the price at which the cost of the *next* item to be produced and sold is exactly equal to the revenue which it produces. The marginal revenue is

$$\frac{dR}{dp} = 10,000 - 100p$$

The marginal cost is

$$\frac{d(TC)}{dp} = -3,500$$

Equating these two quantities, we find that marginal revenue and marginal cost are equivalent if p is $135. Therefore, this is the profit-maximizing price.

Figure 8-5 shows the equivalence of these two approaches. In Figure 8-5a revenue and total cost are shown as functions of price. The vertical distance between revenue and total cost is profit (or loss) as labeled. The graph shows that this vertical distance is maximum at $p = \$135$. Figure 8-5b is a plot of marginal cost and marginal revenue. Whatever the price, marginal cost is unaffected, a constant negative $3,500. Marginal revenue declines as price increases. The marginal cost and marginal revenue are equal at the point of intersection of the two curves, at a price of $135.

Operational Pricing

After exploring the basic pricing ideas of economics and their use in the determination of the best price to charge in an imperfectly competitive market environment, it is well to consider the pricing practices which are in *operational* use among marketing managers. In doing so, we investigate the limits of current knowledge and illuminate some of the difficulties involved in pricing decisions.

FACTORS AFFECTING PRICING DECISIONS

If we recognize that the basic model and the analysis built about it are abstractions of the real world, we realize that there must be factors which exist in the real world but are not included in the model, for such is the case with all useful models. The basic pricing model is perhaps less descriptive of the real world than many of the other models for decision problems which we shall discuss because there are so many factors affecting pricing decisions which are insufficiently understood and therefore unquantifiable. Such aspects are typically omitted from quantitative models.

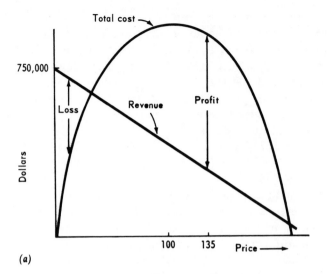

(a)

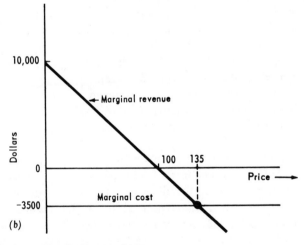

(b)

FIGURE 8-5   (a) REVENUE AND COST
            (b) MARGINAL REVENUE AND MARGINAL COST

The most significant category of intangibles affecting the pricing decision encompasses the many objectives the organization seeks. If its objective is a simple profit-maximizing one, the model is adequate on this count; however, this is seldom the case. First, one has the incompatibility of short-run profits and long-term objectives. Often, a company is willing to forgo the former for the latter. A somewhat diabolical example might be a well-financed organization which decides to ruin its less stable competitors by charging a lower price than anyone can sell at and still make a profit. In incurring the loss, the stable organization hopes to force out competitors, leaving the market to itself alone. Alternatively, a monopoly protected by a patent might choose to charge high prices for its products in the hope that the immediate high profits will not be offset by losses of prestige when others are free to enter the field. Of course, in many markets an organization's natural objectives are to prevent the demise of competition, as might be General Motors' natural objective in its bus lines, and *not* to entice others to enter the field, as might be Polaroid's objective in the home photography field.

Another natural objective which can be of great significance in pricing decisions has to do with price stability. Change for its own sake is unwelcome in most businesses, especially in areas where the fundamental records and operations of the enterprise are concerned. The inexperienced analyst often rebels against the way things are being done because he believes that there is a better way. What he sometimes fails to realize is that although his way might be best for a business which is just beginning, instituting it would require substantial changes. The issue is not simply that there is a better way, but whether this better way has enough benefit to warrant *change*. In pricing, the best price and the current price may not be the same, but the primary question is whether the difference is great enough to warrant changes in bookkeeping, price schedules, etc. The question is of undeniable significance in many industries which are characterized by relatively stable prices.

The nature of the distribution channels for the product interrelates with price stability in their effect on pricing decisions. The longer and more complex the distribution channels are, the larger will be the number of changes induced in the chain reaction which results from a price change. Since each such change involves a cost, greater total costs will be necessary to implement a price change in a situation involving long, complex distribution channels. Conversely, long channels may result in a damping of the effect of a price change, since the organizational elements of the system may absorb (at least for a time) portions of the change. For example, a price increase at the manufacturer's level may be borne by the wholesaler until the next normal printing of his catalogs, since the cost of a special printing might outweigh the cost of absorbing the price increase.

Clearly the basic model does account for the extent of consumer demand for the product. It does not, however, consider the *nature* of that demand. Pricing decisions are greatly affected by the "characteristics of the demand"; e.g., is the demand in the form of a well-informed customer who makes a deliberate choice or is it impulse buying? Are there likely to be fluctuations in the attention paid by consumers to the purchasing decision? Are delivery dates firm or flexible? From such questions it is clear that there is more to demand than simply quantity considerations as assumed in the basic model, for varying answers to these few simple questions could greatly affect the price decisions.

The nature of competitors and their policies also affects operational pricing decisions. In its least important role in pricing, competition places constraints on the degree of pricing flexibility which is practical. Generally, competitors tend to match price changes within certain bounds. In many markets a single firm is recognized as the price leader (usually the one with the highest market share) and the others tend to follow the leader. U.S. Steel's relationship with the industry is the best example. If "Big Steel" raises prices, others follow suit because they are too small to fulfill the additional demand which they might attract at lower prices and because the size and status of the leader tend to make such actions appear necessary and prudent. Even if the effect of competitors is not so direct as it is in the case of steel, the firm which changes prices without giving due consideration to the possible and probable counteractions of its competitors is operating in the dark. In the cold light of day it may find that the neglect of competitors and their possible reactions has caused irreparable harm.

Probably the most significant factor which affects pricing decisions, and yet is omitted from the basic model, is the *nature of the product* itself. We have already argued that products themselves and the factors affecting different products are so different as to preclude the possibility of making a generally valid statement concerning the significance of pricing decisions. Consumer product pricing decisions differ from pricing decisions for industrial products because of the basic "nature of the demand," since customers for industrial goods are generally better informed than are those for consumer goods. However, other distinct differences bear on the pricing decision. The potential customers for industrial goods are generally fewer in number and more homogeneous in makeup than in the consumer case. This permits greater emphasis on substantive product features (such as price and quality) and less appeal to emotion. There is greater reliance on custom-made goods in industrial marketing, with consequent effect on pricing. In many cases bidding for contracts is the established industrial procedure. This situation puts great burdens on pricing decision makers, since they must "price in advance" on goods which are often one of a kind. (We shall discuss aspects of bidding for contracts in a later section of this chapter.)

Another product characteristic which reflects on the basic nature of the pricing decision is the degree of obsolescence associated with the product. Perishable items, such as fresh fruit, become "obsolete" very quickly, as do jet airplanes and women's dresses. We have discussed the purchasing problems involved with such goods in the previous chapter. Here, we are concerned with the other side of the coin: What in-season and postseason price combination will maximize the utility of the organization? If the product is a durable good, the same kinds of problems exist in a less dynamic environment.

Specialty products—those which are distinctive and can command premium prices—involve pricing decision components which are unique. The basic objectives of the enterprise become of overriding significance if the product is of the specialty variety. The retail pricing policy of Volkswagen, a specialty product because of unique features and a distinctiveness established in the minds of the consuming public, is an example of what can be done in pricing specialty products. Virtually all other autos are sold in America on an informal bargaining basis, the price which is printed on the window sticker on each car being the maximum feasible price and the dealer's carefully determined break-even-for-minimum-acceptable-return price (usually unknown to the unsophisticated buyer) being the minimum feasible price. The actual selling price is determined through a complex bargaining process involving the salesman, sales manager, and potential buyer. Volkswagen, on the other hand, can sell at list prices with minimal bargaining because the buyer cannot go down the street to a Chevrolet dealer and get a better price on a car which is, for all practical purposes, the same. He can do this if the dealer is a Ford dealer who does not wish to bargain. For such relatively undifferentiated products as basic Fords and Chevrolets, a list pricing policy cannot easily be maintained.

Not the least significant factor which affects pricing decisions is *tradition*. The best examples of this factor's importance are products which are historically associated with an easy-to-pay price, such as candy bars, chewing gum, and soda pop. It is interesting to note what has occurred as rising labor and material costs have forced price changes on such products. Coke, for example, raised the price from a nickel to a dime with no initial change in the traditional quantity.[5] Hershey chocolate bars, on the other hand, continued to be sold at the nickel price, with quantity reduced to make the price profitable. Neither firm was willing to establish

---

[5] Subsequently, of course, Coke was marketed in bottles and cans containing larger quantities, usually at a premium price.

a nationwide price of 7 or 8 cents because of the resulting inconvenience in making change and in vending-machine sales.

OPERATIONAL PRICING METHODS

It is a truism that *businessmen tend to know more about product costs than they know about the other factors affecting pricing decisions.* Moreover, they typically think that they know more about the "other factors," as discussed in the previous section, than they know about consumers' demand. Consequently it is not surprising to find that the methods used for making operational pricing decisions are oriented toward costs, other factors, and demand—in that order of significance.

In practice, most pricing decisions are made on a complex subjective basis rather than according to a particular method. At best, the technique used is a combination of several methods rather than a specific one. It is useful, however, to delineate several basic operational pricing methods, if only because they serve as inputs to actual pricing decisions.

Cost Pricing. Since cost is the best-understood element of the pricing decision, one would expect cost pricing to be a pervasive pricing method. Of course, there are many different operational pricing methods, each based on cost in some way, which fall into this category.

The policy which is generally referred to as "cost-plus" pricing implies that the price is set at a level which reflects the *average* total cost of each unit of output plus a fixed profit margin or percentage profit. As such, cost-plus pricing (on this full-cost basis) reflects historical costs and past output levels. Of course, ideally one would like to deal with future costs when dealing with prices which will apply to the future. The prediction of future costs, however, is not so easily accomplished from information available in traditional accounting records.

Another problem which is not solved by accounting procedures has to do with joint costs and their allocation. The best example of this is the refinement of crude oil, which produces a multiplicity of end products. The training cost for multiple-line insurance agents is another such situation. In such instances the "fair" proportion of these costs to be charged against each product (whether heating oil, kerosene, gasoline, etc., or life insurance, fire insurance, casualty insurance, etc.) is not at all clearly defined by accounting procedures. One major insurance company has undertaken a large study of several years' duration with the primary objective of determining cost allocations. When this is done, the "simple" cost-plus pricing method, which is a necessity in some lines of insurance, will become operationally useful.

Other cost-based pricing methods might also be called "cost-plus," although it is the "full cost plus a fixed margin" method which is normally accorded that title. Since the selling price must be greater than unit cost (in some sense) for a sale to be profitable, all cost methods have the characteristic of making an addition to cost—thus the term "cost-plus." An incremental cost method focuses on the extra variable cost ("out-of-pocket cost") associated with an additional unit of output. Of course, this is a short-run costing procedure, since a seller who uses incremental costs to "bid" for additional sales will ultimately find that other customers will insist on the lower price given to the last customer. Thus, an ever-growing group of customers would be supplied at a price based on incremental cost, leaving an ever-diminishing group to pay prices which include fixed costs—an extremely unstable situation.

Rate of Return Pricing.   In rate of return pricing, the margin to be added to an average cost is determined from a desirable rate of return to be achieved on the organization's investment. Total cost for the average production of a time period is determined, and the ratio of invested capital to total cost is computed. The product of this ratio and the desired rate of return yields the *markup* (the amount to be added to the average cost) which is necessary to achieve the desired rate of return. This method obviously aims at averaging out over the long term, and it therefore breeds price stability.

Competitive Pricing.   Wherever the primary factor in setting a price is the price already charged by a competitor, the method implicitly used is competitive pricing. An administered pricing situation, in which a small number of sellers of an undifferentiated product allow a price leader to dominate prices, is at one end of the competitive pricing spectrum. Competition, in such instances, tends to be on a nonprice basis, e.g., quality control (though not always quality itself), services provided along with the product, the quality of such services, etc. A gas war in which service stations compete almost solely on a price basis, with no natural leader being recognized, is another competitive pricing situation, as is the supermarket that must carry (and advertise) "loss leaders" at price reductions which are comparable to those of competitors.

Analytical Pricing Models.   It would not be valid to suggest that analytical methods are in great use in operational pricing decisions. Yet it is true that a simple model, which we have already discussed, is often used to provide a basis for sound pricing decisions. This model is based on break-even analysis as discussed in Chapter 5.

The reader will recall that the break-even approach can be used to determine the break-even sales quantity or break-even-for-minimum-acceptable-return quantity as a function of decision variables whose values have not been set. In Table 5-17 price was used as one of these decision variables. In fact, just such a procedure is often used in pricing decisions, since *it does not require any explicit delineation of the demand curve.* After the break-even quantities are determined, the pricing decision is judgmental; i.e., "can we achieve the break-even sales quantity at this price?"

In fact, the most pervasive operational pricing methods are subjective and judgmental, and hence, they defy categorization. These intuitive methods may depend on a feel for the market, a knowledge of the customers' ability to pay, or other factors. The significant point concerning such methods, however, is that they are based almost entirely on the "factors affecting pricing decisions" which were previously discussed. Therefore, *all of the operational pricing methods discussed here, and their various combinations, depend on costs and these "other factors."* Demand enters into operational pricing only marginally, and then at a completely subjective level. Paradoxically, the demand factor, which is given so little logical consideration in operational pricing, is of central importance to pricing theory and the basic model.

The history of science is replete with the successful blending of theory and practice to achieve results. It is likely that this will be the case with the scientific analysis of pricing decisions, in which the theory is based on demand considerations and current practice largely avoids such considerations. In fact, one can see the juncture being hastened by the progress which has been made in the area which is the cause for discrepancies between theory and practice— the estimation of demand.

Demand Estimation

The operational pricing methods just discussed are largely subjective; yet they could easily be put on an objective basis if the marketing manager knew the *demand function* which applied to his product. In fact, most of the problem areas which are dealt with in the other chapters of this book would be more easily approachable if this function were known, because the demand function incorporates information on the *response* (of demand) to changes in each of the variables which affect demand. As might therefore be expected, the methods and techniques for estimating demand functions are not so well developed as many of the other techniques dealt with elsewhere in this book. Of course, other avenues are open to us. If one is to analyze the pricing decision outside the systems context which is implied by the demand function, one is required to have some knowledge of the demand curve. The least information which will prove useful is an estimate of relative demand elasticity. At best, one can "know" the entire demand curve, although these are fragile and temporal entities. Most often the marketing manager is satisfied with some moderate amount of information concerning the general shape of the demand curve. He realizes that a true demand curve applies at a point in time and that it can easily be shifted by other factors. Yet, if he is able to learn something about today's demand curve, while the uncertainties are great, this little knowledge is better than none at all. In fact, a meager knowledge of the demand curve is often sufficient to allow the calculation of ball-park estimates of "best" prices, which turn out to produce very profitable results. In this respect, pricing decisions are much like the decisions involved in the selection of complex military weapons systems. As Defense Secretary McNamara proved in an area in which all previous analysis was subjective, a little objective analysis can result in outcomes which are, perhaps, not the best which could be achieved, but which represent tremendous improvements over those outcomes previously attained. The fact that a manager does not (and never will) *know* the demand curve may prevent him from determining the true optimal price, but estimated demand curves can lead to "very good" prices more consistently than can blue-sky guesses.

BASIC TECHNIQUES OF DEMAND ESTIMATION

Three basic classes of techniques prove useful in estimating demand. We may characterize these classes as interview, experimental, and statistical.

Interview. The simplest and most straightforward approaches to demand estimation are personal *interview* techniques. Yet, as in the case of all easily applied methods, the analysis of the results of such techniques is not so simple. In interviewing to estimate demand, the interviewer simply selects a consumer, say at a supermarket where he has just observed her making a purchase of his or a competitor's food product, and asks her what quantity of the product she would buy at various price levels. Alternatively, she can be faced with a hypothetical price decrease and asked if it would induce her to switch brands. Of course, this naïve approach is not likely to produce totally reliable results, since people have simply not thought out in advance what they would do *if*—and they are usually unable to do so on the spur of the moment. Even if they are given an opportunity to weigh and consider the problem, they are not likely to answer a hypothetical question exactly as they would act in the real world. The key to interviews is to *have the subject convey the desired information by his actions in a real choice situation* rather than simply by an answer to a contrived question. In laboratory situations, perhaps with a consumer panel, one can offer real choices, e.g., a product or a given amount of cash. A series of such choices may reveal the implicit values placed on the product by the consumer. Alternatively, a group may be given money in a simulated marketing situation and told to purchase a package of detergent, or cereal, or whatever product is in question. A series of such "experiments" can provide insight into price-demand relationships.

Experimental. The *experimental* approach to demand estimation differs from the simple "laboratory experiments," which are really interviews of a sort, in that no direct contact with individual consumers is made. The approach is the direct estimation of the demand curve—estimating the aggregate demand at each of a number of price levels. To do this, the price which is charged in actual purchasing situations is varied; the price charged at a particular department store might be reduced and the resulting increase in demand estimated, for instance. Several significant aspects of such experiments should be considered. First, to measure an increase in demand, one must know what the demand would have been had the price change not been made. This requires the forecasting of sales. Secondly, in manipulating prices on certain levels, legal difficulties may arise.[6] Moreover, price experiments, particularly those involving price increases, are considered to be potentially dangerous by many marketing managers since they may induce permanent changes in the complexion of the market. For instance, a price increase might drive some consumers to other brands and they may never return, regardless of subsequent decreases. If this occurs, it is possible for the little learning about the demand curve to be a truly dangerous thing.

Statistical. The *statistical* estimation of demand is not necessarily different from the interview and experimental approaches, since statistical methods may be used in analyzing the results of both interviews and experiments. However, statistical analysis is

[6] See the Howard reference.

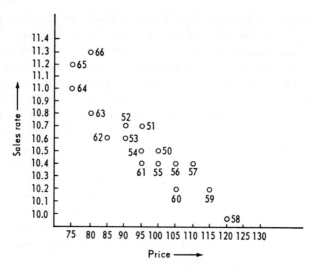

FIGURE 8-6   **SCATTER DIAGRAM OF PRICE-SALES DATA**

the only way to obtain demand information on an after-the-fact basis. Let us suppose, for example, that a history of price changes and corresponding sales levels exists for a product. Suppose these are the data in Table 8-1. A scatter diagram of these data is shown as Figure 8-6.[7]

TABLE 8-1   **PRICE-SALES DATA**

Year	1950	1951	1952	1953	1954	1955	1956	1957	1958	1959	1960	1961	1962	1963	1964	1965	1966
Price, dollars per unit	$100	$95	$90	$90	$95	$100	$105	$110	$120	$115	$105	$95	$85	$80	$75	$75	$80
Sales rate, millions of dollars per year	10.5	10.7	10.7	10.6	10.5	10.4	10.4	10.4	10.0	10.2	10.2	10.4	10.6	10.8	11.0	11.2	11.3

To permit utilization of these price-sales data, we might simply sketch in a continuous curve which appears to fit the points of the scatter diagram. Since our eyes notoriously "see" what they wish to see, however, a more objective approach is desirable. The least-squares technique, previously used in Chapter 6 in drawing inferences from test areas to potential market areas, is the simplest and most direct statistical approach. The reader will recall that the objective of the least-squares approach is to fit a straight line to

[7] The reader should note that we have followed the conventional procedure of plotting the controllable variable—price—on the horizontal axis rather than the traditional economist's method of plotting a demand curve, in which price is shown on the vertical axis.

the points of the scatter diagram in a fashion which *minimizes the sum of squared vertical deviations between the points and the fitted line.*[8]

In fitting a straight line to these data, we are using the model

$$R = ap + b \qquad\qquad (8\text{-}7)$$

where R is sales revenue in millions of dollars per year, p is unit price, and a and b are constants to be determined by the least-squares procedure. The Appendix gives the calculational details relevant to the estimation of a and b, using least-squares techniques.

After the parameters a and b in expression (8-7) have been estimated by using least-squares, the revenues to be derived from various price levels may be predicted from the equation. Of course, the information represented by expression (8-7) also makes it possible to apply the basic pricing decision model.

Both of these uses depend on a belief in the validity of the demand relationship represented by expression (8-7). However, there are a number of factors concerning this relationship which might lead to doubts about its applicability. First of all, the relationship is linear and the real-world demand relationship is probably nonlinear. There may be nothing wrong with the use of a linear approximation over some restricted range of the controllable variable, price. On the other hand, the convenient linear assumption may be a very bad one. In any case, one step which the manager should take is to analyze what is already known concerning the demand characteristics of the particular product in question. He may do this by reviewing previous work done *at the industry level.* Such studies for various industries pervade the literature of economics, and some of them may be useful in analyzing the worth of a linear approximation.

There is one aspect of least-squares analysis which may permit its use even though a strict linear relationship may not appear to be a valid approximation. The idea of linearity in this circumstance refers to the way in which the coefficients enter into the equation. For example, the expression

$$R = a \log (p) + b$$

is a linear relationship for these purposes because the coefficients enter linearly. The equation

$$R = ap^2 + b$$

[8] The reader who is not conversant with least-squares ideas should reread the appropriate section of Chap. 6, which gives the reasoning for choosing this criterion.

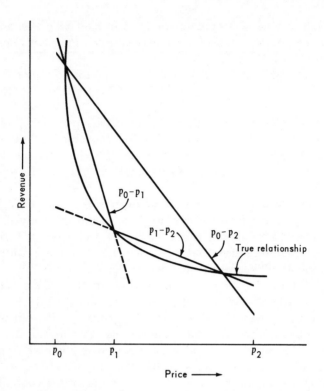

FIGURE 8-7   LINEAR RELATIONSHIP AS APPROXIMATION TO AN UNDERLYING NON-LINEAR RELATIONSHIP

is also considered to be of linear form. Therefore, a wide range of relationships is available for use. There is no difference in the calculations which are necessary except that the log p or p^2 quantity would be tabulated in the same fashion as p would be if expression (8-7) were the basic equation. In effect, we can think of a general equation

$$R = ax + b$$

into which we substitute x equals p, log p, p^2, or some other function of p according to our view of how such a relationship best describes the form of the real-world demand environment.

A further aspect of linearity which should be noted has to do with the range and distribution of the controllable variable, price. Figure 8-7, which shows price and revenue data, may be used to demonstrate this point. The true (but unknown) price-revenue relationship is indicated by the curved line. Although this is the "true" relationship, the observed revenues will probably not fall exactly on this line.[9]

If the range of price which had been observed in the past spanned only the interval from price p_0 to p_1, the linear function labeled "p_0–p_1" might have resulted from least-squares analysis. If the observed range had been from p_1 to p_2, the line labeled "p_1–p_2" could have resulted. It is easy to see that these three linear relationships are quite different, and that one should be careful in interpreting them. Either of the lines estimated by using "partial" data might be a useful approximation *over the appropriate range,* for, as we have previously pointed out, the only validity of any such expression is over the range of the data which were used to estimate it. The overall linear approximation might be a rather bad one, however, since the true curve is not linear. In other words, a linear approximation may be valid over some restricted portion of the price-revenue relationship even if it were a poor approximation to the entire curve.

The second important factor which may bring the validity of simple statistical demand estimates into doubt has to do with other variables which influence demand. The demand *function* relates many variables (advertising, consumer income, etc.) to demand. Price is just one of the factors which influence the demand level. In attempting to relate price and revenue, we have completely omitted the other factors from consideration.

To illustrate how such omissions may affect our results, let us suppose that revenue is *solely* determined according to the relationship

$$R = 10,000 - 20p + 10A \qquad\qquad (8\text{-}8)$$

in which p is price, A is advertising expenditure, and R is revenue. This expression, although unrealistic, is adequate to illustrate our point. Suppose that the several observed combinations of revenue, price, and advertising expenditure, which exactly satisfy expression (8-8), are those given in Table 8-2.

If one were not cognizant of the influence of advertising, or if for some other reason advertising were completely neglected,

---

[9] There may have been recording errors which contribute to these discrepancies, or sales revenue attributable to one time period may be credited to another. Further, stock restrictions may have prevented actual attainment of the revenue prescribed for a particular price. In general, the world is just never observed to operate according to the smooth curves which the manager must use for analysis. As long as such errors or other variations are confined to the revenue variable, there is no cause for concern because our analysis is capable of dealing with them adequately.

TABLE 8-2   OBSERVED DATA

Year	Revenue	Price	Advertising expenditure
1962	9,000	100	100
1963	8,000	150	100
1964	10,000	100	200
1965	11,000	100	300

an attempt to estimate an expression such as (8-7) would determine that the best such equation is

$$R = -40p + 14{,}000 \tag{8-9}$$

This equation seems to imply that each unit increase in price will result in a 40-unit decrease in predicted revenue, whereas the "true" equation (8-8) shows that the real effect is only one-half that amount.[10] Moreover, if we use expression (8-9) to predict the revenue which would be achieved at a price level of 100, we find that the predicted revenue is 10,000. Looking back to Table 8-2, we can see that revenues of 9,000, 10,000, and 11,000 were actually observed at this price level. The reason that the price coefficient has been inflated in our estimation procedure is, of course, that the effect of advertising has been completely neglected. To circumvent this difficulty, we need simply to consider the effect of advertising on demand. In the next section we shall illustrate ways of doing this.

STATISTICAL DEMAND ESTIMATION

In the previous section we pointed out that least-squares techniques may be used in estimating demand. In this section we shall elaborate on the points made there and introduce other statistical ideas which may be employed in demand estimation.

A straightforward extension of the ideas of least-squares regression analysis leads to *multiple regression analysis*. In multiple regression analysis, an equation of the form

$$Y = a_1x_1 + a_2x_2 + \cdots + a_nx_n + b \tag{8-10}$$

in which the x's are independent variables to be used in predicting the dependent variable Y, is estimated from sample data. When

---

[10] We shall say more later concerning the interpretation of this coefficient.

the parameters $a_1, a_2, \ldots, a_n$, and b have been estimated, known values of the x's may be used to predict Y. The least-squares analysis applied to this case is in every way analogous to that described earlier for the case of a single independent variable.

One of the primary advantages of multiple least-squares analysis in estimating demand is that it may obviate the difficulty, which was illustrated earlier, caused by the omission of relevant variables. If, for example, the manager faced with the situation depicted by the data in Table 8-2 felt that revenue was dependent on both price and advertising expenditure [as we, who are privy to the information of expression (8-8), know], he might hypothesize a relationship of the form

$$R = a_1 p + a_2 A + b \qquad (8\text{-}11)$$

If the least-squares regression techniques which are discussed in the Appendix are applied to the estimation of a_1, a_2, and b, the resulting expression (using the data of Table 8-2) is

$$R = -20p + 10A + 10,000 \qquad (8\text{-}12)$$

which is identical to the underlying relationship (8-8). This is because the "observed data" in Table 8-2 are determined directly from expression (8-8). Since all points in the table lie exactly on the same straight line, the best least-squares fit is the line itself (with a corresponding sum of squared deviations of zero). Of course, this is a contrived situation which will not normally be encountered in the real world, for real observed data are subject to the influence of random errors and random fluctuations.

However, one does not gain one's goal simply by introducing more variables into the least-squares analysis and conducting a multiple regression analysis rather than a univariate one. If a dozen variables are used and a thirteenth important factor is omitted, we have the same difficulty as we encountered when only a single predictor variable was used. Each time we introduce another variable, we significantly increase the quantity of data which are required to do a reasonably good job of estimation and prediction. In many cases our data are limited. If the data sources are past time periods, we have an obvious constraint on the number of data points on a scatter diagram such as Figure 8-6. Even if it were possible to do so, we would probably not wish to include pre-Civil War time periods, or probably, pre-World War II time periods, because of a general belief that the underlying structure and "rule of operation" of today's economy are vastly different from those which operated in those eras. As in much of the rest of marketing analysis, we are faced with data shortages and the need to do the best possible analytical job in the light of such shortages. In the case of demand estimation this means that we must consider a compromise between the omission of variables and limited data.

UTILIZING DEMAND ESTIMATES

There are two general ways in which statistical demand estimates may be used in pricing decisions. The distinction between the two may, at first, appear to be insignificant to the novice statistician, and yet, from the viewpoint of statistical theory, the difference is great.

In the first case, estimates of the relevant parameters are made and used in conjunction with hypothetical values of price (and perhaps other controllable variables) to *predict* the demand level. In the case of expression (8-9), for example, the manager might plug in various prices and calculate the revenues which would be predicted to evolve from them.

In the second case, termed *structural analysis,* the coefficients of the independent variables are analyzed as to their effect. Doing this is an attempt to infer *causal* links between variables. For instance, if one were to infer that a unit increase in price in the system relevant to expression (8-9) will *cause* a 4-unit decrease in revenue, that conclusion would be the result of a structural analysis. The objective of focusing attention on the coefficient rather than the predicted value of the dependent variable (revenue) might be a worthy one—to test out a hypothesis or hunch about how the system operates—but we may get into further difficulty in doing so.

A reasonable retort to this proposition might be a hearty "What's the difference?" In truth, the difference is not at all obvious, but rather it has to do with the statistical properties which define a "good" estimate and a "good" prediction.[11] *It is possible to have good predictions of revenue* in cases such as we have discussed, *and at the same time not have good estimates of the coefficients which multiply the independent variables.* In other words, there are conditions under which we may feel confident about utilizing predictions which emanate from one or more estimated coefficients in no one of which we have great confidence. While this may appear to be paradoxical, it is true, and we should be careful to realize that *more stringent conditions are necessary for structural analysis than for prediction.*[12]

[11] See the Spurr et al. reference.
[12] See the Tintner reference for an extensive discussion of the relevant properties and conditions.

Strategic Pricing Analysis

In many pricing situations no objective historical data are available to utilize as a basis for making demand estimates. In such cases it is often deemed best to rely on the judgments of those who are experienced in the marketing of similar products. If such judgments can be combined with objective analysis, basic pricing strategies can frequently be more readily evaluated.

The subjective judgments of experts are most often utilized to assess the *likelihoods* of the various uncontrollable aspects of decision problems—the environmental factors and the actions of competitors. Of course, the use of subjective likelihood judgments is not restricted to pricing decisions, and yet in the pricing area the need, and hence, the worth, of judgmentally developed likelihood estimates is apparent.

Often, subjective likelihood estimates are made for competitive actions and for outcomes (rather than for states of nature). For example, although a pricing decision maker usually does not know which specific action a competitor will take in response to a price change, the range of alternatives which are open to a competitor is generally restricted, and one can often obtain good estimates of the likelihood that he will adopt each of the actions open to him. Although such estimates usually cannot be based on objective historical data, the experience of executives in dealing with similar products and in interacting with their contemporaries in other enterprises provides a sound basis for judgment. Similarly, if no objective historical data are available, the best estimates of the likelihood of outcomes—in terms of levels of market penetration or profitability—may well be obtained through the subjective processes of the experts who are most familiar with the product and the results of the market analyses which have been conducted.

In making use of the subjective likelihood estimates of executives, we are here following a path like that used in the analysis of product decisions in Chapter 5. Pricing decisions are closely related to product decisions in two areas. First, both are major aspects of the decision process necessary to introduce a new product. Most other decisions, such as advertising and promotional outlays, distribution modes, etc., are more transient in the sense that actions can rather easily be altered without great long-term effect on the product's sales. The selection of a product and the price charged for the product are more basic in that these actions are generally less easily revoked and involve more significant long-term effect on profit results than does the adoption of advertising budgets or promotional media.

The other major similarity between product and pricing decisions is the great likelihood that meager historical data will be available to the decision maker in both circumstances. Hence, both areas become prime areas for analyses which make use of subjective estimates to complement objective methods.

Another important factor in pricing decisions which is not incorporated into the previously discussed basic models is the time element. A general pricing strategy[13] is perhaps best thought of as a series of actions to be adopted at various points in time. Each of the individual actions comprising the strategy may be dependent on events which occur in the outside world, such as price reactions by competitors. Or other more complex happenings may be incorporated into a general pricing strategy. For instance, the adoption of a pricing action may be made contingent on a specified informational input such as the result of a demand-estimation experiment.

An illustration of a complex strategy of this kind might serve to make the point more satisfactorily. A *single* pricing strategy might be described in a number of steps.

1. Conduct pricing experiment.
2. If result of pricing experiment is *a*, set price at $1.95; if *b*, set price at $2.39; if *c*, set price at $2.69.
3. If price is $1.95 and competitor reacts with price cut, maintain $1.95 price.
4. If price is $2.39 and competitor reacts with price cut, match his price (down to $1.95).
5. If price is $2.69 and competitor reacts with price cut, act to undersell him by $0.05 (down to $1.95).

In this complex strategy *a*, *b*, and *c* represent precisely defined experimental outcomes related to the estimation of the demand curve.

It is clear that the potential complexity of general strategies of this form is virtually limitless. Also, the number of possible strategies of this kind which need to be considered grows very large as the number of elements in each strategy increases. Practically speaking, it is usually not feasible to consider strategies which are extremely complex—either from the analytic viewpoint or in terms of the degree to which it is possible to delineate *meaningfully* the possible combinations of circumstances and events which may ensue. Thus, most complex strategies are limited to four or five essential elements.

The advantages of analyses which incorporate both complex pric-

[13] As opposed to the simple courses of action which we have called, and will continue to call, "strategies" throughout the text.

ing strategies and subjective likelihood judgments are clear. One is able to bring an effective combination of subjective inputs and objective analysis to bear on a decision problem. Additionally, it is possible to conduct prospective analysis and at the same time to permit the adaptation of actions to events which may occur as time passes. This flexibility can be introduced on a basis which is much superior to the sort of panic decision making which so often takes place in situations involving complexities such as competitive responses and information which is received after actions have already been implemented. Thus, by utilizing subjective likelihood judgments and complex strategies, it is possible to plan ahead—through prior analysis of the alternative strategies—and at the same time to respond effectively to happenings in the marketplace which occur as time passes. The simultaneous feasibility of these two aspects of analysis is the essential characteristic of what we shall refer to as "strategic pricing analysis."

DECISION TREES IN PRICING ANALYSIS

One of the best means of analyzing pricing strategies in this expanded context is the decision *tree*. A decision tree is simply a diagram of the series of choices, events, and alternatives which make up a complex pricing problem.[14]

Figure 8-8 depicts a simple pricing decision problem in the form of a tree. The tree outlines a series of actions and events, beginning at the left and moving toward the right. The leftmost node indicates that the decision maker has two price alternatives for his new product—89 cents and 79 cents.[15] The two alternatives are for a new product which is to be marketed two months before a competitor markets a similar product. The company feels that the two-month lead time should be used to advantage, but it does not feel that the lead time is sufficient to make any meaningful evaluation of the product's sales performance. This is the case because distribution of the product is to be accomplished on a piecemeal basis from market to market and no reliable data will be available for about four months. This means that the competitor will introduce his product before our company has much information on the degree of market penetration.

[14] Of course, like the majority of the analytic methods which are presented in this book in the context of specific marketing decision problems, decision trees are not limited to pricing analysis.

[15] As in all other aspects of the example to be discussed, the number of alternatives is restricted only for ease of presentation. The methods of analysis used are equally applicable to situations involving larger numbers of alternatives, nodes, etc.

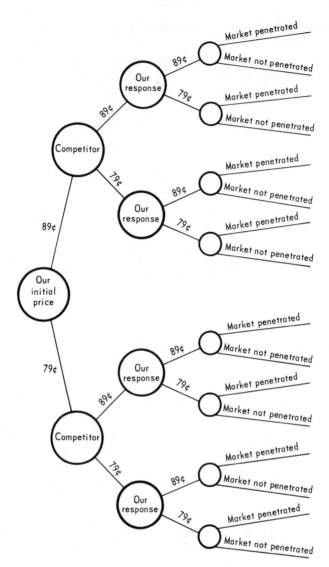

FIGURE 8-8   PRICE DECISION TREE

The two nodes labeled "competitor" in Figure 8-8 indicate that the next meaningful event is the competitor's introduction of his product. We assume that he too has only the same two price alternatives—89 cents and 79 cents. The positions of these two nodes indicate that the tree represents a tracing out of each of the possible combinations of actions, events, and circumstances which *might* occur. For instance, the topmost branch indicates that we set an initial price of 89 cents, our competitor establishes a price of 89 cents, etc.

The third set of four nodes indicates that we may wish to take a price action in response to our competitor's price. We still allow only the 89-cent and 79-cent alternatives. The last set of nodes indicates that after four months have elapsed, we will obtain some information concerning the degree of market penetration, both in the introductory period and in the period subsequent to competitive introduction. For simplicity, we have summed up the information which we might receive into two disjoint and exhaustive states—"market penetrated" and "market not penetrated." In any real situation, of course, it would be unlikely that two states would be adequate to describe outcomes of this nature meaningfully.

Clearly, each branch of the tree describes a series of events which *may* ensue. If we follow the topmost branch from beginning to end, for example, we trace out the events: "we set initial price of 89 cents"; "competitor uses price of 89 cents"; "we continue selling at 89 cents"; "market penetrated." Every other branch describes a similar sequence of happenings; i.e., *each branch describes a complex decision problem which takes place over time in terms of a series of events and a single outcome.*

Each possible outcome depends on three factors—our strategy, our competitor's strategy, and the degree of market penetration achieved. This illustrates the direct applicability of the conceptual framework as described in Chapter 3, for the strategy is made up of our controllable price actions, the state of nature determines the level of market penetration, and the other uncontrollable element is the competitive price reaction.

PRICING STRATEGIES

The strategies which are available to the decision maker in a pricing situation such as that described by the tree of Figure 8-8 involve two elements—the initial price and the price to be set subsequently in response to the competitor's price. One strategy which might be considered would be a simple specification of the two prices. For example, the strategy

(89¢,79¢)

indicates that we set an initial price of 89 cents and then reduce the price to 79 cents after our competitor has entered the market. The difficulty with such a strategy, however, is that it ignores the price our competitor sets. It would certainly be preferable to make the second element of our strategy—the "response" price—dependent on the price chosen by our competitor, since we know our competitor's price when we are called on to render a response price, and it is illogical to act as though we did not have that information. In fact, the feasibility of an adaptive response to happenings is one of the bases which we have used to argue for the use of complex strategies.

It is quite simple for us to make our response price contingent on the price set by our competitor. We may do this by including three elements in each strategy rather than two. The first element is the initial price. The other two elements are the price actions which we would take in response to an 89-cent and a 79-cent competitive price respectively. Hence, the strategy

$(89\cancel{c};79\cancel{c},89\cancel{c})$

indicates that we set an initial price of 89 cents and if the competitor introduces his product as 89 cents, we cut our price to 79 cents. If he introduces at 79 cents, we remain at 89 cents. Thus, the two prices appearing after the semicolon in such a strategy indicate our price *if* the competitor's price is 89 cents or 79 cents respectively.

The strategy

$(89\cancel{c};89\cancel{c},79\cancel{c})$

is perhaps more intuitively appealing than the one previously described. It involves an initial price of 89 cents and then a price-matching policy; i.e., if the competitor's price is 89 cents, ours is maintained at 89 cents; if the competitor's price is 79 cents, our price is reduced to 79 cents.

Strategies of this kind have two principal advantages. First, since each strategy incorporates the action to be taken at a number of decision points which occur at different times, one is able to make an analysis of an entire sequence of decisions on a prospective basis, rather than simply facing each decision individually. The advantage of this in terms of the systems approach should be clear. Secondly, by specifying future actions in terms which hinge upon the course of future events, one retains flexibility in being able to respond to the developing situation. A simple strategy involving only two prices in this illustration would require that the response price be set before we know the price to be charged by our competitor. By using strategies of the more complex form described for this example, we are combining "the best of two worlds." We are able to make a complete analysis *before the fact* and to respond to happenings which take place after our analysis.

The number of possible strategies of this kind which must be considered is determined by the number of elements of the strategy and the number of alternatives available. Here, since there are three elements and two alternatives for each, there are 2^3, or 8, strategies. These are outlined in Table 8-3 in the same form as

TABLE 8-3 PRICE STRATEGIES

Strategy	Initial price	Response price if competitive price is 89¢	Response price if competitive price is 79¢
S_1	89¢	89¢	89¢
S_2	89¢	89¢	79¢
S_3	89¢	79¢	89¢
S_4	89¢	79¢	79¢
S_5	79¢	89¢	89¢
S_6	79¢	89¢	79¢
S_7	79¢	79¢	89¢
S_8	79¢	79¢	79¢

previously described. It is easy to see that an exhaustive list of strategies consists of all possible combinations of initial prices and the prices which can be set in response to either of the two possible competitive prices.

SOLVING PRICING PROBLEMS BY USING DECISION TREES

To select the best of the eight strategies which are available in the example discussed in the previous section requires two kinds of *predictions*. The first—the prediction of the *worth of each outcome*—is the more susceptible of objective determination. For instance, if forecasts of the size of the total market for the product are available and if we quantitatively define the meaning of "market penetrated" and "market not penetrated," it should not be difficult to predict the general profit levels which would be achieved (over some planning period) for each of the outcomes represented by the rightmost ends of the tree branches in Figure 8-8. In any real situation, of course, we would undoubtedly need to consider a larger number of descriptions of the degree of market penetration in order to make such predictions accurately. Here, we shall assume that the profit entries at the extreme right of Figure 8-9 represent our predictions of the profit consequences for each outcome.

The second kind of prediction which we find necessary involves the *likelihoods with which the various levels of market penetration and the various competitive actions may occur*. We have previously argued that numerical probability estimates of these likelihoods are probably best made subjectively by executives who are familiar with the market analyses which have been conducted, the competitor and his objectives, and other pertinent factors which have not previously been explicitly considered. The likelihood estimates may be made in the form of betting odds; e.g., "the odds are 2 to 1 that our competitor's price will be 89 cents if ours is 89 cents." When translated into numerical probabilities—in this case, a probability of $\frac{2}{3}$ for a price of 89 cents and $\frac{1}{3}$ for a price of 79 cents if our price is 89 cents—such likelihood estimates may be used in scientific problem analysis.

The probabilities in parentheses in Figure 8-9 represent those which we shall assume have been generated by querying executives and other experts in the manner described. We should be careful to note exactly what these probabilities represent. The 0.70 entry on the 89-cent competitive price branch indicates that there we assess a 70 percent chance that our competitor will use an 89-cent price *if we set an initial price of* 89 *cents.* This is so because this probability appears on the major branch corresponding to an 89-cent initial price set by our firm.

This 0.70 *probability is not the probability that our competitor will set a price of* 89 *cents.* It is the probability that he will do so *if* we use an initial price of 89 cents. This is easily verified by viewing the lower major branch of the tree and noting that the probability that he will use an 89-cent price if we set an initial price of 79 cents is assessed as 0.10. The reader should note that because of this, each of the probability pairs sums to unity, e.g., 0.70 and 0.30, 0.10 and 0.90, etc. This is so because of the *if* statement included in each probability statement and because the relevant events have been defined to be mutually exclusive and exhaustive; e.g., our competitor sets a price of *either* 89 cents *or* 79 cents *but not both,* and he can set no other price.[16]

Similarly, the 0.50 probability at the upper right of the tree indicates that we assess as 0.50 the probability that the market will be penetrated *if* we set an initial price of 89 cents, our competitor uses a price of 89 cents, and we thereafter continue using the same price throughout the planning period.

To determine the best of the eight available strategies, we need only to follow the path on which each leads us through the tree describing the decision problem, note the probability and profit outcome associated with each, and calculate the *expected profit* for each strategy. Strategy S_1, for example, tells us to set an initial

---

[16] The reader who is conversant with probability theory will note that these are conditional probabilities.

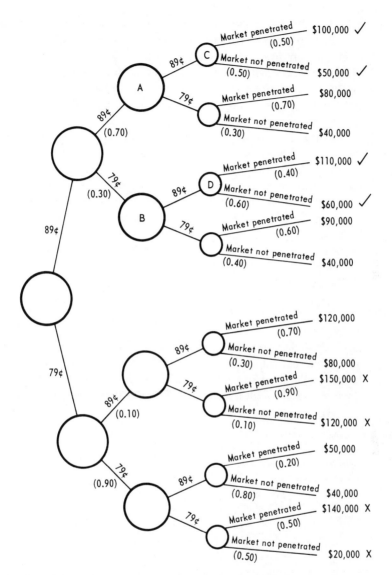

FIGURE 8-9   PRICING DECISION TREE WITH PROFITS AND PROBABILITIES

price of 89 cents and retain that price regardless of the price set by our competitor. This strategy leads us along any one of four branches of the tree, depending on the set of uncontrollable circumstances which occur. The outcomes associated with these four branches are indicated by check marks in Figure 8-9. To trace them out, we note that the first element of S_1 tells us to use an initial price of 89 cents. This takes us along the upper main branch of the tree. Then, depending on the competitor's price, we go to either node *A* or node *B*. These two nodes represent the point at which we must set our response price. Still following S_1, we maintain our price at 89 cents in either case, continuing to either of the nodes labeled *C* and *D*. From there we proceed to one of the four check-marked outcomes, depending on whether or not the market is penetrated. (The reader should follow through this process for the other seven strategies to make certain that he understands the relationship between the strategies and the problem description which the tree entails.)

The probability of our arriving at any of these four outcomes by using S_1 may be determined as the product of the probabilities of the constituent uncontrollable aspects. Hence, the $100,000 profit outcome will be achieved under S_1 with probability 0.70 times 0.50, or 0.35. In the same fashion we determine that S_1 may lead to an outcome of $100,000, $50,000, $110,000 or $60,000 with probability 0.35, 0.35, 0.12, or 0.18 respectively.

The expected profit associated with S_1 is therefore

$$E(S_1) = 0.35(\$100,000) + 0.35(\$50,000)$$
$$+ 0.12(\$110,000) + 0.18(\$60,000)$$

or $76,500. The expected profits associated with the other strategies are given in Table 8-4. The reader should verify each of these for himself, using the strategy descriptions given in Table 8-3 and tracing each through the decision tree in Figure 8-9.

TABLE 8-4   EXPECTED PROFITS FOR VARIOUS PRICING STRATEGIES

Strategy	Expected profit
S_1	$76,500
S_2	73,500
S_3	69,600
S_4	66,600
S_5	48,600
S_6	82,800
S_7	52,500
S_8	86,700

The strategy with the greatest expected profit is S_8, with its expectation of $86,700. Thus, following the optimal strategy, the decision maker should set an initial price of 79 cents and maintain that price regardless of the price action taken by competition. This is the case because, as shown in Table 8-3, S_8 involves a 79-cent price in the event of either an 89-cent or a 79-cent competitive price. With this strategy the actual path traced out through the decision tree will be one of those leading to the outcomes marked with crosses in Figure 8-9.[17]

In viewing the pricing decision problem in this way, the decision maker has introduced several useful details into his decision making procedure. First, his use of subjectively estimated probabilities allows him to take advantage of the knowledge and experience of experts in a fashion which is completely compatible with scientific methods. Next, he has considered complex strategies which involve a considerable greater degree of perception and sophistication than do the simple pricing strategies which he might previously have been able to consider. These strategies are more complex in two major respects. First, they permit simultaneous consideration of a series of individual pricing actions over time rather than just the independent establishment of single prices at moments in time. Moreover, they allow the price that will be set in each case to be contingent on happenings in the marketplace. Hence, even though the decision maker is able to plan ahead with the analysis of an entire series of price actions, he is not tied to an inflexible price level which may not be suitable at the time when it needs to be implemented.

Of course, an alternative method of achieving price flexibility would be simply to treat each price action independently or, alternatively, to develop a complex pricing strategy and then ignore it if conditions seem to dictate a different price at the time a particular price is to be put into effect. Clearly neither of these alternatives is very attractive, since the first does not permit simultaneous consideration of the sequence of price decisions and the latter results in the kind of quick and dirty decisions which, in the panic of the moment, so often neglect salient aspects of the problem situation.

Competitive Bidding

Many organizations do not deal with large numbers of consumers, but rather have a relatively small number of potential customers and opportunities to sell their product or services. Contractors who

[17] This optimal strategy vividly illustrates one of the difficulties which can be encountered if, in fact, no careful consideration has been given to the degree to which profit is a good approximation to utility. With the optimal strategy S_8, it is rather likely (0.45) that a profit of only \$20,000 will be realized. In some circumstances a very cogent argument for the complete avoidance of this poor profit outcome could be presented. This, of course, would be equivalent to saying that profit is not a valid approximation to utility, since seeking complete avoidance of that outcome imputes a very small relative utility to it.

deal with the Defense Department are probably the extreme example of this. Some undiversified aerospace companies realize virtually all of their annual revenue from a small number of government contracts for which they and their competitors bid. Others, which are diversified, still rely heavily on such contracts for revenues. Much of industrial marketing takes place in a similar environment. Companies frequently purchase machinery, vehicles, and other equipment on a bidding basis which may be formal or informal. It should be emphasized, however, that the large size of contracts for military weapons systems (which with increasing frequency total in the billions) and relatively large industrial situations involving the purchase of equipment should not lead us to the idea that all bidding situations are on a large scale. A great number of small service organizations, from janitorial service to plumbing, acquire business in a similar fashion.

In bidding for contracts, an organization is effectively setting the sales price for its product or service. In preparing the bid, the enterprise faces some of the same problems as price decision makers in conventional pricing situations, and many different ones. Probably the two most important aspects of bidding situations are the necessity to price in advance, without opportunity for revisions or second thoughts, and the overwhelming significance of competition.

The most important bidding situation is the one in which the details of the service to be performed or of the product to be purchased are published among interested organizations along with an invitation to bid. The firms which choose to bid determine a "price" for the desired product and submit their bids to the buyer. If the process is a formal one, the competing bids may be opened simultaneously at a scheduled ceremony. The procedure may be completely informal, however. In any case, there is usually only one opportunity to bid,[18] and the lowest bid is usually the one accepted (subject perhaps to some constraints).

Whatever may be the particular bidding rules, the organization's pricing decision is to establish a bid which will be high enough to provide a profit and low enough to be accepted. The firm's objectives obviously have a great effect on the price which will be established. As we have discussed, the bidding strategy will be affected by the current economic status of the company, order backlog, etc. A large aerospace contractor, in the face of a small backlog of contracts, might be willing to sacrifice immediate profit for the sake of employment and organizational stability. His reasoning might be that failure to obtain a certain contract would force

[18] Some bidding situations on an auction basis are part of the marketing environment. However, they are less significant than the closed bid type.

him to lay off employees and otherwise decrease his overall capability. Reduced capability, in turn, might lead the government to believe that he could not perform adequately on later contracts and might decrease his chances of obtaining them. If he will simply break even on the present contract, he will be able to maintain his organization and chances for future contracts. All of these factors would, of course, be incorporated into the utility measure which is applied to the possible outcomes of the decision problem.

THE BIDDING PROBLEM

The "bidding problem" entails the determination of the best bid (price) to set for a specified contract. We shall attack the problem as though we are certain that a bid is to be placed and will therefore concentrate attention on the amount of the bid itself. Subsequently, we will discuss the "problem of bidding," i.e., the problem of whether or not to place any bid at all. If the cost of performing the contract is fixed, it seems to be to the organization's advantage to place a bid which is as high as possible. If the bid is R and the cost of performing the contract is known (estimated) to be C, the difference

$$R - C$$

or profit, should be large. Since C is not generally open to a significant degree of control, the obvious way to do this is to make the bid as large as possible. However, this argument is totally predicated on an *if*—if the contract is awarded to the firm being considered. If the award is made to someone else, the profit is not $R - C$, but zero.[19] Hence, we must deal with the likelihood that the award will be made and, consequently, with *expected profit*.

To incorporate likelihoods into the analysis of this decision problem (under risk), we need to have information concerning the probability of being awarded the contract. Since this probability depends on the bids which competitors submit, we would like to have data on their bids. However, they are undoubtedly not going to be so congenial as to provide such information, and so we must garner data from historical records. But, of course, even our access to the historical data of our competitors is limited. It is not uncommon to find that the only available information concerning competition is the bids which are announced at the past contract awards. Nothing is usually known, for instance, about the rela-

---

[19] If the "problem of bidding" has already been resolved, all costs involved in the preparation of bids are *sunk costs*, which are unaffected by (and have no effect on) the "bidding problem."

tionship of each bid to the cost estimate on which the competitor's bid was based.

Bidding against a Single Competitor. If the situation involves bidding against a single competitor, we should search into past records of his bids in similar situations. We may then relate each of *his bids* to *our cost estimates* for the same contracts. The reason for doing this is simple: we do not have his cost data, but our cost data provide a consistent basis for comparison. Also, our cost is a known quantity to us in the current decision problem, whereas our bid is the decision variable. The only relevant basis for comparison is, therefore, our cost.

Suppose we found that our competitor bid (in relation to our costs) with the frequency shown in Table 8-5. This table illustrates that of the 50 times both our firm and our competitor bid in the past, his bid was 80 percent of our cost estimate twice—a relative frequency of $\frac{2}{50}$ or 0.04. Also, his bid was 90 percent of our costs three times—a relative frequency of 0.06. The other tabulated figures have similar interpretations. For purposes of illustration, we have used large intervals of the ratio of our competitor's bid to our cost in Table 8-5. In practice, we would use a much smaller interval and perhaps attempt to fit a continuous function to this empirical distribution.

TABLE 8-5   PAST HISTORY OF COMPETITOR'S BIDS

Ratio of competitor's bid to our cost	0.8	0.9	1.0	1.1	1.2	1.2	1.4	1.5
Frequency	2	3	8	14	10	7	4	2
Relative frequency	0.04	0.06	0.16	0.28	0.20	0.14	0.08	0.04

With these data the outcome array may be developed. The strategies are the various bids which can be entered. It is simplest to express the potential bids as multiples of the cost estimate C. Because actual bids are placed in dollars and cents, the possibility of tie bids is remote. To forestall that possibility in this simple illustration, we shall choose for strategies multiples which are slightly less than those for which we have obtained competitor's data. In other words, rather than considering bids of $0.8C$, $0.9C$, etc., we shall use the quantities $0.75C$, $0.85C$, etc. This will preclude the possibility of ties and at the same time permit an illustration which is not unwieldy.

The competitor's various bids are the determinants of the (generalized) states of nature. In fact, the competitive actions are probably the only uncontrollable elements which need to be considered. The relevant outcome array is given in terms of profit $(R - C)$ as Table 8-6. Competitive bids are given with the same degree of accuracy as that used in collecting the historical data of Table 8-5.

TABLE 8-6 OUTCOME ARRAY (PROFIT)

Strategy	Competitor's bid							
	$0.8C$	$0.9C$	$1.0C$	$1.1C$	$1.2C$	$1.3C$	$1.4C$	$1.5C$
S_1: bid $0.75C$	$-0.25C$	$-0.25C$	$-0.25C$	$-0.25C$	$-0.25C$	$-0.25C$	$-0.25C$	$-0.25C$
S_2: bid $0.85C$	0	$-0.15C$	$-0.15C$	$-0.15C$	$-0.15C$	$-0.15C$	$-0.15C$	$-0.15C$
S_3: bid $0.95C$	0	0	$-0.05C$	$-0.05C$	$-0.05C$	$-0.05C$	$-0.05C$	$-0.05C$
S_4: bid $1.05C$	0	0	0	$0.05C$	$0.05C$	$0.05C$	$0.05C$	$0.05C$
S_5: bid $1.15C$	0	0	0	0	$0.15C$	$0.15C$	$0.15C$	$0.15C$
S_6: bid $1.25C$	0	0	0	0	0	$0.25C$	$0.25C$	$0.25C$
S_7: bid $1.35C$	0	0	0	0	0	0	$0.35C$	$0.35C$
S_8: bid $1.45C$	0	0	0	0	0	0	0	$0.45C$
S_9: bid $1.55C$	0	0	0	0	0	0	0	0

Note that the lower left of the table has zero entries, indicating that these are the outcomes on which the contract award goes to the competitor. In the upper right the award is won and the entries are the indicated bid less C (the cost of fulfilling the contract). The nonzero entries are the same for each strategy because the total revenue derived is exactly the amount of the bid if the award is won (regardless of the competing bid) and zero if the award is lost.

The historical data of Table 8-5 provide insight into the likelihood of occurrence of each of the states of nature. The relative frequencies given there might be adjusted to represent probabilities more adequately if some pattern has also been detected in the past behavior of the competitor. If, for example, the competitor never bid below 125 percent of our costs when he had a large backlog of orders, we might adjust the relative frequencies to conform to the current situation. In other words, if we know that our competitor now has a large order backlog,[20] we may not use the relative frequencies as direct estimates of the probability that the competitor will make each of the respective bids. Rather, we might inflate the probabilities which are applicable to the higher bids ($1.3C$, $1.4C$, and $1.5C$) to account for our added knowledge.

---

[20] Such information is readily available from quarterly and annual corporate reports.

Assuming that no such adjustments are necessary in this case, we may simply assign the relative frequencies of Table 8-5 as the probability for each of the competitive bids in the outcome array of Table 8-6. If *maximization of expected profit* is used as a -proxy for maximization of expected utility, the decision maker needs only to sum the products of probabilities and profits across all competitive bids for each strategy and choose as best that strategy which has the highest expected profit. The expected profit for S_1 (a bid of $0.75C$) is

$$E(S_1) = 0.04(-0.25C) + 0.06(-0.25C) + 0.16(-0.25C)$$
$$+ \cdots + 0.04(-0.25C)$$

which, since the probabilities must sum to unity, is just $(-0.25C)$. In each of the expected profit calculations for S_2 through S_8, there need to be only two terms, since there are really only two *different* outcomes. For S_2, for example, we have

$$E(S_2) = 0.04(0) + 0.96(-0.15C) = -0.144C$$

and for S_3

$$E(S_3) = 0.10(0) + 0.90(-0.05C) = -0.045C$$

The reason for simplification to two terms is that, for any strategy, the same profit is associated with any competitive bid which results in winning the contract and the same profit is associated with any competitive bid resulting in the loss of the award. This means that we may simply collect all of the terms involving the same profit for each strategy and display the expected profit in a simple two-term expression.

The expected profits for all strategies are given in Table 8-7. The highest expected profit results from S_5. Hence, the bid which

TABLE 8-7   EXPECTED PROFITS

Strategy	Expected profit
S_1	$-0.25C$
S_2	$-0.144C$
S_3	$-0.045C$
S_4	$0.037C$
S_5	$0.069C$
S_6	$0.065C$
S_7	$0.042C$
S_8	$0.018C$
S_9	0.0

should be entered to maximize expected profit is one which is 115 percent of our cost estimate (S_5 is $1.15C$). If our competitor bids according to his past behavior with respect to our costs, we will win the award with probability 0.46 and lose with probability 0.54. These quantities are simply the sum of the probabilities associated with the highest four and lowest four possible competitive strategies which have been considered. (See Table 8-6.)

The phrase "which have been considered" should be emphasized in the preceding sentence, for it is clear that both our competitor and we *could* bid something which is different from the strategies in Table 8-6. What we have assumed there is that we have rounded off all possible bids to the values shown, using some consistent rule, and have therefore considered *all possible* competitive bids and strategies. Of course, our estimates of the likelihoods involved are based on those bids which have actually been made by our competitor in the past.

Bidding against Several Competitors. If the contract in question is to be bid on by several known competitors, the same basic method of analysis can be applied. The only significant difference is that the bit of historical information which must be known is the ratio of the *lowest* competitive bid to the decision maker's costs. The data should be obtained from historical records, and a single distribution such as that in Table 8-5 can be compiled. If, for example, competitor A had once been the lowest competitive bidder with a bid of 80 percent of our cost and competitor Z had twice been the lowest competitive bidder with a bid of 80 percent of our cost (with no other instances of a lowest competitive bidder placing a bid of 80 percent of our cost), these three occurrences should be entered as a frequency of 3 under 0.8 in that table. *The resulting distribution should then be treated as though it represented a single competitor even though data from a number of different competitors are included in it.*

The reason for this is that we are concerned only with the lowest competitive bid on each contract. We win if our bid is lower than the lowest competitive bid, and we lose if it is higher. As we saw in calculating the expected profits from Table 8-6, the only significant thing is *whether* we win and by *what bid* (if we win) and not how badly we lose (or, indeed, not even how well we play the game). The determination of the best bid in the case of several competitors is identical to the procedure in the case of a single competitor provided that the historical bidding data (such as that in Table 8-5) represent the lowest bids of competitors.

THE PROBLEM OF BIDDING

The "problem of bidding" (as differentiated from the "bidding problem") is the determination of *whether* or not to place a bid on a particular contract. In effect, there appear to be only two strategies in such a problem—to place a bid or to refrain from bidding.

The primary reason for the significance of this problem is the *cost* which is associated with the bidding process. In order to place a bid, some costs must be incurred by the bidder—if only those involved in the perusal of the contract details and bidding rules. In most cases, if nonstandard items are involved, significant costs of research and development may be necessary before a meaningful bid can be made. Again, the best example is the aerospace industry, in which millions of dollars may be spent by several companies to develop a basic airplane which will be the vehicle of competition for an Air Force contract. Obviously only one company is success-ful, and the losers derive no immediate monetary returns to com-pensate for the costs they incur.[21] In almost all cases involving non-standard items, the cost of establishing and analyzing the cost data which may form the basis of a price bid is substantial. Because of this, it is often not possible to follow a strategy of bidding on all available contracts.

A little thought will reveal that the problem of bidding is not just a simple bistrategy situation which may be attacked indepen-dently of the bidding problem. The reason for this is that the outcomes of the strategy "place bid" depend on the value of the bid. Hence, the two problems are intrinsically interrelated.

The simplest way to approach the combination of the two prob-lems is to *add an additional strategy* to those shown in Table 8-6 and to *adjust the outcomes for the cost of bidding*. The addi-tional strategy is "do not bid," and the cost adjustment applies to all strategies except the nonbidding one. If the cost of bidding is known to be B dollars, the outcome array would be identical to Table 8-6 except that every entry would be reduced by B dollars and a new row would be added for the strategy "do not bid." The outcome for every entry in this row would be zero. The reader

[21] Recent Defense Department practices provide some compensation in such instances. However, the amount is typically far short of covering actual costs. The companies derive benefit in the knowledge gained in producing the bid, and they are often rewarded through the use of this knowledge on other civilian or military contracts. Such is the case with the huge C-5A transport airplane and the SST (supersonic transport), both of which will almost certainly be produced by more than one firm even though the govern-ment contract goes to only one.

should construct this array and note that S_9, which had all zero entries in Table 8-6, now has entries of $-B$, since the bid will never result in the award of the contract, but the cost of bidding has been incurred. All aspects of the solution using this revised table are identical to those demonstrated for Table 8-6. The expected profit for the "do not bid" strategy will be zero. Hence, some bid will be entered if any strategy has positive expected profit.

BIDDING ON SEVERAL CONTRACTS

If several contracts are available (or will be available) for bidding, the manager should consider the opportunities presented by each. If there is no restriction on the number of contracts which may be bid on, the analysis of the previous section may be carried out and the "best" bid entered on each (assuming, of course, that the best bid results in positive expected profit).

It is often not possible, however, to bid on every available contract even though each may have a "best" bid which entails positive expected profit. This constraint may come about because of budget limitations or other restrictions imposed by the organizational policy of the bidder or buyer. For instance, a contract preparation budget may curtail the number of bids which may be placed; or organizational objectives may determine that the total cost estimates of contracts on which bids are entered may not exceed a given amount. Similarly the purchaser (the Federal government, for instance) may decree that no single company can bid on (or be awarded) more than a fixed number or volume of contracts. In such instances, the trade-offs between the various contract opportunities should be considered.

Bidding on Several Contracts at Different Times. If several contracts will be available for bidding at different points in time, the *ultimate outcome* of all of those contract opportunities for which information is available should be considered. To demonstrate this, let us consider a simple case of two bidding opportunities. Contract 1 is to be available first, and then contract 2 is to come up for bids. Our manager is faced with a total bid preparation budget of $20,000, and the preparation of bids for contracts 1 and 2 is expected to cost $10,000 and $15,000 respectively. For simplicity, let us consider two possible bids by our firm on each contract ("high" and "low") and three possible bids by our opponent ("low," "medium," and "high"). The dollar amounts of the bids for us and our competitor are different, so that the winner is clear for every possible combination. The profit outcomes (neglecting the cost of bidding) are given in Table 8-8*a* and *b*. Suppose further that past behavior of the competitor indicates that he bids "low," "medium," and "high" respectively 20, 40, and 40 percent of the time in situations like contract 1, and 30, 40, and 20 percent of the time in situations like contract 2. (These probabilities are also entered in Table 8-8.) It is clear that our bid preparation budget is not adequate to permit us to bid on both contracts, but suppose that we have been given permission to exceed our budget if high revenues are gained from the first contract; i.e., if we are awarded contract 1 with a "high" bid, we may spend some of its revenue in preparing a later bid for contract 2. If we win with a "low" bid, we are still not permitted to exceed our budget.

TABLE 8-8   OUTCOMES FOR TWO BIDDING SITUATIONS

Strategy	Probability and competitor's bids		
	(0.20) Low	(0.40) Medium	(0.40) High
S_1: low	0	$100,000	$100,000
S_2: high	0	0	$150,000

(a)   *Contract 1*

Strategy	Probability and competitor's bids		
	(0.30) Low	(0.50) Medium	(0.20) High
S_1: low	0	$200,000	$200,000
S_2: high	0	0	$250,000

(b)   *Contract 2*

We may analyze contract 1 by using the diagram of Figure 8-10. The three lines originating from the left circle in this figure indicate that we have three alternatives: "bid low," "bid high," or "do not bid." If we choose the last, the only possible outcome is the loss of the contract as indicated at the lower right. The 1.0 there indicates the contract will be lost with probability 1 (certainty) if no bid is entered. Similarly, if the "low" bid is selected, the probability of winning is 0.80 because our opponent bids "medium" or "high" with this relative frequency and in these situations we win with a "low" bid. Hence, the probability of losing with a "low" bid is 0.20. (These probabilities may be seen in Table 8-8*a*.) The analogous probabilities for a "high" bid are 0.40 and

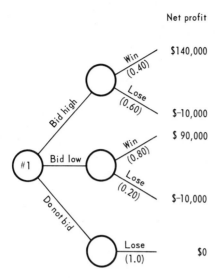

Net profit

FIGURE 8-10   CONTRACT 1

0.60, since we win with a "high" bid only if the opponent also bids "high." These probabilities are entered on the appropriate lines in Figure 8-10.

We can describe the various possible combinations of events which may occur by tracing all paths from left to right through the tree. For instance, the lower path ("do not bid—lose") results in a zero net profit and the upper path ("bid high—win") results in a net profit of $140,000. This $140,000 is the profit indicated in Table 8-8*a* minus the $10,000 cost of bid preparation. The other outcomes are specified at the right in the figure. Each should be traced through by the reader.

The best strategy for contract 1 (independent of contract 2) can be found by calculating the expected profit for each strategy—"bid high," "bid low," and "do not bid"—just as was done previously. The expected profit for "bid high" can be seen from Figure 8-10 to be

$$E \text{ (bid high)} = 0.40(\$140,000) + 0.60(-\$10,000)$$

or $50,000. The expected profit for "bid low" is

$$E \text{ (bid low)} = 0.80(\$90,000) + 0.20(-\$10,000)$$

or $70,000. Since the expected profit for "do not bid" is zero, *the low bid is the best strategy if only contract 1 is considered.*

Considering contract 2 independently in the same fashion, we obtain Figure 8-11. The three possible strategies ("high," "low," "do not bid") have expected net profits of $35,000, $125,000, and

Net profit

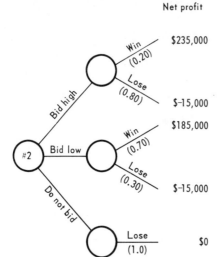

FIGURE 8-11   CONTRACT 2

$0 respectively, as can be verified by the reader. This means that *the best strategy is the "low" bid if only contract 2 is considered.*

However, because the choice which is made on contract 1 can affect the choice on contract 2 (i.e., if we bid on contract 1 and lose, we cannot bid on contract 2), *both* should be viewed together in terms of the ultimate outcome. Figure 8-12 repeats the network of Figure 8-10, to which the various possible situations related to contract 2 are connected at the right. Following the lower branch from the left, we note that if we do not bid on contract 1, we lose it with certainty and are left with sufficient budget to bid on contract 2. If we bid "high" on contract 2, we win with probability 0.20, and if we bid "low," we win with probability 0.70 (from Table 8-8*b*). The total net profit (from both contracts) resulting from each complete path through the network from left to right is indicated at the right. For example, the second entry from the top is $125,000. It indicates that we have bid "high" and won on contract 1 and bid "high" and lost on 2. The gain on contract 1 was $150,000, and the costs of bidding were $10,000 and $15,000 respectively, for a net profit of $125,000. The reader should follow each path through the network and note three factors concerning each:

1. The reason for the existence of the path in terms of the bidding rules which have been described.
2. The total net profit associated with the path.
3. The probability associated with each outcome.

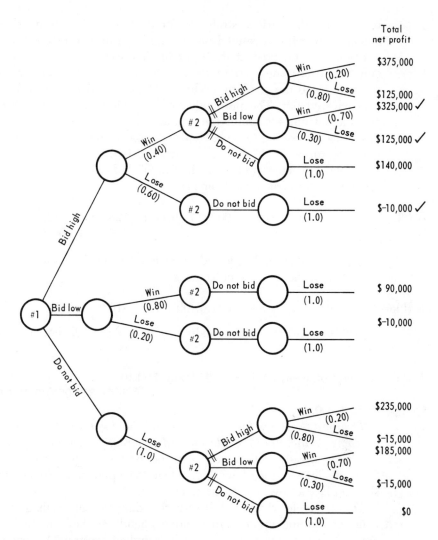

Total
net profit

$375,000

$125,000
$325,000 ✓

$125,000 ✓

$140,000

$-10,000 ✓

$ 90,000

$-10,000

$235,000

$-15,000
$185,000

$-15,000

$0

FIGURE 8-12 CONTRACTS 1 AND 2

Now, we have already analyzed contract 2 independently and found that *if we are given the opportunity* (according to the prescribed rules) *to bid on contract 2, we shall bid "low."* At the lowest of the decision points labeled 2 in Figure 8-12, we know that if we actually arrive at this situation, we shall place a "low" bid on contract 2. Similarly, at the topmost of the decision points labeled 2, we shall do the same. This has been indicated by the placement of double lines across the alternatives which we shall never choose if we are in either of these two situations.

Having made this analysis of contract 2 first, we may proceed to evaluate the various strategies for contract 1. There are *three* outcomes which can result from a "high" bid on contract 1. These outcomes are checked at the right in Figure 8-12. We know that only three outcomes can result because of our initial analysis of contract 2 and the resulting exclusion of the two paths which have been crossed out. The probability of achieving each of the three possible outcomes is simply the product of the respective probabilities of events leading to each. Hence, the probability of the $325,000 outcome for a "high" bid on contract 1 is 0.28 (0.7 times 0.4). The total expected profit for the "high" bid on this contract in the light of the *ultimate* outcome is

$$E \text{ (``high'' on contract 1)} = \$325,000(0.7)(0.4) + \$125,000(0.3)(0.4) \\ + (-\$10,000)(1.0)(0.6)$$

or $100,000. The total expected profit for the "low" bid on contract 1 in terms of the ultimate outcome is

$$E \text{ (``low'' on contract 1)} = \$90,000(1.0)(0.8) + (-\$10,000)(1.0)(0.2)$$

or $70,000, and the corresponding value for the "do not bid" alternative is

$$E \text{ (``do not bid'' on contract 1)} = \$185,000(0.7)(1.0) \\ + (-\$15,000)(0.3)(1.0)$$

or $125,000.

The best strategy for contract 1, in the light of the ultimate outcome, is therefore the "do not bid" alternative. Of course, since we have already determined that the best strategy for contract 2 taken alone is the "low" bid, when the opportunity arises, we should place a "low" bid on contract 2.

This example vividly illustrates the difficulty of dealing with contract bidding opportunities independently if they are tied together by some constraint (such as our budget constraint). Looking back to Figure 8-10 and its related comments, we see that the best strategy for contract 1, taken alone, is a "low" bid, whereas when the two opportunities are combined, we find that a higher *total* expected profit can be achieved by passing up the first contract and placing a "low" bid on the second. This is somewhat contrary to one's immediate intuition; yet a little reflection results in an understanding of how this can be so (indeed, we have shown that it is so in this case). Think, for instance, of contract 2 as being very lucrative with virtual certainty. If there is some chance that a bid on a very ordinary prior contract can preclude a bid on the later contract, we should pass up the first, *even though the opportunity itself may be a good one*. It is simply not "good enough" to be worth the risk of being unable to bid on the later contract.

The other important analytic principle which is illustrated by this simple example is that *the order of evaluating the contracts should be the reverse of that in which they will occur.* By analyzing contract 2 first, we were able to eliminate all nonoptimal alternatives for contract 2 for each of the states we might be in when faced with a decision on contract 2. The basis for this approach has been formally stated by Bellman (see references) as the "principle of optimality" for multistage decision problems of the kind we are discussing here.

Simultaneous Bidding on Several Contracts. If several contracts must be bid on simultaneously, each should be analyzed separately in the fashion described in the "Bidding Problem" section. If there are no constraints on the amounts which may be bid, the overall best strategy is simply composed of each of the individual best bids (assuming that each has positive expected profit).

If, however, there is some restriction on the total amount which may be bid, it may be impossible to bid each best amount and to satisfy the restriction simultaneously. Suppose, for example, that simultaneous bids may be placed on three contracts and that each has been independently analyzed as described in the "Bidding Problem" section. Suppose that the *expected profits* associated with the various bids are those given in Table 8-9. (The contracts have been numbered 3 through 5 to distinguish them from those of the previous section.) The best bid on each of the three contracts is $125, $300, and $400 respectively, because these are the bids

TABLE 8-9   EXPECTED PROFIT DATA FOR BIDS ON THREE CONTRACTS[22]

Contract 3		Contract 4		Contract 5	
Bid	Expected profit	Bid	Expected profit	Bid	Expected profit
$ 0 (do not bid)	$ 0	$ 0 (do not bid)	$ 0	$ 0 (do not bid)	$ 0
100	60	100	100	100	100
125*	62.5	200	150	200	200
150	60	300*	165	300	225
200	40	400	100	400*	300
300	0	500	0	500	250
				600	250
				700	0

[22] The asterisks indicate the best bid on each contract. The determination of these best bids is explained in the text.

which have maximum expected profit. (These bids are indicated by asterisks in Table 8-9.) If all three contracts are bid on simultaneously and there is no constraint on the bidding, the best overall strategy is simply composed of these three individually best bids.

Suppose, however, that a restriction has been placed on the total amount which may be bid. This may be the result of a corporate policy aimed at preventing the overextension of resources that might occur if large total bids were placed with the expectation that some awards would not be won, and an unexpectedly high number were actually won. If the restriction is that the total bids may not exceed $700, it is easily seen that the strategy which is composed of the three best bids is an unacceptable one, because the total bid required is $825. Even though this strategy is the *unconstrained optimal strategy,* when the restriction is imposed, it cannot be undertaken because it violates the constraint. What is needed is a *constrained optimal strategy,* one which is the best of all those which satisfy the constraint.

We may determine the constrained optimal strategy by beginning with the unconstrained optimal strategy and analyzing the changes which are introduced by changes in each contract bid. In particular, we are concerned with the *unit change in expected profit which is introduced by a change of bid.* If we suppose that no bids other than those listed in Table 8-9 may be entered, the analysis is not difficult. In bidding for contract 3, for example, we note that a change of bid from $125 to $100 results in a decrease in expected profit of $2.50 (from $62.50 to $60). We consider a *reduction* of the bid because the total bid for the unconstrained optimal strategy was larger than the maximum allowable under the constraint. Similarly, a $100 reduction from the best bid on contract 4 produces a $15 loss of expected profit (from $165 to $150), and a $100 reduction from the best bid on contract 5 causes a $75 loss of expected profit (from $300 to $225). The change in expected profit *per dollar decrease in bid* for the smallest downward possible changes from each best bid is shown for each contract in Table 8-10.

Table 8-10 shows that the least loss in expected profit per dollar decrease in bid occurs for contract 3, where the loss is 0.10 (it

TABLE 8-10   CHANGE IN EXPECTED PROFIT PER DOLLAR DECREASE IN BID

Contract	
3	$-2.50/25 = -0.10$
4	$-15/100 = -0.15$
5	$-75/100 = -0.75$

is 0.15 and 0.75 respectively for contracts 4 and 5). This involves a change in bid from $125 to $100 on contract 3. If this change is made and the other two best bids are retained, the total bid quantity would be $800, i.e., $100, $300, and $400 on the three respective contracts. Since this total does not satisfy the $700 constraint, we must investigate other changes which can also be made. It is important to note that since the change from a bid of $125 to $100 on contract 3 is already indicated to be the best possible adjustment, we shall consider that it will be made and analyze *additional* bid changes.

Given that the indicated change has been made on contract 3, the respective changes in expected profit per dollar decrease in bid for each of the downward adjustments which may be further considered are shown in Table 8-11. Note that since no change has previously been made on contracts 4 and 5, their ratios are unchanged. Now we are considering the *next* change which may be made to bring the total amount bid closer to (or within) the bounds of the constraint. As such, we are comparing a *further decrease* in contract 3 with the previously considered decreases in contracts 4 and 5.

TABLE 8-11   CHANGE IN EXPECTED PROFIT PER DOLLAR DECREASE IN BID

Contract	
3	$-\,{}^{60}\!/_{100} = -0.60$
4	$-\,{}^{15}\!/_{100} = -0.15$
5	$-\,{}^{75}\!/_{100} = -0.75$

Since the next possible decrease in bid on contract 3 is from $100 to $0 (no bid), with a loss in expected profit of $60, its ratio is —0.60. The least loss in expected profit per dollar decrease in bid occurs for contract 4, where a loss of $15 expected profit from a $100 decrease in bid results in a ratio of —0.15. Hence, this change will be selected as the second to be made. The total bid amount on the three contracts is then $700 ($100, $200, and $400). Since this sum satisfies the $700 constraint, the best strategy is the one composed of bids of $100, $200, and $400 on contracts 3, 4, and 5 respectively.

Of course, if the last change had not exactly satisfied the maximum total under the constraint, the *total expected profit* would need to be analyzed to determine whether the reduction which has the least *per dollar* loss is actually the best to make. It is possible to find a situation in which the tentative bid reduction which for the first time satisfies the constraint at the lowest expected loss in profit per dollar of bid, reduces the total amount bid by such a quantity that a reduction which involves a higher per dollar cost, but a smaller reduction in the amount bid, will have a lesser effect on total expected profit. Therefore, *when the increments by which bids may be reduced are not the same for all contracts, the reduction giving the lowest loss per dollar should be chosen until such a reduction meets the constraint for the first time. Then, the total effect on expected profit should be analyzed to determine which of the possible reductions will permit the constraint to be satisfied at the least loss in expected profit.*

If the constraint to be considered is of an entirely different kind, the same general method of approach may be useful. For example, if the government were awarding the three previously discussed contracts, they might place a limit on the *number* of contracts on which bids may be placed. If the maximum number were two, the unconstrained optimal strategy for three bids could be altered in a fashion similar to that for the constraint on the amount of the bids. The difference, of course, would be that the change would require a shift to the $0 (no bid) strategy on at least one contract rather than a simple reduction in the bid.

Pricing and Purchasing

A basic objective of scientific analysis is to obtain an overall *systems* viewpoint of organizations and the decision situations which determine the future of organizations. Of course, it is not always possible to consider all decision problems in their broadest context, for to do so might require infinite analytic capability. Usually, the natural and traditional decision problems are initially treated somewhat independently of one another, with the objectives and needs of other organizational elements treated as constraints. For example, the "best" possible new product, in and of itself, might not be the one chosen because it would require expensive retooling or because it would not fit into the existing distribution network. Or, at the very least, a good new product idea might have to be so good that, even before its selection, it would be obvious that any such inherent difficulties would be offset by its profitability.

One of the natural ways to approach a systems viewpoint more closely is through the combination of related decision problems and models which have previously been treated independently. A combination of pricing and purchasing decisions is a natural step along these lines. The significant element which is common to both pricing and purchasing decisions is *demand*. In the purchasing models of Chapter 7, demand (the output of the system) was treated as known, either deterministically or probabilistically. In pricing decisions, demand is treated as a *known* (estimated) *function of price*. The discrepancy between the two approaches lies in the words "known function of price." The purchasing models implicitly assumed that demand was either fixed, as in the case of a contracted demand level, or so stable that it could be approximated as a fixed known quantity or probability distribution. The possible dependence of demand on another of the organization's controllable variables, price, was ignored.[23] In pricing decisions this dependence of demand on price is of paramount importance. Hence, it seems natural to combine the two decision problems by way of their common element, demand.

Again, to avoid the unnecessary complexities of transforming from one dimension to another, we shall consider the decision problem of a processor who buys units of raw material, processes them, and sells them to consumers. His basic decision variables are those of the purchasing and pricing decision problems—the purchase quantity and frequency (q and t respectively) and the price to charge (p).

The total relevant cost expression may be written, in the fashion of Chapter 7, as[24]

$$TRC = C_H \frac{q}{2} T + C_0 \frac{D}{q} + Dk \qquad (8\text{-}13)$$

where C_H is the cost of holding a unit in stock for a time period, q is the purchase quantity, T is the time period under consideration, C_0 is the fixed order cost, D is the demand quantity during T, and k is the unit cost (purchase cost plus unit variable processing cost). The first two terms of this expression are identical with those of expression (7-6). The last term becomes an element of total relevant cost here because the demand quantity D will not be considered as fixed. In Chapter 7, D was considered fixed, and therefore the latter quantity was unaffected by the decision variables; hence, it was considered irrelevant. Here, this cost element is relevant since a change in the sales price affects D (and therefore the product D times k).

---

[23] Note that the situations treated in the "Purchasing with Variable Prices" section of Chap. 7 referred to *purchase prices*. Here, we are dealing with the controllable variable *sales price*.

[24] The assumptions made for this expression are the same as those discussed for the basic model of Chap. 7, except for those dealing with demand. Here, demand for the period T is known only as a function of price in terms of the demand curve.

Differentiating TRC with respect to q to minimize total relevant cost, in the fashion of Chapter 7, we obtain

$$q_0 = \sqrt{\frac{2C_0 D}{C_H T}} \tag{8-14}$$

which is identical with expression (7-8). Expression (8-14) gives the best purchase quantity q_0 in terms of the known quantities C_0, C_H, and T and the unknown D. If we substitute this best purchase quantity into the total relevant cost expression (8-13), we obtain the expression for the minimum total relevant cost (TRC_0), which after some simplification is

$$TRC_0 = \sqrt{2C_H C_0 D T} + kD \tag{8-15}$$

We shall assume that demand is known only through its dependence on price, i.e., the demand curve. For simplicity we shall again use a linear curve

$$D = ap + b \tag{8-16}$$

In no way, other than computational tedium, however, does this unrealistic linear demand curve restrict the applicability of this approach.

Profit for the period T may be written as revenue minus fixed costs f and variable costs. If we assume that the purchasing of raw materials is performed in an optimal fashion, the total relevant variable costs will be those given as TRC_0 in expression (8-15). Hence, profit P is

$$P = pD - f - \sqrt{2C_H C_0 D T} - kD \tag{8-17}$$

since pD (unit price times demand quantity) is total revenue.

Since demand is known in terms of price via the demand curve, we may express profit as a function of price alone as

$$P = p(ap + b) - f - \sqrt{2C_H C_0 (ap + b) T} - k(ap + b) \tag{8-18}$$

To determine the profit-maximizing price, we should differentiate expression (8-18) with respect to the decision variable price and equate to zero, i.e.,

$$\frac{dP}{dp} = 0 \tag{8-19}$$

The solution of this expression for p gives the optimal price p_0. Because the algebra involved in this is a bit messy, we shall defer it to the technical note to follow. The important thing to notice is that *we may develop a numerical value for the best price to charge from the solution of expression* (8-19).

The process which we have followed and shall follow in integrating the purchasing and pricing problems is outlined as Figure 8-13. First, the *form* of the total variable (relevant) cost expression and

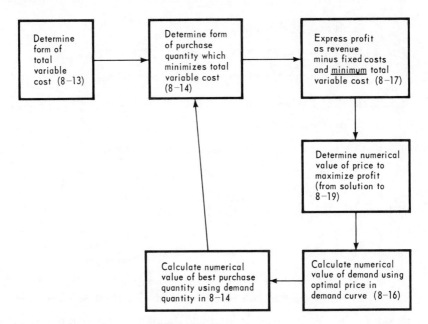

FIGURE 8-13 PRICING AND PURCHASING ANALYSIS

the *form* of the best purchase quantity were determined. We emphasize the word "form" because the determination of these numerical values would require that the numerical value of the demand quantity be known. Next, profit was expressed as revenue minus costs, the total variable cost being assumed to be the minimum which could be obtained, i.e., the cost attained using the optimal purchase quantity. The numerical value of the profit-maximizing price was then determined. By using this price in the demand curve, the numerical value of the demand quantity which the best price would generate could then be determined. Finally, by using this demand quantity, the numerical value of the best purchase quantity could be found. In this fashion the problems of determining a best purchase quantity and best price were solved simultaneously.

In the technical note that follows, we shall demonstrate the algebraic steps which are necessary for the determination of the best price.

***Technical Note on the Solution of Expression* (8-19).**[25] The profit P in the foregoing pricing and purchasing model is given in terms of price p in expression (8-18) as

$$P = p(ap + b) - f - \sqrt{2C_H C_0(ap + b)T} - k(ap + b)$$

[25] This technical note requires only algebra and elementary calculus.

This may be expanded as

$$P = ap^2 + bp - f - (2C_H C_0 apT + 2C_H C_0 bT)^{1/2} - kap - kb$$

Differentiating with respect to p and equating to zero, we obtain

$$\frac{dP}{dp} = 2ap + b - \tfrac{1}{2}(2C_H C_0 apT + 2C_H C_0 bT)^{-1/2}(2C_H C_0 aT) - ka = 0$$

This is equivalent to

$$2ap + b - ka = C_H C_0 aT(2C_H C_0 apT + 2C_H C_0 bT)^{-1/2}$$

Squaring both sides, we obtain

$$(2ap + b - ka)^2 = \frac{C_H{}^2 C_0{}^2 a^2 T^2}{2C_H C_0 apT + 2C_H C_0 bT}$$

or

$$4a^2 p^2 + b^2 + k^2 a^2 + 4apb - 4ka^2 p - 2kab = \frac{a^2 C_H C_0 T}{2ap + 2b}$$

Multiplying both sides by the denominator of the right side and then subtracting the numerator of the right side, we obtain

$$8a^3 p^3 + (16a^2 b - 8ka^3)p^2 + (10ab^2 - 12ka^2 b + 2k^2 a^3)p + 2b^3$$
$$- 4kab^2 + 2k^2 a^2 b - a^2 C_H C_0 T = 0 \quad \textbf{(8-20)}$$

which is a cubic equation which must be solved for p. We shall not proceed with the mechanics of solving this equation since any college algebra text discusses techniques for determining the rational roots of polynomials such as this. The important thing for our purpose here is that at least one rational root exists; hence, the optimal value of p can be determined numerically.

Summary

The price to be charged for a product is a basic component of an overall marketing strategy. The elementary concepts and models of pricing are an intrinsic part of traditional economic analysis; yet, in no other area of marketing is so little formal analysis utilized by decision makers.

One of the problem areas in analytic pricing is the determination of demand. Because this function is so complex, it is often best to treat pricing decisions in a sequential manner. In "strategic pricing decisions" subjective likelihood and utility judgments are used for initial price actions. Then, happenings in the marketplace (perhaps in test markets, for instance) are viewed and the price level is modified on the basis of this information. Since each of the possible happenings in the marketplace can be outlined before the initial price action, it is possible to develop an optimal pricing strategy which permits the adaptation of future pricing actions to future happenings. Such a strategy can be determined on a completely objective systems basis prior to the adoption of an initial pricing action.

Since many organizations and industries effectively set their prices on a competitive bidding basis, this topic is included as a part of pricing analysis. The problem of price setting in a competitive environment differs in structure from most other pricing decisions, because the most significant single uncontrollable factor which needs to be considered is the conscious choice of an action by a rational competitor.

The other major components of an overall product strategy involve promotion and distribution. In the next chapter promotional decisions in the context of advertising are discussed. Subsequent chapters deal with other promotional decisions and distribution decisions.

EXERCISES

1. Using expression (8-1), show that if the elasticity of demand is unity, a change in price from p_1 to p_2 will have no effect on revenue.
2. What is the fallacy in the consistent use of an "incremental cost" approach to pricing?
3. In what respect might a typical demand curve such as that shown in Figure 8-1 be descriptive of the behavior of a single individual? How would it describe the behavior of aggregates of consumers? What special circumstances might exist for a single individual which would indicate that such a curve would be meaningless in describing his behavior?
4. Give an argument, based on a trial-and-error approach by producers and consumers, as to why an equilibrium price should prevail for a product.
5. How would a prolonged period of adverse weather affect the supply curve for a farm product? What happens to the equilibrium price in this case? Why?
6. One of the difficulties involved in scientific marketing analysis is that the marketing system often cannot be experimented on, as can many physical systems. As a result, the marketing scientist must simply *observe* the system. The difficulties inherent in this can be demonstrated by considering the observations which might be made before and after a shift in demand. To show this, consider the shift described by Figure 8-4. If the demand curve shifts up to BB' in good times and down to AA' in bad times, consider what the analyst would record during two successive observations on the system—the first in good times and the second in bad. What might his conclusion be? Does this observation invalidate the basic model? Why or why not?

7. The quantity of a product which is bought by consumers and that which is sold by producers is always equal. How, then, can we say that the equality of supply and demand determines an equilibrium price?

8. How closely do you think an organized trading institution such as the New York Stock Exchange approximates a situation of perfect competition? Why?

9. The following is a fictitious press release (from the Samuelson reference). Comment on its validity.

   The effect of a tax on a commodity might seem at first sight to be an advance in price to the consumer. But an advance in price will diminish the demand. And a reduced demand will send the price down again. Therefore it is not certain, after all, that the tax will really raise the price.

10. Publications listing wholesale auto prices are becoming available to the average auto buyer. What effect do you think this has on the pricing policies of dealers?

11. Advertising agencies generally price their services at 15 percent of the total media expenditures of the advertiser. What negative effect on the agency might this have in the case of a new product? How might this be remedied?

12. Verify expression (8-9), using the data of Table 8-1 and the least-squares approach of the Appendix. Sketch in the estimated linear function in Figure 8-6. How good a fit does it represent?

13. Verify expression (8-12) in the same way as (8-9). Comment on the fit obtained.

14. What is the difference between a demand curve and a demand function?

15. Comment on the validity of the statistical estimation of demand in light of the situation described in exercise 6.

16. Suppose that the sequence of events in the decision problem described by the tree of Figure 8-8 is changed so that first we set a price, then the degree of market penetration accomplished *prior* to competitive introduction is reported, then our competitor's price becomes known, and finally we set a response price. Sketch the decision tree. What strategies are available to us? How does this problem differ in structure from the example given in the text?

17. Make certain that the decision tree from exercise 16 has been constructed so that 89-cent alternatives and "market penetrated" events always lie on branches which are above 79-cent alternatives and "market not penetrated" events respectively. In such a tree, associate probabilities of 0.50 and 0.70 to the "market penetrated" event for initial prices of 89 cents and 79 cents respectively. Also, assume that the probability for a competitive price of 89 cents is 0.10, 0.70, 0.10, or 0.40 in the event that the initial price is 89 cents and the market is penetrated; the initial price is 89 cents and the market is not penetrated; the initial price is 79 cents and the market is penetrated; or the initial price is 79 cents and the market is not penetrated, respectively. Label the 16 profit outcomes at

the right of the tree 200, 180, 160, 150, 140, 170, 100, 140, 120, 180, 150, 140, 100, 120, 50, and 100 thousands of dollars respectively. Which is the best strategy?

18. Distinguish between the "bidding problem" and the "problem of bidding." (You should recognize that although this particular terminology has no great significance, the distinction between the two problems is of importance.)

19. The bidding history of three competitors is compiled below. The data given represent each competitor's bid and our cost estimate for a number of contracts. What bid should we place on the next contract for which we will compete with these organizations?

Competitor A	30	140	150	100	110	70	30	10	100	80	170	30	15	40	20	120	100
Competitor B	30	150	140	100	100	60	20	10	70	80	160	20	20	30	20	120	70
Competitor C	40	110	110	100	70	70	20	15	60	80	150	50	20	30	30	130	90
Our cost	20	100	100	80	80	50	20	10	70	80	120	10	10	20	10	100	60

Competitor A	100	100	70	10	60	70	10	40	80	80	150	70	80	150	90	220	60
Competitor B	70	110	90	10	60	80	20	30	70	40	100	100	70	200	120	210	40
Competitor C	90	110	60	10	50	100	10	20	70	60	90	100	50	120	100	150	40
Our cost	70	120	30	20	40	90	10	20	60	80	90	50	40	100	120	110	20

20. Solve the "problem of bidding," using the situation described in Table 8-4 if the cost of bidding is 10 percent of our project cost for bids which are 25 percent or more above cost and 20 percent of cost for bids which are less than 25 percent above cost.

21. Adopt the viewpoint of the competitor in the situation of Table 8-6. Suppose that he attributes equal likelihood to each of our strategies. What should he do?

22. Consider the bidding situation of Table 8-4. Suppose that our cost estimate is $9,000. What bid will we place? If the actual cost turns out to be $10,000, what is our expected profit? Suppose that our competitor's past bidding behavior is based on *actual* costs. What effect does this have?

23. What is the significant efficiency which is realized through working backwards in analyzing a bidding situation such as that described in Figure 8-10?

24. Bellman has stated the "principle of optimality" for multistage decision problems as:

An optimal policy has the property that, whatever the initial state and initial decision are, the remainder decisions must constitute an optimal policy with regard to the state resulting from the first decision.

A "policy" is an entire sequence of decisions from the beginning to the end of a network such as that in Figure 8-10. Interpret this formal principle in terms of this example.

25. In what circumstances may the determination of an unconstrained optimal strategy be sufficient even though constraints, such as on the total amount bid on several contracts, must be met? What does this suggest about a reasonable way to go about solving a problem involving constraints?
26. Suppose that the constraint imposed on the bidding situation described in Table 8-7 were that bids could be entered on two contracts at most. What would be the best bidding strategy?
27. If the constraint imposed on the bidding situation described in Table 8-7 were that the total amount bid could not be more than $750, what would be the optimal bidding strategy?

REFERENCES

Bellman, R.: *Dynamic Programming*, Princeton University Press, Princeton, N.J., 1957.

Churchman, C. W., R. L. Ackoff, and E. L. Arnoff: *Introduction to Operations Research*, John Wiley & Sons, Inc., New York, 1957, chap. 19.

Friedman, L.: "A Competitive Bidding Strategy," *Operations Research*, vol. 4, no. 1 (February, 1956), pp. 104–112.

Green, P. E.: "Bayesian Decision Theory in Pricing Strategy," *Journal of Marketing*, vol. 27 (January, 1963), pp. 5–14.

Howard, M. C.: *Legal Aspects of Marketing*, McGraw-Hill Book Company, New York, 1964.

Roberts, H. V.: "Bayesian Statistics in Marketing," *Journal of Marketing*, vol. 27 (January, 1963), pp. 1–4.

Samuelson, P. A.: *Economics: An Introductory Analysis*, 7th ed., McGraw-Hill Book Company, New York, 1967.

Sasieni, M. W., A. Yaspan, and L. Friedman: *Operations Research: Methods and Problems*, John Wiley & Sons, Inc., New York, 1959, pp. 172–174.

Savage, L. J.: *The Foundations of Statistics*, John Wiley & Sons, Inc., New York, 1954.

Schlaifer, R.: *Probability and Statistics for Business Decisions*, McGraw-Hill Book Company, New York, 1959, chap. 2.

Spurr, W. A., L. S. Kellogg, and J. H. Smith: *Business and Economic Statistics*, rev. ed., Richard D. Irwin, Inc., Homewood, Ill., 1961.

Tintner, G. A.: *Econometrics*, John Wiley & Sons, Inc., New York, 1952.

CHAPTER 9

Advertising Decisions

Decisions related to advertising are unquestionably among the most significant problems which the marketing manager faces. This importance is made apparent by the simple magnitude of advertising expenditures, which annually represents over 2 percent of the gross national product and is expected to approach $20 billion per year in 1970. However, the real importance of advertising decisions to marketing management lies in the prevalent belief that *significant portions of this huge advertising expenditure are wasted*. One of the best-known anecdotes of the advertising world, which is so often repeated that it is part of the folklore of marketing, is about the marketing manager who expressed the conviction that over half of his firm's advertising outlays were wasted. "The only trouble," the story has him say, "is that we don't know which half!"

It is clear that such inefficient use of funds would never be tolerated in other segments of the business enterprise. The production manager whose machines were only 50 percent utilized or the foreman whose crew worked only half the time would undoubtedly become familiar with the advertising world through the classified employment section of the newspaper. Yet, the marketing manager can announce to all that advertising expenditures are only 50 percent effective and not feel uncomfortable in his job. Paradoxically he can at the same time expound on the virtually unlimited advertising opportunities which he believes to exist for the same product and still be considered a perfectly rational and competent creature.

The reason for these peculiar circumstances, of course, is that so *little is known about the effects of advertising.* To ascertain that certain advertising is effective is not difficult; one needs only to view the thousands of new consumer products which have been successful and recognize that their success is to a large degree due to the information communicated by advertising. Who would ever have purchased a cold-water detergent or a green fluoride toothpaste from the shelf of a supermarket or from a door-to-door salesman without the information that some New Jersey housewives had found the former was satisfactory and that "Jimmy's group had 34 percent fewer cavities"? And, or course, it was advertising that informed the consuming public of these tests and thus made their success possible (if it did not ensure success). Moreover, if it were not for the general credibility which national advertising lends to a product, would the information concerning these tests have been sufficient to lure consumers from their perfectly satisfactory hot-water detergents and non-fluoride toothpastes? The answer is, of course, that no one knows with any degree of certainty. Yet, there is a strong body of opinion which says that advertising plays a significant role and performs a useful purpose in such instances. This opinion is descriptive of the current state of "knowledge" with respect to advertising effectiveness.

But advertising is not meant to be only a means of conveying information about new products. Advertising, in the final analysis, is meant to *sell,* and to sell both new and old products. How is the maker of Budweiser beer, or Hershey chocolate bars, or Morton salt—all well-established brands of stable products which hold a prominent position in their markets—to view advertising, its purpose, its media, and its effectiveness? Again, the answer is that no one truly knows what view he *should* take.

This is not to say that nothing is known in the advertising field. There is available a good deal of information which deals with specific products or media or with specific products advertised in a particular medium, but very little knowledge of any general applicability is available to advertising managers and analysts. However, a methodology has developed (and is still developing) which deals with how one should make optimal advertising decisions in the light of the kind of information which is only infrequently available. The value of such a methodology lies in the way in which it directs thinking into logical channels, forces one to view the significant (and complex) problems of advertising rather than the simple (and meaningless) ones, and makes explicit the information which is needed to increase the 50 percent effectiveness to 51, 60, or perhaps even 75 percent. Further, the methodology encompasses even the process of gathering this necessary information.

It is this methodology to which we shall address ourselves in this chapter. To do so in the best fashion would require us to consider advertising decisions in the context of the overall promotional program of an organization. Media advertising, personal selling, point of purchase, and other forms of sales promotion make up the total promotional strategy of a firm, and any treatment of one aspect while ignoring the others must necessarily be fragmentary and incomplete. However, because of our inadequacies and a desire for simplicity which will facilitate pedagogy, we shall speak only in the language of advertising in this chapter. As a first approach to promotional decisions, this is probably best. In the next chapter the interactions with other promotional forms will be investigated.

Advertising Objectives

In any endeavor in which resources are to be expended, as time, money, and personnel are "spent" in advertising, some consideration should be given to the objectives or goals of the enterprise. Indeed, for the scientific analyst, the objectives of a decision maker form the very basis of his decision problem.

A great deal is said about a wide range of possible objectives of advertising, but the single basic objective, which cannot be ignored, is that *advertising is undertaken in order to sell products.* Although subobjectives such as the building of goodwill and the communication of information may be relevant, in all respects they are either incidental to the primary objective or they are not really objectives at all. For instance, one phase which is frequently used in this sense—"to convey information"—is not an objective of advertising (unless the firm's business is information), but rather it is a *means of achieving* the primary sales objective. Indeed, to convey information is not even a subobjective of most advertising, if we interpret the word "information" with its usual connotation of objectivity. Most advertisers propagandize rather than inform, and it is probably valid to say that this is exactly what they should do.

Of course, there is a kind of advertising which is directed toward such long-term sales goals that the short-run objective may be considered communicative. Public opinion analyst Elmo Roper has commented on the role of this kind of advertising in the 1970s:

> There will be more advertising devoted to the transmission of ideas rather than simply the selling of products. There will, in other words, be more corporate advertising, more advertising concerned with improving public relations. We exist in a society where public relations— human relations— are increasingly important, and there will be increasing

awareness that the ideas and feelings and attitudes that form the substance of these relations are as crucial to the well-being of many corporations as the movement of products.[1]

However correct Mr. Roper's prognostications may be, the same *Saturday Review* article estimates that "idea" advertising represents barely 1.5 percent of today's total advertising expenditures. And, of course, we can argue that this sort of advertising has a sales objective, although it is admittedly long-range and once-removed from the hard sell with which we are all familiar. In any case, advertising which has a direct sales objective is of such overwhelming importance in today's economy that we may well devote greatest attention to it, even though some of those who are charged with achieving sales goals by spending the money which goes into advertising often seem all too prone to forget the basic objective— sales—and to focus undue attention on some subobjective or irrelevancy.

Of course, the basic sales objective is a complex one, and it is necessarily so because of the many elements of communication, persuasion, consumer awareness, etc., which are an intrinsic part of selling via advertising.

Advertising deals with the psychology and social behavior of human beings. The study of human psychology and sociology has not progressed to such a point that use of the results of behavioral studies in the pursuit of desired objectives has been made on any wide scale.[2] As a result, understanding of advertising and its proper role in society is not great. (And it probably will not be until further behavioral results are available.) Yet, advertising is big business and clearly should be analyzed by whatever methods and tools are available.

The concentration of advertisers and advertising professionals on the subobjectives and alternative means of achieving them is not entirely invalid, of course. In special situations the subobjectives should often be emphasized. In the case of a new product, as we have already noted, great attention must be paid to information concerning its existence, uses, availability, special features, etc. Although the long-range sales objective may initially be neglected in favor of near-term necessities, it should be kept foremost in mind. This point, which will undoubtedly seem obvious and overplayed to some, is not, in the author's experience, unfailingly followed by managers and advertisers. Too often has the quest for more and more "exposures" led an advertiser to spend great sums of money in ways which were warranted neither by the subse-

---

[1] *Saturday Review*, Apr. 9, 1966, pp. 70–77.
[2] One possible exception to this is the field of motivational research.

quent sales nor by the sales which could be predicted at the time of the expenditure decision. With due respect to the members of the advertising profession, who have unquestionably contributed greatly to the economic growth of this nation, it should always be borne in mind that no advertising agency executive has ever, in good conscience, promised a potential advertiser that advertising *causes* sales. This causal relationship, which most people believe exists in some degree, is basic to the advertising environment; yet no one has proved or disproved it with any degree of generality.

It is the author's belief that a large percentage of the total advertising expenditures made in the United States in any given year is motivated by fear. In interviews with corporate personnel at all levels from president to clerk, the author has seldom found the element of fear absent. The fear motivation is the *dread of the outcomes which would result from a failure to advertise or from a decrease in the level of advertising.* If one examines the historical expenditures of these corporations, one finds that few have ever significantly reduced advertising expenditures from one year to the next and that virtually none have ever completely discontinued advertising expenditures. Typically, for reasons we shall discuss later, the successful corporation makes successive moderate year-to-year increases in its advertising budget. In fact, then, there seems to be little factual basis for the fear motivation of advertisers. If one could point out numerous instances of dire events following reductions in advertising expenditure, there would be a factual base. However, this is not the case. The fear of advertisers is largely the fear of the unknown.

To argue that a fear of the unknown is always groundless would be fallacious, however, for it may well be that distinctly negative results would be produced by cuts in the advertising budget for many products. One is led to the belief that there are many well-established, heavily advertised products that could well benefit from less advertising. Here, the word "benefit" is intended to refer to the overall profit result rather than to imply that sales increases would result from advertising decreases.[3] Hershey Chocolate, the prime example of the advertising antagonists, is the only major U.S. corporation which does not advertise.[4] Of course, Hershey

[3] The possibility that this inverse relationship exists in some cases is not so remote as it might, at first, appear. The author knows of a situation in which significant sales increases occurred immediately after substantial decreases in advertising. The reason for this was presumed to be that personal sales *effort* (but not expenditure) was increased to compensate for what was expected to be a disastrous effect of the advertising reduction.

[4] Hershey has begun at least two promotional campaigns in recent years. In one, the beginning of marketing in hitherto-unpenetrated Canada, in

got into the chocolate bar field during its infancy and has risen to heights where its brand name virtually replaces the generic name for the product. Most products do not enjoy such recognition. However, the suspicion that many mature products possess in large measure the same attributes which contribute to Hershey's non-advertising success leads to the natural conclusion that such products might well be advertised less heavily.

There is also an ethical question concerning the overadvertising of physically undifferentiated products in order to create perceptible differences in the minds of consumers. For instance, two brands of detergent sold in Great Britain are almost chemically identical. The companies which produce these detergents have been involved in a "spending war" for almost two decades. Payment for all of these promotions is, of course, ultimately made by the consumer in the form of higher prices. The ethical question is: "Is this detrimental to the public interest?"[5]

Of course, it is almost certainly equally true that some corporations do not spend enough on advertising and that some which might profitably cut back expenditures in their current advertising program might be able to achieve significant gains from advertising in different ways. It is these decision problems of advertising which we shall initially address ourselves to in the next sections of this chapter, "how much to spend" and "how to spend it."

The Advertising Budget

An *advertising budget*, like any budget, is an organized plan for the expenditure of funds. Most firms use advertising budgets to plan and control the nature and timing of advertising expenditures.

---

the face of entrenched competition, was accompanied by advertising. Similarly, a promotional campaign, involving coupons good toward part of the purchase price of its cocoa products, was instituted in the United States.
[5] The British Monopolies Commission has put forth a tentative *yes* answer to this question. *Business Week* (Aug. 20, 1966, pp. 34–35) in an article entitled "Can You Spend Too Much for Ads?" reports that the Commission held that the two companies in question were "wasting money on advertising at the public's expense. Nearly 25% of the final retail price . . . is the result of 'selling expenses'—primarily advertising, sales promotion, and market research." To charges that the advertising and marketing power of large companies serves to restrict competition, Dr. Jules Backman of New York University replies that his study "Advertising and Competition" "demonstrates that there has been no relationship between the extent of economic concentration and the intensity of advertising" (quoted in *Business Week*, April 8, 1967). Campbell Soup Company's Red Kettle brand of dry soup, which was withdrawn from the market after a $10 million advertising expenditure, is given as an illustration that heavy advertising does not guarantee product success.

Of course, the elements of the advertising budget vary greatly from product to product and company to company. Some organizations might include expenditures for dealer catalogs, house organs, market research, corporate annual reports, etc., in their advertising budgets, and others might not. Almost all companies would include media charges, advertising research, direct mail costs, artwork, etc., in their budgets. This, then, is the first major point to be considered when one analyzes an advertising budget: the determination of the constituent elements of expenditures which are included in the budget. Once this has been done, the determination of the size of the advertising budget may be begun.

OPERATIONAL BUDGETING

Like pricing decisions, advertising budgeting has its theory and its practice, and, in truth, in some companies the two have not yet met. In most organizations which advertise heavily, however, a great deal of attention is given to the analytic determination of advertising budgets. Before discussing budget analysis and the objective determination of advertising budgets, we should review the traditional methods of budgeting which are still the only ones used by many organizations. As in the case of operational pricing methods, there is usually no single method which is always applied by an organization (the closest thing to such a universal method being the first one to be discussed). Most organizations make use of complex combinations of these methods and analytic results to determine advertising budgets.

Spending Related to Sales. The most common simple method of budgeting advertising is to relate the advertising expenditure to sales revenue on a fixed percentage basis. Thus, the simple rule of thumb that "it is necessary to spend at least x percent of sales revenue on advertising in the widget industry" is often heard and often applied. At its simplest, such a rule is based on last year's sales revenue. In a more sophisticated form it is based on forecasts of future sales during the period in which the expenditure will be made. In any event it is surely a safe and simple approach to budgeting.

The method has great weaknesses, however, in its theoretical foundation and its operational implications. If advertising is to be viewed as causally related to sales, there is little justification for tying the advertising expenditure to sales already achieved or forecast. Because the method is so widely used in one form or another, the operational effects tend to be confusing. Anyone who attempts to measure the *effect of advertising* on *sales,* in cases where such a method has been used to determine expenditure levels, will find an extremely high correlation between advertising expenditure and sales revenue. However, the real causality is the exact opposite of what is hypothesized: it is sales that have caused advertising expenditures (via the rule of thumb which fixes advertising as a percentage of sales) and not advertising that has caused sales.

There seem to be few justifications for budgeting advertising as a fixed percentage of sales other than simplicity and safety. It is a safe policy in the sense that it is traditional and can therefore be defended as an element of corporate "policy." To set advertising budgets by tossing coins would also have the virtue of simplicity, but it could not so easily be defended as a rational policy formulated on the basis of years of practical experience. Nonetheless, this is unquestionably the most pervasive method of advertising budgeting and its use clearly indicates why most successful companies do not significantly reduce advertising expenditures. In our expanding economy successful companies generally experience sales increases from year to year, and therefore the fixed-percentage advertising budget also increases from year to year in an unending (but sometimes temporarily interrupted) spiral.

Task Approach.   One of the most common approaches to the budgeting of advertising depends on the definition of the job to be done and the resources which will be needed to accomplish that job. In theory, to use this approach, one first specifies the advertising objectives, the alternatives which are available to accomplish the goals (media, copy, etc.), and the costs of each alternative. The advertising budget to be used is then chosen as the minimum cost alternative which will accomplish the objectives, or some reasonable value which will closely approximate the achievement of the objectives. This procedure, which incorporates some of the basic ideas of the conceptual framework and the "principle of bounded rationality," has considerable theoretical justification. In practice, however, the concept is often applied at a completely subjective level, and it sometimes deteriorates into an attempt to justify advertising expenditures by relating them to some specific goal which is not necessarily the basic objective of the organization. Most often the objectives which are considered are stated in terms of exposures to advertising messages and personal response measures rather than sales. In a competitive advertising environment which is much impressed with the development of quantitative analysis and mathematical tools, the author has seen this method applied to the task of measuring advertising effectiveness. In other words, the "job" is defined as the measurement of the effectiveness of advertising and the cost of doing it is defined as y additional dollars of expenditures. Thus the budget is set to perform a data collection task. There is nothing necessarily wrong with such an approach if the basic objective—sales—is not neglected. But if such a task is paramount, the costs of achieving the task goal should be carefully investigated and the task evaluated in terms of them.

In particular, the task approach is often useful for new products. In such cases attention must necessarily be paid to the subobjectives of communication, exposures, etc. A primary requisite for successful application in the case of new products is that the systems view be taken of the promotional program so that all of the promotion which is undertaken will be coordinated and directed toward the achievement of the same basic goals. To have a media campaign directed toward maximum exposure frequency (the *number* of exposures regardless of duplications) while sales promotions were being carried on to increase market penetration (the total number who have used the product at least once) might not be consistent, since the former objective has no concern for the *reach* (or total unduplicated exposures) and the latter directs itself toward the pursuit of triers.

Competitive Budgeting. In many instances the actions of competitors are of paramount importance in determining advertising budgets. An unaggressive approach is often taken for mature products when the primary goal is retentive—to maintain the current sales level. A basic assumption in many uses of this method is that sales of a particular brand are more dependent on the *proportion* of advertising which it expends in a particular market than on the absolute amount of such expenditures. Hence, the advertising share and market share of the brand often are about equal. In some cases, of course, the advertising shares of the various competitors are relatively stable over time, but unequal to their respective market shares. In such instances, one product is presumed to require a disproportionate advertising share to compensate for some other deficiency.

This approach has certain similarities to relating spending to sales since it basically involves the measurement of some performance level which has occurred, or the prediction of what will occur, and the establishment of a budget on the basis of this determination. In the case of spending in relation to sales, the item which is measured or predicted is sales of the brand in question. In the case of competitive budgeting, it is the advertising expenditure level of competitors.

Maximum Spending.   Another approach to budgeting involves the spending of whatever amount is available or can be afforded. The determination of the amount which can be afforded is somewhat arbitrary, of course. However it is determined, if calculations of "necessary" expenditures are made and "excess" funds are budgeted for advertising, this is the method which is being used.

The difficulties inherent in this method are rather obvious. It assumes that if advertising is good, then more advertising is better. Even if this is true in some instance, it is almost certainly valid only up to a point. Advertising expenditure, like almost any return-oriented activity in this world, is probably subject to *diminishing returns;* above some point, each increased dollar of expenditure yields less and less until the incremental return is less than the dollar expended. To spend to the maximum as a rule of thumb in any endeavor which is subject to diminishing returns is not a rational approach.

Investment Budgeting.   Many individuals and organizations view advertising as a capital investment which produces two kinds of returns—current sales and long-term goodwill. This viewpoint (for it is really that rather than a method) equates dollars expended in advertising with those spent in any other internal capital investment venture. As such, the argument goes, advertising should compete with these alternatives for funds.

The problems involved in putting this view into practice are great. One does not know how to estimate the size or timing of the residual effect, for instance. Nonetheless, while no one would claim to actually use the investment approach to determine numerically the size of a particular budget, the investment viewpoint is held by many in business and advertising. If one listens carefully to the views of advertising agency executives when budgets are being discussed, reference to the investment viewpoint is almost invariably implicitly made.

ANALYTICAL BUDGETING

The determination of the total advertising budget by using only the operational methods discussed in the previous section has little basis in logic. Each of these methods has serious flaws which cannot easily be obviated. The alternative to these methods, which in practice is usually a complementary approach to the operational methods, is the budgeting of advertising according to an analytic advertising model.

A Basic Static Budgeting Model.   The question of how large the advertising budget should be is intrinsically related to the deter-

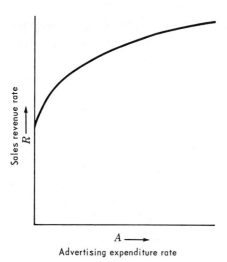

FIGURE 9-1   SALES RESPONSE CURVE

mination of the *sales response to advertising*. We may hypothesize the existence of a static sales response function such as that shown in Figure 9-1. The curve sketched there is one which exhibits the *decreasing marginal return* property which is almost universally accepted as characteristic of advertising. In the figure, R is sales revenue rate (in dollars per year) and A is advertising expenditure rate (in dollars per year).

The difficulty in operationally estimating sales response curves such as that shown in Figure 9-1 is the most significant drawback to objective analysis of advertising decisions. Just as the demand curve is critical to pricing, so is the sales response curve necessary to advertising analysis. In fact, both are really reflections of the underlying basic relationship, the demand function. We shall say a great deal more about the determination of advertising effectiveness later. Here, we desire only to demonstrate its role in simple budgeting analysis. For these purposes we may think of the response curve as depicting the sales revenue generated in a particular medium for expenditures in that medium. Later we shall enlarge on this interpretation.

If s units of the product are produced and distributed during the year, the total relevant cost TRC, exclusive of the cost represented by the advertising expenditures which we seek to determine, may be represented as

$$TRC(s) = f + cs \qquad\qquad (9\text{-}1)$$

where c is a unit variable relevant cost, f is total relevant fixed costs, and $TRC(s)$ denotes that the total relevant cost expression is in terms of the sales quantity s. This function may be written in terms of R, recognizing that total revenue (for the year) is simply s times the unit sales price p, or

$$s = \frac{R}{p} \tag{9-2}$$

Hence,

$$TRC(R) = f + c\frac{R}{p} \tag{9-3}$$

i.e., total relevant cost as a function of the revenue rate R is given by expression (9-3).

Profit (for the year) may then be written as revenue minus total cost, where total cost is composed of the total relevant cost (which excludes the advertising expenditure which we seek to determine) and the advertising expenditure A. Hence,

$$\text{Profit} = R - TRC - A = R - f - c\frac{R}{p} - A$$

which is equivalent to

$$P = R\left(1 - \frac{c}{p}\right) - f - A \tag{9-4}$$

To maximize profit, we set the derivative of this function with respect to the decision variable A equal to zero.

$$\frac{dP}{dA} = \left(1 - \frac{c}{p}\right)\frac{dR}{dA} - 1 = 0$$

or,[6]

$$\frac{dR}{dA} = \frac{1}{1 - \dfrac{c}{p}} \tag{9-5}$$

The solution, as represented by expression (9-5), gives the *slope of the sales response function* dR/dA in terms of quantities which are assumed to be known, the unit sales price and the unit variable relevant cost.

The right side of expression (9-5) deserves further attention. Writing it as

$$\frac{1}{(p - c)/p}$$

we can see that it is the *reciprocal of incremental profit as a fraction of the unit selling price*. Since the derivative dR/dA is the *slope*

---

[6] The reader should note that R is given as a function of A by the sales response function of Figure 9-1. Hence, dR/dA is the slope of that function.

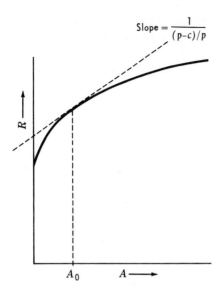

Slope $= \dfrac{1}{(p-c)/p}$

FIGURE 9-2   DETERMINATION OF OPTIMAL ADVERTISING EXPENDITURE

of the sales response curve, the solution given by expression (9-5) tells us that *we should increase the expenditure level A to the point on the sales response function where the slope of the curve is 1 divided by the incremental profit as a fraction of the selling price*. The optimal value of advertising expenditure A_0 determined in this fashion is shown in Figure 9-2. The slope of the sales response curve for all points to the left of A_0 is greater than this critical value, and the slope for all points to the right of A_0 is less than this value. At A_0 the next dollar spent on advertising returns

$$\frac{1}{(p-c)/p}$$

dollars in revenue.

The logic of this solution is made clear by a numerical example. If the unit selling price is $6 ($p$) and the unit variable relevant cost is $4 ($c$), the unit profit is $2. The critical slope of the sales response curve is therefore 3; i.e., we should continue expenditures up to the point where dR/dA is 3. There, to spend an additional advertising dollar produces $3 in revenue, $2 of which is accounted for by c (which is $4 per $6 unit, or $2 per $3 in revenue) and $1 of which is the advertising dollar, for a net of zero. At any expenditure level to the left of A_0, the slope of the response curve is greater than 3. Hence, a positive net marginal profit will result. At any point to the right of A_0, the slope is less than 3 and a negative net marginal profit will result. Therefore, A_0 is the advertising expenditure level which will maximize total profit.

A Basic Dynamic Budgeting Model. One of the significant omissions of the basic static model is its neglect of the time dimension. A model developed by Vidale and Wolfe (see references) which includes the element of time may be considered a basic dynamic model of advertising. This model assumes that the behavioral patterns which reflect advertising response can be represented by three basic parameters: a sales decay constant, a sales saturation level, and a sales response constant.

Sales decay constant. The sales decay constant is a feature of the model which describes the behavior of sales revenue *in the absence of advertising.* If no advertising is undertaken, it is assumed that the product loses sales through customers who switch to other brands and customers who use the product less frequently. The hypothesis is that these customers forget the product and are lured away by competitors who do advertise. Empirical evidence given by Vidale and Wolfe indicates that the temporal pattern of this decay is described by

$$R(t) = R(0)e^{-\lambda t} \tag{9-6}$$

where $R(t)$ is the sales revenue rate (in dollars per unit time), $R(0)$ is its value at some starting point (zero), λ is the sales decay constant, and t represents time.[7] A graph of expression (9-6) is shown as Figure 9-3. It is important to note that this exponential decay pattern is specifically hypothesized for an *unadvertised* product.

Sales saturation level. A practical limit of sales which can be generated by advertising is the second important element of the dynamic model. The sales saturation level M is the known sales rate which will not be exceeded regardless of the advertising level. Of course, this presumes that the advertising media are fixed, and its intuitive justification may be partially based on the limited reach of any particular media combination. If other media were tried, the saturation level might change because potential new customers could be attracted.

Sales response constant. The basic concept of sales response to advertising in the model is built around a sales response constant *r. The sales response constant represents the sales revenue generated by a dollar of advertising expenditure when sales are zero.* In other words, if a firm with no sales spent $5,000 per month on advertising and the advertising caused the sales rate to become

---

[7] The symbol *e*, of course, is the frequently occurring mathematical constant which is equal to about 2.718.

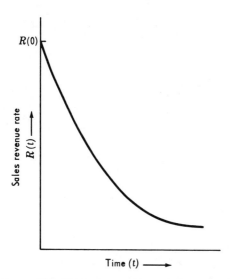

FIGURE 9-3   SALES DECAY OF AN UNADVERTISED PRODUCT

$10,000 per month, the sales response constant would be 2; i.e., each dollar of advertising produced $2 in sales revenue.

However, since the number of potential customers that can be affected by advertising decreases as sales approach the saturation level, the net effect of a dollar of advertising should become less and less. This assumes that the only effect on advertising is on nonconsumers of the brand and that advertising which is used up on current customers is wasted. The degree of effect of advertising on nonconsumers is assumed to be constant (per advertising dollar); i.e., the sales per dollar of advertising which results from advertising undertaken at a specific sales level R (below the sales saturation level M) is

$$r \frac{(M - R)}{M} \tag{9-7}$$

Hence, if r is 2 and the current sales level is at a rate which is 70 percent of the saturation level, each dollar of advertising will produce $0.6 of revenue, and at a zero sales level each dollar will produce $2 in revenue. This is accounted for by the fact that, although each advertising dollar has a constant effectiveness on nonconsumers, in this case there are only 30 percent of the potential consumers who are not already customers, and in the case of zero sales there are no customers. Thus, the total observed sales increase per advertising dollar becomes smaller as the saturation level is approached because there are fewer and fewer noncustomers with whom advertising can be effective.

The actual total response per advertising dollar is given by expression (9-7), which may be written

$$r - \frac{rR}{M}$$

to show that the per dollar response at a sales level of zero ($R = 0$) is r and that this per dollar response decreases at the rate r/M for every dollar increase in the sales rate R. At a sales rate M, the response is zero.

The basic dynamic model which incorporates the sales decay constant, the sales saturation level, and the sales response constant is

$$\frac{dR(t)}{dt} = r\frac{M-R}{M}A(t) - \lambda R(t) \qquad (9\text{-}8)$$

where $R(t)$ is the sales revenue rate at time t, $A(t)$ is the advertising expenditure rate at time t, and the parameters are as previously defined. $dR(t)/dt$ is the instantaneous rate of change of $R(t)$.

To utilize this model for advertising budgeting, one needs only to answer the question: "What level of advertising is necessary to *maintain the sales revenue rate at a constant level?*" This level of advertising is the solution to the equation

$$\frac{dR(t)}{dt} = 0 \qquad (9\text{-}9)$$

since maintenance of $R(t)$ at a constant level implies that the rate of change is zero. The solution [obtained by equating expression (9-8) to zero and solving for A] is

$$A = \frac{\lambda}{r}\frac{RM}{M-R} \qquad (9\text{-}9a)$$

Hence, the closer sales are to the saturation level, the more costly it is to maintain the level (since $M - R$ is small and the product RM is large), and the larger the ratio of the decay constant to the response constant, the more costly is maintaining the sales level. If the sales saturation level is $40,000 per month, the current level of sales is $30,000 per month, the sales response constant is estimated to be 2 and the decay constant 0.4, then the advertising level needed to maintain sales at the $30,000 per month level will be

$$A = \frac{0.4}{2} \cdot \frac{30{,}000 \cdot 40{,}000}{40{,}000 - 30{,}000}$$

or $24,000. Hence, to maintain a sales revenue rate of $30,000 per month, $288,000 (12 times $24,000) should be budgeted for the coming year.

Although there is some empirical evidence that this model for budgeting advertising is a reasonable one[8] and there is considerable intuitive and logical justification for the three salient aspects of the model—sales decay, sales response, and a sales saturation level—one of the primary faults of the model is its concentration on potential customers and consequent neglect of current customers. As we illustrated in the discussion of test marketing, long-term product success is guaranteed only through repeat sales, and one of the fundamental aims of advertisers is to appeal to current customers to consume at a faster rate—to have two cars or TV's instead of one, or to buy beer in eight-packs. To certain products which have a rather stable usage rate (such as soap or toothpaste) or have large bodies of loyal customers (such as cigarettes), the model may be directly applicable. For others, where purchase quantity affects usage (as in home consumption of beer) or where customers skip from brand to brand, the model may require modification. In any event, its intuitive appeal and empirical justification qualify it as a basic dynamic model of advertising budgeting.

Like any mathematical model, this basic dynamic advertising model incorporates parameters whose numerical values must be estimated before the model can be applied. In this instance the parameters are the sales decay constant, the sales saturation level, and the sales response constant. Since the estimation of these quantities is peripheral to the main body of methodology, we shall defer discussion of this topic to a technical note at the end of the chapter.

Allocation of Advertising Expenditures

Problems involving the allocation of advertising expenditures among media, market areas, consumer groups, or time periods constitute one of the most significant classes of advertising decisions. The importance of such problems is made apparent by the declining-marginal-return form which is hypothesized as being descriptive of advertising sales response. If the portion of the total advertising budget which is being expended in television spots is so great that the "flat" upper portion of the response curve for that medium (such as that in Figure 9-1) is in evidence, it is likely that additional benefit can be obtained by reducing the TV spot outlay and spending the money in another medium. But how much should the television expenditure be cut, and what other media should be considered? Such questions will be dealt with in this

[8] See the Vidale and Wolfe and the Benjamin, Jolly, and Maitland papers, for example.

section. Primarily we shall discuss allocations among media, since this type of allocation problem has received most attention. The methodology is equally applicable to other allocation problems, however. In particular, the specialized media allocation model which is discussed incorporates aspects of allocation to media and to consumer groups.

The models dealing with media allocation problems are primarily of the decision-making-under-*certainty* kind. The known state of nature is taken to be the sales response elements and/or the characteristics of the media audiences. The strategies are the various expenditures of money which can be made for each medium.

THE BASIC STATIC UNCONSTRAINED MODEL

The basic static model which was developed in the previous section for the determination of an overall budget or the budget for a particular medium may be easily extended to the case of various media. The basic requirement for this extension is a knowledge of the sales response curve for each medium. Since it cannot be pointed out too frequently, we shall again indicate that this is a rather restrictive assumption. Nonetheless, the basic approach serves two vital purposes. First, it illustrates what we must know in order to conduct the analysis. Secondly, it presents in theoretical finery that logical process which we may follow to utilize the rough information of the real world in making good advertising expenditure allocations.

We may symbolize the sales response curve for n media (or market areas, consumer groups, etc.) as

$$R = R(A_1, A_2, \ldots, A_i, \ldots, A_n)$$

i.e., the sales revenue rate generated by an expenditure of A_1 dollars in medium 1, A_2 dollars in medium 2, etc. This is the multidimensional function which corresponds to the single-dimensional curve of Figure 9-1. There, A was taken to be the expenditure in one medium. Here, we are concerned with the *allocation* of advertising dollars rather than with the setting of an overall budget. Therefore here, we must begin with a knowledge of the contributions of the various media to sales.

In the same way we used to develop expression (9-4), we may write the profit equation in terms of the advertising expenditures in the several media as

$$\text{Profit} = R\left(1 - \frac{c}{p}\right) - f - \sum_{i=1}^{n} A_i \qquad \textbf{(9-10)}$$

where R is sales revenue rate (dollars per year) as a function of the A_i's, c is unit variable relevant cost, p is unit selling price, f is fixed cost, and the A_i quantities are the advertising expenditures in the various fixed-cost media under consideration. The term "fixed cost" as applied to the media indicates that the particular cost is not a function of the sales generated, as it would be if we were discussing the cost of salesmen's commissions, for example.

To determine the best allocation to each of the media, we must equate the partial derivatives of profit with respect to the various media expenditure levels equal to zero, i.e.,

$$\frac{\partial P}{\partial A_i} = 0 \qquad\qquad (9\text{-}11)$$

for each A_i.[9] This is equivalent to

$$\frac{\partial R}{\partial A_i}\left(1 - \frac{c}{p}\right) - 1 = 0$$

for all i, or

$$\frac{\partial R}{\partial A_i} = \frac{1}{1 - c/p} \qquad\qquad (9\text{-}12)$$

for all i. This expression is identical to that which represents the solution to the basic static budgeting model (9-5) except for the appearance of the i subscript and the partial derivative. The interpretation is completely analogous. Expression (9-12) says that the optimal advertising expenditure *in each medium* is that at which the slope of the sales response curve in the direction of increasing values of each A_i (with all other A_i values fixed) is 1 divided by the incremental profit as a fraction of the unit selling price. In other words, the overall optimum allocation is achieved at a point where *a dollar spent in any of the media will produce the same sales return*—the quantity

$$\frac{1}{1 - \dfrac{c}{p}}$$

For example, if this quantity is 2.5 and the revenue function is

$$R = 2 \ln A_1 + 4 \ln A_2 \qquad\qquad (9\text{-}13)$$

the partial derivatives equated according to (9-12) are[10]

$$\frac{\partial R}{\partial A_1} = \frac{2}{A_1} = 2.5$$

[9] This is a necessary but not sufficient condition for the existence of maximum point of the profit function. See any calculus test for a full discussion.
[10] The symbol "ln" denotes the natural logarithm, or logarithm to the base e.

and

$$\frac{\partial R}{\partial A_2} = \frac{4}{A_2} = 2.5$$

The solutions of these equations give

$A_1 = 0.8 \qquad A_2 = 1.6$

which, if all quantities were expressed in millions of dollars per year, would say to allocate \$0.8 million to medium 1 and \$1.6 million to medium 2.

However, a simple additive response function across the media ignores any interactions which might occur. For example, if each medium is directed toward a different consumer group, expression (9-13) might well be applicable. If the media overlap in their coverage, a more complex function might be required, e.g.,

$$R = 2A_1{}^2 + 4A_2{}^2 - 0.2A_1A_2$$

The last term in this expression accounts for the interaction of the two media.[11]

Applying (9-12) here, we have

$$\frac{\partial R}{\partial A_1} = 4A_1 - 0.2A_2 = 2.5$$

$$\frac{\partial R}{\partial A_2} = 8A_2 - 0.2A_1 = 2.5$$

where the right side of expression (9-12) is still regarded as being 2.5. Note here that *each of the equations involves both A_1 and A_2*. This demonstrates the general case in which *expression (9-12) represents a set of n simultaneous equations in n unknowns.* Solving simultaneously for A_1 and A_2 in this case, we find the best expenditure rate A_1 to be about 0.64 million and the best expenditure rate A_2 to be about 0.33 million.

The best total advertising budget in this case is simply the sum of the best allocations to each medium. In these two examples, total budgets of \$2.4 million and \$0.97 million should be set for the year for the two media considered. In the next section we shall deal with the problem of allocating advertising expenditures when the total budget is determined in some other fashion.

THE BASIC MODEL WITH A BUDGET CONSTRAINT

The unconstrained problem of allocating advertising expenditures to media in the best possible way, regardless of the total amount

[11] Later in this chapter we shall say more about the general forms of sales response functions.

spent, is not very descriptive of the realities of advertising decision situations. Most often the advertising manager is faced with the problem of *allocating a fixed total budget* which has already been set in some manner (the manner usually being one of the operational budgeting methods which we have discussed). Proceeding in this two-step fashion, involving the arbitrary determination of the total budget and then the optimal spending of this fixed amount, must necessarily lead to suboptimization, but as a practical procedure it may not be a bad one. The reason for this is the large uncertainty inherent in any "knowledge" of sales response. In such cases a "marriage" of subjectivity and objectivity often produces good results. In any case, a better overall result will probably be produced by suboptimization than by a purely subjective approach to both the setting of the total budget and its proportional allocation among media. Since we shall undoubtedly have to live with arbitrarily determined total budgets for some time, extensive analysis of the constrained allocation problem appears to be justified on pragmatic grounds.

If the fixed budget to be allocated is G (dollars per year), the constrained advertising allocation problem is to allocate advertising dollars among media in a way which maximizes profit:

$$P = R \left(1 - \frac{c}{p} \right) - f - \sum_{i=1}^{n} A_i \qquad (9\text{-}14)$$

Profit is to be maximized subject to the constraint that the total of all allocations should equal the budget G, i.e.,

$$\sum_{i=1}^{n} A_i = G \qquad (9\text{-}15)$$

Of course, the decision variables are the A_i—the advertising allocations—and we should remember that the sales revenue R in expression (9-14) is presumed to be known as a function of the A_i.

The method of solution for the mathematical problem represented by expressions (9-14) and (9-15) involves the use of the technique of "Lagrange multipliers." The basic notion of Lagrange multipliers may be simply stated. It says that the conditions necessary for the maximization of the function P (profit) subject to a constraint which may be written as

$$\sum_{i=1}^{n} A_i - G = 0$$

are the same conditions as those necessary for the maximization of the function

$$P + d \left(\sum_{i=1}^{n} A_i - G \right) \qquad (9\text{-}16)$$

subject to no constraints.[12] The quantity d is the Lagrange multiplier to be determined, and the key gain made by its introduction is the transformation of a constrained model into an unconstrained mathematical problem.

Therefore, we are formally faced with the problem of determining the values of the A_i which maximize expression (9-16). These will be the same values which maximize (9-14) subject to constraint of (9-15). The entire expression written in terms of the A_i [from expression (9-14)] is

$$R(1 - c/p) - f - \Sigma A_i + d(\Sigma A_i - G)$$

To maximize this quantity, we set the partial derivatives with respect to the A_i equal to zero, i.e.,

$$\frac{\partial R}{\partial A_i}\left(1 - \frac{c}{p}\right) - 1 + d = 0$$

or,

$$\frac{\partial R}{\partial A_i} = \frac{1 - d}{1 - c/p} \qquad (9\text{-}17)$$

for all i. Expression (9-17) represents n equations in $n + 1$ unknowns—n quantities represented by the A_i's and d—which may be supplemented by the basic constraint (9-15) to obtain a solution.

Suppose, for example, that the same response function used previously

$$R = 2 \ln A_1 + 4 \ln A_2$$

is applicable and that the quantity

$$1 - c/p$$

is $\frac{2}{5}$. If the problem is to determine A_1 and A_2 to satisfy a total budget of 2 million, then

$$A_1 + A_2 = 2.0$$

Expression (9-17) gives

$$\frac{2}{A_1} = \frac{1 - d}{\frac{2}{5}} \qquad (9\text{-}18)$$

and

$$\frac{4}{A_2} = \frac{1 - d}{\frac{2}{5}} \qquad (9\text{-}19)$$

and the budget constraint is

$$A_1 + A_2 = 2.0 \qquad (9\text{-}20)$$

[12] See the Hildebrand text in the references.

Solving these three equations simultaneously for A_1, A_2, and d, we obtain

$$A_1 = \tfrac{2}{3}$$
$$A_2 = 1\tfrac{1}{3}$$
$$d = -\tfrac{1}{5}$$

The solution to the constrained problem is to allocate $\tfrac{2}{3}$ million to medium 1 and $1\tfrac{1}{3}$ million to medium 2.[13]

To investigate one further important aspect of the constrained problem, let us suppose that the fixed budget is 2.5 million instead of 2.0 million. Expressions (9-18) and (9-19) remain the same and (9-20) becomes

$$A_1 + A_2 = 2.5$$

These three equations have the solution

$$A_1 = \tfrac{5}{6}$$
$$A_2 = 1\tfrac{2}{3}$$
$$d = \tfrac{1}{25}$$

which indicates that $\$\tfrac{5}{6}$ million should be allocated to medium 1 and $\$1\tfrac{2}{3}$ million to medium 2.

We should now recall that the best allocation for the unconstrained problem, as determined in the previous section, had a total allocation of 2.4 million. Hence, the allocation of 2.5 million which is imposed here *must necessarily result in a lower total profit.* This indicates that, unless there is some outside reason for requiring that *exactly* 2.5 million be spent, *it would be better simply to spend the 2.4 million required by the optimal unconstrained solution.* The reader should verify this by calculating the total profit derived by the optimal allocation of the 2.4 million (0.8 million to medium 1 and 1.6 million to medium 2) and the optimal allocation of the 2.5 million ($\tfrac{5}{6}$ million to medium 1 and $1\tfrac{2}{3}$ million to medium 2).

The general conclusion which we may draw from this is that if the true constraint is the inequality

$$\sum_{i=1}^{n} A_i \leq G \tag{9-21}$$

rather than the equality (9-15), *the unconstrained problem should be solved first.* If the solution to the unconstrained problem happens to satisfy the constraint (9-21), the unconstrained solution will produce greater profit than the constrained solution obtained

---

[13] The significance of the unique numerical value for d is not germane to the understanding of this model and may therefore be thought of as simply a mathematical convenience. The interested reader may see the Allen text in the references.

by using the Lagrange technique, and it should therefore be the one which is used. If the unconstrained solution violates expression (9-21), the constrained problem should then be solved in the fashion shown and its solution should be implemented.

A SPECIALIZED MEDIA ALLOCATION MODEL[14]

In this section we treat a specialized media allocation model. It is "specialized" in the sense that it does not deal with the basic strategic objectives of the organization, but rather with the tactical objectives of the media specialist. The emphasis is on reaching the appropriate audience rather than on direct profit maximization. To do this, the model addresses itself to the role of advertising media as *communication channels* (rather than as sales producers) and seeks to provide an allocation which produces effective communication.

As we have previously noted, the emphasis on a subobjective of advertising is pervasive in the advertising world, but in general it is to be avoided. However, one of the basic virtues of quantitative analysis is that it may permit one to suboptimize effectively, and such suboptimization may produce outcomes superior to those which could have been achieved without analysis. On this basis, then, we proceed to develop a specialized model which seeks suboptimization rather than the direct achievement of the "true" objectives of the organization. In a subsection we shall deal with the relationship between the specialized model and the sales objectives of the organization.

The model and method of approach are best illustrated in terms of an example. Suppose that the decision maker has a choice of two media vehicles[15] and that his decision variables are the *number of messages* to be carried in each medium, a_1 being the number in the first and a_2 the number in the second. For example, if the first medium were a magazine, a_1 might be the number of full-page insertions during a particular planning period. If the second medium were a newspaper, a_2 might be the number of insertions of a specified size. If one is considering insertions of various size (or duration, in the case of radio and television), *each specific message can be treated as though it were a separate medium.* Thus, a full-page ad in the morning newspaper would be a separate alternative to a half-page entry in the same paper.

---

[14] The model in this section is adapted from the Brown and Warshaw paper in the references.

[15] Neither the method nor computational algorithms for the method are limited to two media. Only two alternatives are used here because the underlying ideas can be illustrated graphically for a problem of this dimension.

Constraints. A basic set of *constraints* on the values which a_1 and a_2 may take on can be readily developed. The most obvious of these is a budgetary constraint. If the unit costs of a message in the two media are $1,000 and $500 respectively and the total media budget is $20,000, the budget constraint is

$$1,000a_1 + 500a_2 \leq 20,000 \qquad (9\text{-}22)$$

which reflects the fact that the total expenditure may not exceed the budget. However, as indicated by the inequality, it is not necessary that all of the budgeted amount actually be expended.

Certain technical constraints on the values which a_1 and a_2 may take on are also apparent. If the first medium is a monthly magazine and the planning period is a year, no more than one message per issue would probably be desired. Thus,

$$a_1 \leq 12 \qquad (9\text{-}23)$$

would be a technical constraint. The constraint is termed "technical" because it emanates from the environment rather than from the organization.[16]

Nontechnical constraints—those which emanate from within the organization—are often imposed on the advertising allocation problem. For example, if it is decreed that at least six messages should appear in the second medium, the resulting constraint is

$$a_2 \geq 6 \qquad (9\text{-}24)$$

Of course, the number of nontechnical constraints, each of which usually expresses a subobjective generated by an organizational element, could become quite large. In general, the limits on the media program which these subjectively based constraints entail should be kept as loose as possible, for, as we shall see, the placing of too severe a set of restrictions on the media program can hinder the effective achievement of the communications goals of the organization.

The Feasibility Space. The set of all points (a_1, a_1) which simultaneously satisfy *all* of the constraints will be termed the *feasibility space*. This space is graphically illustrated in Figure 9-4 for the constraints previously illustrated. The three lines plotted in the figure represent expressions (9-22), (9-23), and (9-24), with the inequalities converted to equalities. Thus, all points on the horizontal line have a_2 *equal to* 6 and all points on the vertical line have a_1 *equal to* 12. Since the constraints are in the form of in-

---

[16] See the King reference for a discussion of technical and nontechnical constraints and their role in production decision problems.

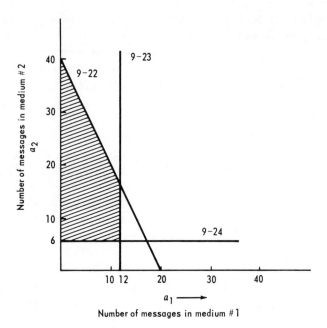

FIGURE 9-4   ADVERTISING FEASIBILITY SPACE

equalities, we must determine "which side" of these lines represents the points satisfying the strict inequality. In the case of the horizontal line, it is all of the points above the line, since these points all have a_2 greater than 6 and expression (9-24) is in the "greater than 6" form. Similarly all points satisfying expression (9-23) lie to the left of the vertical line. The budget constraint is represented by the slanted line. All points to the lower left of this line satisfy the constraint. This can be demonstrated easily by allowing both a_1 and a_2 to equal zero and determining whether the constraint is satisfied. If it is, this implies that the origin is one of the points satisfying the constraint and hence, that all of the points on the "side" on which the origin lies also satisfy it. If it does not, the "other side" of the line encompasses the points satisfying the constraint. In this case, if a_1 and a_2 are both zero, expression (9-22) becomes

$$0 \leq 20,000$$

which is valid. Hence, the origin lies in the region which satisfies the constraint.

The points in the feasibility space must satisfy all three of these constraints in addition to the obvious constraints

$$a_1 \geq 0 \qquad a_2 \geq 0 \tag{9-25}$$

which simply indicate that a negative number of messages is not meaningful. The feasibility space is indicated by the crosshatched area of Figure 9-4, which incorporates all of the points which simultaneously satisfy the five constraints represented by expressions (9-22) to (9-25). It should be noted that the points lying on the lines which bound the feasibility space are included in the feasibility space. This is because none of the constraints is in the form of a strict inequality.

The feasibility space shown in Figure 9-4 is therefore the graphical representation of all combinations of the number of messages to appear in the two media which satisfy the cost, technical, and nontechnical constraints imposed on the decision maker. It thereby represents all of the feasible strategies.

The Objective. The communications objective can be expressed as the maximization of the number of effective exposures which can be obtained within the limits of the constraints. The word "effective" is the tie which binds this specialized model to the overall sales objectives of the organization. First of all, we should make clear that an "exposure" is an awareness of a message by a single consumer. An "effective exposure," however, is an awareness by a *potential customer* or other specific consumer type. The general characteristics of potential customers are presumed to be known. For instance, bald men are consumers but they are not potential buyers of hair tonic. Moreover, families subsisting on unemployment compensation are probably not potential customers for luxurious pleasure boats. In general, market analysis can identify general characteristics of potential customers, and it is the job of the media decision maker to communicate with them, not just to maximize total exposure without attention to those who are exposed.

Suppose that approaches such as those described for carrying on market analysis in Chapter 6 have been used to characterize the potential customers for the product according to those factors shown in Table 9-1, and that weights which indicate the relative importance of those factors have been developed in a fashion similar to that used for product search in Chapter 5. These weights are given in the last column of the table as "importance weights."

TABLE 9-1 POTENTIAL CUSTOMERS

Characteristic	Importance weight
Age: 25 to 45	0.60
Income: over $10,000	0.20
Home location: urban	0.20

These data indicate that the age range of potential customers is from 25 to 45 and that this characteristic is three times as important (0.6 versus 0.2) in the description of the "market" as either the income or the home location characteristic. Although a first reaction to a characteristic such as income level might indicate that it is superfluous, this is not actually the case. Although a mail-order house might *hope* that its *customers* were all wealthy, so that they could immediately purchase whatever they desire, it *knows* from the results of market analyses and warranty card responses that its primary *potential customer* group does not have the highest income level. If it feels that its future customers will be much the same as its past customers, its primary desire is to communicate with potential customers; therefore, an exposure of a potential customer (as defined by the characteristics in Table 9-1) is what is meant by an effective exposure. Exposures of anyone else are considered incidental to the objective of maximizing total *effective* exposures.

The idea of effective exposures is not limited to treating as potential customers only those consumers with characteristics similar to those of past customers. The organization may specify a new target consumer group to which it wishes to appeal. One of the motivations for doing this might be the conversion of an entirely new type of consumers into customers. Alternatively, the appeal might be directed toward a status group of consumers (such as physicians or lawyers) whose purchases can be expected to influence other consumers.[17]

We shall assume that the audience makeup of each of the media under consideration is known in the same terms as in Table 9-1. Suppose that the proportion of the audience of each medium which falls into our definition of a "potential customer" is given by Table 9-2. These differences between the two media are of great significance because of the desire to communicate effectively with potential customers.

TABLE 9-2   MEDIA AUDIENCE CHARACTERISTICS

Characteristic	Medium 1	Medium 2
Age: 25 to 45	60%	30%
Income: above $10,000	40%	70%
Home location: urban	50%	80%

[17] The similarity between this approach and the focusing of attention on a particular group by the basic dynamic model should be noted. There, the "effective exposures" represented all exposures by noncustomers.

The effective audience for each medium may be calculated as the product of the medium's gross audience, the medium's *average noting score*, and an *effectiveness coefficient*. The noting score is the percentage of the audience that "note" the advertisement. This is a standard measure used by the Starch Readership Service[18] to denote the proportion of magazine readers that indicate recognition of an advertisement when they are interviewed after having "read" a magazine. The effectiveness coefficient sums up the audience objective, the weights assigned to various audience characteristics, and the proportional composition of each medium in terms of these characteristics. In the case of the first medium, the effectiveness coefficient is calculated as

$$0.6(0.6) + 0.4(0.2) + 0.5(0.2) = 0.54$$

i.e., the sum of the products of the proportion of the medium's readership which possess each characteristic and the weight assigned to that characteristic in the audience objective as determined from the data of Tables 9-1 and 9-2. The corresponding calculation for the second medium is

$$0.3(0.6) + 0.7(0.2) + 0.8(0.2) = 0.48$$

Suppose that the gross audiences of the two media are 5 million and 3 million respectively and that the average noting scores are 0.3 and 0.2 respectively. If we interpret the average noting score as a probability—the likelihood that the message will be noted—and the percentage composition of the audience is fixed, the *expected effective audience* for the first medium is

$$5,000,000(0.54)(0.3)$$

or 810,000, and the expected effective audience for the second medium is

$$3,000,000(0.48)(0.2)$$

or 288,000.

Solving the Model. It is the desire of the decision maker to *maximize expected effective exposures subject to the constraints which make up the feasibility space*. Figure 9-5 depicts the feasibility space with the extraneous portions of the bounding lines removed. Superimposed on it are various values of the total effective exposure relationship; i.e., the total number of effective exposures *TEE* is

$$TEE = 810,000a_1 + 288,000a_2 \qquad (9\text{-}26)$$

[18] See Lucas and Britt, chap. 3, for a complete discussion.

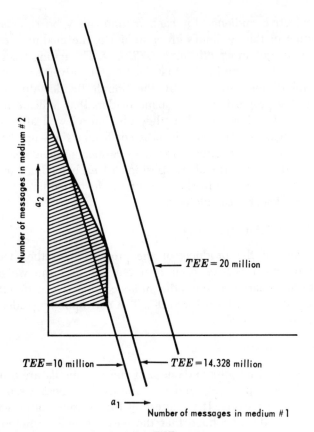

FIGURE 9-5   FEASIBILITY SPACE WITH *TEE* LINES

For example, all combinations of points (a_1, a_2) for which the total number of effective exposures is 10 million, 14.328 million, and 20 million are graphed in the figure. It is easy to see that greater expected effective exposure totals are represented by *TEE* lines which are farther to the upper right and lesser *TEE* values are represented by lines farther to the lower left. Since we seek to maximize this value subject to meeting all of the constraints, we should select a *TEE* line which has as great a value as possible and yet has at least one point in the feasibility space. Such a line is represented by

$$TEE = 810,000a_1 + 288,000a_2 = 14,328,000$$

since the point $(a_1 = 12, a_2 = 16)$ lies on that line and is also a point in the feasibility space. Hence, to maximize total expected effective exposures subject to the constraints, we should use 12 messages in the first medium and 16 in the second. This will give a total of 14,328,000 expected effective exposures and will cost exactly $20,000, the budget limitation.

We should note how simple this determination was. We found it necessary only to graph the feasibility space and several lines representing various *TEE* values. The *TEE* lines are parallel to one another, so that by "moving" in the direction of increasing *TEE* (in this case, toward the upper right), we must finally arrive at a *TEE* line which has at least one point in the feasibility space and for which a further small move would give a line which has a greater *TEE* value, but no points in the feasibility space. The line having maximum *TEE* and still having at least one point in the feasibility space is the best we can do while subject to the constraints. Hence, the point (or points) in the feasibility space has coordinates which represent the best media allocation (subject to the constraints). Of course, this simple graphical analysis is possible only for the case of two alternative media. If a greater number of media are available, the graphical approach cannot be used, but efficient computational devices (called algorithms) have been devised to accomplish the analogous optimization task in a greater number of dimensions. This sort of analysis is part of the subject matter of *linear programming*, and the primary computational device is the *simplex method*.[19]

The simplex method is based on a fundamental theorem, which we shall not attempt to prove; we shall state it informally because of its usefulness in graphical analysis such as that just shown. The theorem says that the optimal solution to a problem of this type lies at one of the "corners" of the feasibility space. There can be alternative optimal solutions (those having different values of a_1 and a_2 but the same value of *TEE*) which can lie on the bounding lines, but one of the optimal solutions will also always be at a corner point. The optimal solution will never lie at an interior point of the feasibility space. The implication of this for graphical analysis is clear. We can omit the awkward procedure of drawing lines which have various *TEE* values and trying to determine the one which has the greatest value by moving these lines in the direction of increasing *TEE*; instead, we can simply sketch the feasibility space, note which constraints are relevant to each corner point, determine the coordinates of the corner points, and determine the *TEE* value associated with each point. The corner point with the greatest *TEE* value is the best allocation strategy.

[19] See the Garvin reference for a simple discussion of the basic ideas of linear programming and the simplex method. Any other introductory linear programming text will contain similar material.

For example, the corner points of the feasibility space in Figure 9-5 are

1: $(a_1 = 0, a_2 = 6)$
2: $(a_1 = 12, a_2 = 6)$
3: $(a_1 = 12, a_2 = 16)$
4: $(a_1 = 0, a_2 = 40)$

These are determined simply by equating the symbolic representations of the two constraints which combine to produce the corner. Corner point 3 is at the intersection of expressions (9-22) and (9-23), for example. The equality corresponding to (9-22), in terms of a_1, is

$$a_1 = \frac{20,000 - 500a_2}{1,000}$$

and for (9-23) it is

$$a_1 = 12$$

Equating these, we have

$$\frac{20,000 - 500a_2}{1,000} = 12$$

or,

$$a_2 = 16$$

Therefore, the corner point is $(a_1 = 12, a_2 = 16)$.

The value of *TEE* for each of the corner points is found simply by inserting the coordinates of the points into expression (9-26). Hence,

$$TEE\ 1 = 810,000(0) + 288,000(6) = 1,728,000$$
$$TEE\ 2 = 810,000(12) + 288,000(6) = 11,448,000$$
$$TEE\ 3 = 810,000(12) + 288,000(16) = 14,328,000$$
$$TEE\ 4 = 810,000(0) + 288,000(40) = 11,520,000$$

Since corner point 3 has the highest *TEE* value, it is the optimal allocation of advertising messages. This was determined without resort to the graphing of various values of *TEE*. All that was necessary was a realization that the best allocation lay at one of the corner points and the calculation of *TEE* values for each corner point.

Extending the Model. A number of possible refinements which serve to make the model more applicable to real-world situations can easily be made. In most media selection situations, there are a number of qualitative factors which should be considered along with the obvious quantitative ones. For example, if both black-and-white and color advertisements in a particular medium are possible, each should be treated separately. In such a case, since the audience and the importance weightings of the audience characteristics will be the same for both color and black-and-white, the effectiveness coefficients for each will be the same. Unless the difference between the two kinds of messages is reflected in the average noting score, some adjustment must be made to differentiate between the two. If black-and-white is deemed to be only 60 percent as effective as color, the effective audience for the black-and-white medium may be multiplied by 0.6 to correct for this. (Of course, depending on the medium, the noting score may already convey this information; in which case no adjustment is necessary.)

Similarly, if placement of ads of different sizes is considered, some adjustment in the effective audience should be made. If a half-page ad is considered 65 percent as effective as a full-page ad, the medium's effective audience (which applies to both sizes of ads) can be adjusted by multiplying by 0.65.

The careful reader will also note that the solution which results from models of this kind is not necessarily restricted to integer values. Thus, the best allocation might be 17.5 messages in the first medium and 14.3 in the second. Of course, one cannot actually purchase partial advertisements. Even the interpretation of restricting the length or duration is inapplicable since we have specified that each message of different size or duration is to be considered as a different medium. In this case, the integer solutions around the optimal should be investigated. For instance, the *TEE* value associated with the strategies

S_1: $(a_1 = 17, a_2 = 14)$
S_2: $(a_1 = 17, a_2 = 15)$
S_3: $(a_1 = 18, a_2 = 14)$
S_4: $(a_1 = 18, a_2 = 15)$

should be compared, since the analytically derived solutions are presumed to be $a_1 = 17.5$ and $a_2 = 14.3$, and the consequences of the possible small violations of constraints should be investigated. In fact, the difference in *TEE* value between a high-valued strategy which slightly violates a constraint and a lower-valued strategy which does not, can be considered to be the cost of viewing the constraint as inviolable. From this standpoint, it is usually easy to make a selection among the near-optimal strategies.

Some Implicit Assumptions. The most significant assumption which is implicitly made by the model has to do with the special-

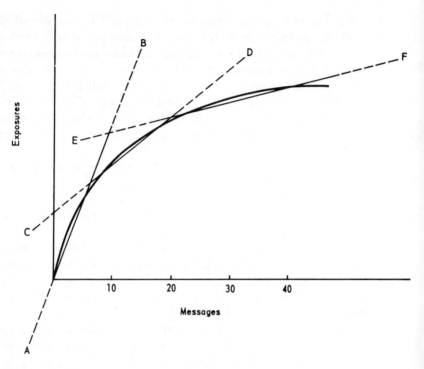

FIGURE 9-6   APPROXIMATING A NONLINEAR RESPONSE CURVE

ized nature of the model—its emphasis on exposures rather than a performance measure such as profit or sales.

Additionally, the model assumes that there is a linear relationship between the number of messages and the number of effective exposures; i.e., each additional message generates the same incremental number of exposures. If the relationship between messages and exposures is actually one which is best described by a function which possesses the decreasing marginal return characteristic, the approximation may not be a valid one. Brown and Warshaw (see references) have proposed a way of handling this difficulty by utilizing linear approximations to segments of the response curve. Figure 9-6 shows how such linear approximations may be made for three segments of a response curve. The segment between 0 and 10 messages (for a particular medium) is approximated by the line *AB*, the segment between 10 and 20 messages by *CD*, and the segment between 20 and 40 by *EF*. We may not be concerned with the response curve beyond 40 messages if one of the constraints limits us to 40 messages in the medium. If we were concerned with the remainder of the curve, we would simply use another line to approximate it over some finite range (say up to

100 messages). The details of the method used to convert the model to account for these approximations are interesting from a technical viewpoint but do not serve to increase. understanding of the basic methodology. Hence, they are given as a technical note below.

Another implicit assumption of the model which is not so easily remedied is that dealing with the valuewise independence of advertisements in the same medium. Since we treat any two different messages (say two of different sizes) as different media in the model, it is not unlikely that a best strategy could be made up of some messages of both kinds. The implicit assumption in this situation is that the net effect of both is the *sum* of their individual effects (if each were used alone). This is clearly a poor assumption in this case, particularly since we have focused on exposures rather than reach (the total number of different people exposed). Of course, audience duplications among the various media are also neglected.

Technical Note on Nonlinear Response Curves in Media Allocation.[20] If the true response curve in Figure 9-6 is approximated by the straight lines over the intervals shown, the effect of the approximation can be included in the model. Suppose that the curve shown is applicable to medium 1. The basic decision variable a_1 can be decomposed into variables a_{11}, a_{12}, and a_{13} defined over the respective ranges as

$0 \leq a_1 \leq 10$ $\quad$ $a_{11} = a_1$
$\qquad\qquad\qquad$ $a_{12} = 0$
$\qquad\qquad\qquad$ $a_{13} = 0$

$10 \leq a_1 \leq 20$ $\quad$ $a_{11} = 10$
$\qquad\qquad\qquad$ $a_{12} = a_1 - 10$
$\qquad\qquad\qquad$ $a_{13} = 0$

$20 \leq a_1 \leq 40$ $\quad$ $a_{11} = 10$
$\qquad\qquad\qquad$ $a_{12} = 10$
$\qquad\qquad\qquad$ $a_{13} = a_1 - 20$

Because of these definitions,

$$a_1 = a_{11} + a_{12} + a_{13}$$

[20] The mechanics of this technical note require only algebra. An understanding of its implications in the solution of programming problems requires a knowledge of linear programming. Its coverage is not necessary to any remaining sections of the Chapter.

If the slopes of the three linear approximations are 800,000, 600,000, and 500,0000, the portion of the objective function relating to the first medium can be changed to

$$810{,}000a_{11} + \frac{600{,}000}{800{,}000} \cdot 810{,}000a_{12} + \frac{500{,}000}{800{,}000} \cdot 810{,}000a_{13}$$

to account for the decreasing effectiveness.

The constraints also must be changed. For instance, the constraint

$$a_1 \leq 12$$

would become

$$0 \leq a_{11} \leq 10$$
$$0 \leq a_{12} \leq 2$$

and a_{13} must necessarily be zero by virtue of this constraint.

If a_1 is changed to the sum of the three double-subscripted a's in all other parts of the model, it may then be solved directly for these new variables.

This procedure is easily generalized so that each medium's response function is approximated in a piecewise linear fashion and each basic decision variable is decomposed into several variables, one for each interval of approximation.

Measuring Advertising Effectiveness

Before proceeding to other decision problems and models of advertising, we shall detour along a path which is perhaps the most critical area of advertising analysis—the measurement of advertising effectiveness. The consideration previously given to the budgeting and allocation problems is sufficient to illustrate the need for discussing this topic. In each model discussed there, assumptions were made concerning the response of sales to advertising. In all cases, these assumptions were critical to the model. In fact, as we shall later see, this is the case for virtually all advertising decision models, so that the degree of our faith in the solutions which are derived from the model can never be greater than the degree of validity of the advertising effectiveness assumptions.

The ramifications and importance of measuring advertising's effect on sales were vividly put by Dr. Alfred N. Watson, vice-president of Alfred Politz Research, Inc., in a talk before the Operations Research Society of America in 1960:

. . . the solution (measurement of the effect of advertising on sales) would offer rewards of fabulous proportions to the successful operations researcher and his employer. The possession of a successful formula in the hands of one company could literally spell corporate death for its competitors. . . . Perhaps no discovery short of that of a formula for determining future stock prices would have greater impact on American business, the course of our national economy or of the economy of the person who first discovers it.[21]

These comments, and all of the foregoing discussion, presume that the effectiveness which we are attempting to measure is on a rather gross basis. We have concentrated our attention on the number of sales dollars which are generated by *advertising dollars*. Advertising effectiveness is presumably related to the number of dollars expended. However, and perhaps more importantly, it is also related to *the content of the advertising message*. In this treatment we shall focus first and primary attention on media effectiveness because, little as we know, more is known in that area than about message effectiveness. Subsequently we shall briefly discuss the latter.

MEDIA EFFECTIVENESS

The effectiveness of media advertising is defined by the degree to which it achieves the goals of the organization. In most cases, the most direct measure of this achievement is sales or profit. As we have previously indicated, the measurement of overall performance is often impossible or impractical; in which case, measurement of the degree of achievement of tactical subgoals is usually attempted. The subgoals are most frequently related to the communication function of advertising.

In some situations the direct measurement of the sales or profits resulting from advertising is particularly easy. For example, direct mail advertising for magazine subscriptions requiring the return of a coupon necessitates only a count of the returns and an adjustment for nonpayers and cancellations in order to estimate the resulting profit. In other cases, such as coupon returns for encyclopedias which elicit informational brochures and a call by a salesman, the separate effects of advertising and personal selling on sales are more difficult to distinguish. In the many situations in which the advertising and sale are not directly connected by any physical means, the determination of the sales response to advertising is even more difficult.

[21] A. N. Watson, "Can Operations Research Measure the Effect of Advertising on Sales?" talk presented to Operations Research Society of America, May 19, 1960. Quoted with permission.

General Forms of the Sales Response Function. We may consider several general forms which the total sales response function may take on, our objective being to clarify our understanding of the complex measurement problems of sales response rather than any dogmatic or exhaustive compilation. If the amounts allocated to n different media are $A_1, A_2, \ldots, A_n$, the general form of the sales response function is

$$R(A_1, A_2, \ldots, A_n)$$

i.e., sales revenue as a function of the various media expenditures.

If different media are directed toward various consumer groups—e.g., *Fortune* to businessmen, *McCalls* to housewives, KDKA to Pittsburgh, and WBZ to Boston—the form of this function might be additive, i.e.,

$$R(A_1, A_2, \ldots, A_n) = R_1(A_1) + R_2(A_2) + \cdots + R_n(A_n) \qquad (9\text{-}27)$$

where $R_1(A_1)$ is the revenue generated by the first medium (which is directed toward a particular consumer group), $R_2(A_2)$ is the revenue generated by the second medium (directed toward another group), etc. This is the form of the function which we used for an illustration in the first allocation model

$$R = 2 \ln A_1 + 4 \ln A_2$$

and the assumption necessary for its use is that A_1 and A_2 have no interactive effect—that the effect of each dollar on medium 1 is the same regardless of the level of expenditure in the second medium. This would not be the case if some of the audience were duplicated and the media reinforced or detracted from one another. Note, however, that we are *not* assuming in such a representation that the effect of each advertising dollar is constant—only that its effect is itself unaffected by the level of expenditures in the second medium and, of course, that the converse lack of effect is also valid.

The second general form of the advertising response function which has both intuitive and logical appeal may be represented as

$$R(A_1, A_2, \ldots, A_n) = R(a_1 A_1 + a_2 A_2 + \cdots + a_n A_n) \qquad (9\text{-}28)$$

i.e., the sales response is some function R of an additive relationship across the various media. This is the case which we illustrated in the specialized model in which the total expected effective exposures were of the form

$$a_1 A_1 + a_2 A_2 + \cdots + a_n A_n$$

where the a's were the effective audiences of the various media. Because we were not concentrating on the organizational performance measures, the form of the R function was unspecified.

This functional form may be interpreted as descriptive of the situation in which the appeals of various media are directed toward the same basic audience. In the form used previously, we were concerned with potential customers, as defined through market analysis, and our communications objective was to achieve the maximum number of *exposures* (regardless of duplications). Hence, the form of the expression for the total exposures in the group is obviously additive across the media. Of course, the form of the R function in (9-28)—the relationship between the total exposures and sales revenue—may take into account the effect of duplicated exposures.

If several media are directed to several groups, the form of the response function might well be a combination of these two. Suppose that two media (1 and 2) are directed at consumer group I and three media (3, 4, and 5) at consumer group II. The resulting form of the function would be

$$R_I(a_1A_1 + a_2A_2) + R_{II}(a_3A_3 + a_4A_4 + a_5A_5)$$

In this relationship each group's sales response has the additive form over the media directed to it and the total response of the two groups is simply the sum of the individual group responses.

Finally, if there is an interdependence which is so great that it must be explicitly accounted for, the general form (for two media) which is applicable is

$$R(A_1, A_2) = R_1(A_1) + R_2(A_2) + R_3(A_1A_2)$$

where the third term (involving the *product* of A_1 and A_2) represents the interactive effect. This effect may be positive if the two media are reinforcing or negative if one detracts from the other. This is the form we used in an earlier illustration of media allocation.

$$R = 2A_1^2 + 4A_2^2 - 0.2A_1A_2$$

There, the effect was negative, indicating that the media detract from each other's effect.

The idea of an *interactive* effect is represented by a product because an interaction means that the effect of a change in one decision variable depends on the absolute level of the other. In the above expression the effect of changing A_1 from 100 to 200 is

$$[2(200)^2 + 4A_2^2 - 0.2(200)A_2] - [2(100)^2 + 4A_2^2 - 0.2(100)A_2]$$

or

$$2(200^2 - 100^2) - 20A_2$$

which clearly does involve the level of A_2. Therefore, the effect of changing A_1 from 100 to 200 depends both on the change itself *and* on the level of A_2; i.e., the two are interactive.

In the case of the function used earlier

$$R = 2 \ln A_1 + 4 \ln A_2$$

the effect of a similar change is

$$(2 \ln 200 + 4 \ln A_2) - (2 \ln 100 + 4 \ln A_2)$$

or

$$(2 \ln 200 - 2 \ln 100)$$

which depends only on the change itself and not on the level of A_2. Hence, this effect is noninteractive.

The implications of these general forms are important from the viewpoint of the manager who is interested in estimating advertising response by some relatively simple means. If the *forms* of the various component functions are not too complex, each may be attacked by using least squares, for example. (The reader will recall that we demonstrated in Chapter 8 that we were not restricted to purely linear relationships with that technique, but only to those in which the *coefficients* entered linearly.) In most cases the general media situation permits one to choose between the two basic additive forms; e.g., if there is separate promotion to various groups or general media promotion of the same group, the choice is clear. Then, the second basic question has to do with interactive effects. If these are believed to exist, consideration should be given to incorporating multiplicative terms. Lastly, the form of the component functions must be specified, e.g., logarithmic, quadratic, etc.

Dynamic sales response. The form of sales response as a function of time is also significant in the determination of media effectiveness. The basic dynamic model of advertising, expression (9-8), may be used to shed light on this aspect. The model is

$$\frac{dR(t)}{dt} = r \frac{M - R}{M} A(t) - \lambda R(t)$$

where $R(t)$ is sales revenue rate, $A(t)$ is the advertising expenditure rate, r is the sales response constant, M is the sales saturation level, and λ is the sales decay constant (all as previously defined).

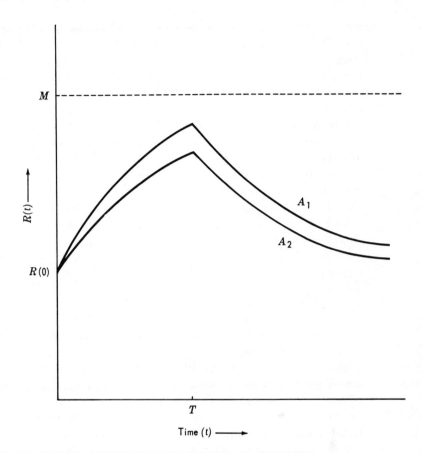

FIGURE 9-7 TEMPORAL PATTERN OF SALES RESPONSE TO ADVERTISING

For a *constant* advertising expenditure rate during the period from $t = 0$ to $t = T$, the sales revenue rate is[22]

$$R(t) = \frac{MrA}{rA + \lambda M}(1 - e^{-(rA/M+\lambda)t}) + R(0)e^{-(rA/M+\lambda)t} \qquad t < T \qquad (9\text{-}29)$$

where $R(0)$ is the sales revenue rate at $t = 0$. If advertising stops at time T, sales revenue decreases exponentially:

$$R(t) = R(T)e^{-\lambda(t-T)} \qquad t > T$$

which is just expression (9-6) with the time origin shifted to T.

The sales response to an advertising campaign of constant expenditure intensity and of duration T is shown graphically in Figure 9-7. Sales revenue rate increases most rapidly at $t = 0$, and as

---

[22] A technical note at the end of the chapter gives details of obtaining this solution.

sales revenue saturation approaches, the increase is progressively less. When advertising is discontinued at time T, sales begin to decay exponentially. [The two curves shown depict $R(t)$ for two different constant expenditure rates, A_1 and A_2, with A_1 greater than A_2.]

Experimental Determination of Media Effectiveness. In discussing demand estimation in Chapter 8, we briefly noted experimental approaches, those in which prices were purposefully varied in order to measure the resulting effect on sales. Then, we discussed the uses of least squares and other statistical ideas in demand estimation. Here, in measuring advertising effect, we are faced with the same kind of problem, but we shall not emphasize the same aspects. In estimating the sales response to advertising, it is conceptually (and sometimes operationally) feasible to use all of the statistical approaches which were discussed in Chapter 8 in the context of demand estimation. Indeed, we have alluded to this in the preceding section of this chapter. However, it is often more practical to emphasize *experimentation* in measuring advertising effectiveness.

The basic distinction between these two approaches should be kept clearly in mind. In applying statistical devices such as least-squares models, we are usually analyzing, *after the fact,* changes which have occurred in some controllable variable (e.g., price or advertising expenditure) in relation to changes in a relevant performance variable (e.g., sales revenue). Although the price or advertising variable is controllable by the decision maker, the analytic process occurs after all controllable aspects have been set and performance results have been determined. The least-squares analyst must take the best of that which is given him, even though he might prefer to have other data to analyze. For instance, if the past changes in the controllable variable for which he has data have been rather small, he might desire to have data covering a greater range of values. Since he is working with past happenings, however, his desire is of no avail.[23]

Experimentation, on the contrary, is *controlled* on an a priori basis. Not only is the advertising expenditure controllable in the usual sense, but it is controllable for purposes of the analysis to follow. In much the same way as we viewed test markets in two ways in Chapter 6—as laboratories for learning about the product and as full-fledged market areas—we view an advertising outlay which is incorporated into a controlled experiment as being di-

---

[23] The statistical analyst will note that we are relegating least-squares analysis to the realm of observational, as opposed to experimental, procedures. Although doing this is not theoretically valid, from an operational point of view, in marketing analysis this is usually the explicit role which it plays.

rected to the basic objectives of the firm *and* to a learning objective. The learning, of course, is about the sales response to advertising. As the reader will note, the ideas of experimental determination of advertising effect and those of test marketing are closely related. Indeed, the ideas to be developed here may be directly applicable to test marketing. We shall discuss this interaction of test marketing and experimentation in a later section.

Experimental objectives. Before determining any experimental plan for analyzing media advertising effectiveness, we should consider our experimental objectives. The overall objective, of course, is to obtain an estimate of the sales response to advertising *which can be used for decision making purposes.* There is no virtue in experimentation for its own sake (at least, there is no profit to the business firm). In some advertising experiments this simple point has not been understood, or at least not applied. In those cases with which the author is familiar, the results of such an oversight were exactly as might be expected. Since the experiments were not set up to account for the use of the possible results in making decisions, the experiments had no discernible impact on decisions. Of course, such experiments, other than exploratory experiments which may be conducted as basic research, should be avoided by business firms and advertising agencies. Nothing will deter the rapidly growing use of science in marketing so much as corporate expenditures on scientific activities which do not produce meaningful and useful results. Therefore, *the prime consideration which should be given to an advertising experiment is its interpretation in terms which are meaningful to decision making.*

Experimental design. Before proceeding with an experiment, it is necessary to consider the design of the experiment. The term "design" is used here in a broad sense which includes both formal techniques of experimental design[24] and the specialized planning and analysis which must be undertaken *before conducting an experiment.* It is this prior analysis which is most often omitted by the advertising experimenter who, in his haste to "learn," forgets what the learning is all about. This omission, together with the neglect of the basic decision objective, is undoubtedly the prime cause of the vast proportion of unsuccessful advertising experiments.

The most significant aspects of the experimental design have to do with *analyzing the accuracy of the experiment in advance and expressing the results in terms which are meaningful to the decision*

---

[24] See the Cochran and Cox text for a comprehensive treatment of experimental design.

maker. We shall demonstrate these aspects in the first illustration, which deals with the estimation of the *slope* of the sales response curve.

Estimation of the slope of the sales response curve. The simplest technical goal of advertising experimentation is to estimate the slope of the sales response curve. This is analogous to the estimation of the price elasticity of demand, as discussed in the previous chapter. The simplest way to estimate this slope experimentally is with a two-level spending experiment. Let us suppose that current advertising expenditures (on a per capita basis) are about equal from area to area. Call this level a_1. The worth of an increase to the level a_2 is to be evaluated experimentally. The emphasis on an *increase* would probably be due to an outside influence such as a proposal to increase advertising. In effect, the experiment would be focused on evaluating the worth of such a proposal.

To perform the experiment, one needs to select two groups of areas—a *control group* and an *experimental group*. The experimental group will be "treated" with the high advertising level a_2, and the control group will remain at the level. a_1. Two bases which determine the composition of these groups are *matching* and *randomizing.*

Matching is a process of selecting *pairs* of areas which are similar with respect to some uncontrollable factor which the experimenter believes might affect the sales response demonstrated in the areas For instance, if the interviewer feels that the income level of an area might have something to do with the response to advertising, he wants to be very sure that he really measures the effect of advertising and not simply the effect of income level. If he were to select market areas randomly, he might find that he has selected all high-income areas for the experimental group and all low-income areas for the control group. If this were so, he would not know whether any measured effect was caused by advertising or income. To avoid this ambiguity, he picks matched pairs of areas— pairs matched according to similar income levels.

It is possible, however, that there are other uncontrollable factors, unknown to the experimenter, which also affect the sales response to advertising. For instance, climate might be such a factor. To deal with this problem, the experimenter *randomizes;* he assigns one member of each pair to the experimental group and one to the control group in a random fashion. For instance, he might just flip a coin for each pair, with heads representing assignment to the control group and tails assignment to the experimental group. In this way he is assured that *differences which exist between the control and experimental groups* (other than the advertising expenditure difference which is to be imposed by the experimenter) *will be due to chance alone.*

The results of the experiment will be determined by measuring the per capita sales revenue achieved in the experimental and control groups over some *predetermined* time interval. Of course, a per capita measure is necessitated by the varying populations of the areas.

The purpose of the experimental group of areas is clear: to produce any change in sales which may be caused by the advertising change. The purpose of the control group is to ensure that it is really the effect of advertising which is being measured. Suppose, for instance, that sales (per capita) increase in the experimental group. It would not be reasonable to attribute the change to the increased advertising if sales also increase by about the same amount in the control group. Probably we are just observing an overall increase which was caused by some other factor. The control group assures us that we know whether there are any factors which affect sales in *all* areas during the experiment.

Suppose that the increased advertising does increase sales ("on the average") in the experimental group. Such a situation is depicted in Figure 9-8, where each area's value of per capita sales revenue is given as a function of the per capita advertising expenditure level in the market. If the average (arithmetic mean) value of per capita sales revenue is $\bar{r}_1$ in the control group and the

FIGURE 9-8   TWO-LEVEL EXPERIMENT

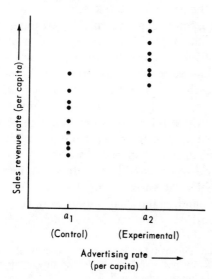

average is $\bar{r}_2$ in the experimental group, the *resulting advertising response slope estimate is*

$$\text{Slope} = \frac{\bar{r}_2 - \bar{r}_1}{a_2 - a_1} \tag{9-30}$$

i.e., the average dollars of added sales per dollar of added spending.

All of the analysis and planning which we have discussed should be conducted *prior to carrying out the experiment.*[25] Also, before the experiment, we should establish a sales response *goal. The goal will enable us to interpret the results of the experiment in a fashion which is meaningful to decision making.* Further, it serves as a criterion for evaluating the experimental results and as a guide for planning for the experimental accuracy which will be required. Such a goal is best set in relation to the decision problems of the organization. For instance, we know from the basic static model that the reciprocal of the *incremental profit as a fraction of the unit selling price* is a significant slope of the sales response function. Expression (9-5), which is the solution of the basic static model, tells us to increase expenditures until this slope is attained. Therefore, we may loosely consider that an experimentally determined slope which is greater than this value indicates that the increase from a_1 to a_2 may be desirable (since we are presumably operating on the steep portion of the curve shown in Figure 9-1) and that a lesser value indicates that the increase may not be warranted (since we might be operating on the flat portion of the curve). Of course, these statements completely neglect the errors involved in experimental determinations (which we shall discuss later), but they do serve to illustrate the general methodology.

The use of the sales response goal for further planning of the experiment requires that the probabilistic concept of a *standard deviation* be understood. The standard deviation of a random variable is simply a numerical measure of the dispersion (or variability) associated with values of the variable. Thus, a random variable which takes on a wide range of values will have a higher value of this measure than one which takes on a smaller range of values (all other things being equal). For instance, if the per capita sales revenue for various market areas had the distribution shown in Figure 9-9a, its standard deviation would be quite small, indicating that the possible values which the variable might assume are closely compacted. On the other hand, if the distribution were that shown in Figure 9-9b, the standard deviation would be much larger, indi-

---

[25] The reader should note that this can be done in the sense that we have not found it necessary to use any numerical results in the foregoing. We are planning, using symbols, for the results which will occur.

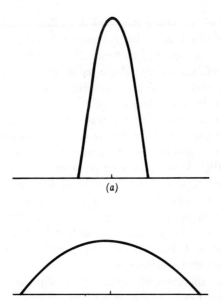

(a)

(b)

FIGURE 9-9   DISTRIBUTIONS OF PER CAPITA SALES REVENUE

cating the relatively wide dispersion of the values which the variable may take on.

When we calculate the slope indicated in expression (9-30), we are really *estimating* some true, but unknown, slope of the sales response curve. In this context, the standard deviation serves as a measure of the amount of random fluctuation which is associated with the particular estimate. The idea of random fluctuation in an estimate is given intuitive meaning when one considers that if other areas had been chosen for the control and experimental groups (or if the same areas had been assigned to the two groups differently), the numerical value of the slope estimate would likely be different. The degree to which slope estimates vary is measured by the *standard deviation of the estimate*, i.e., the standard deviation of the slope estimate.

All of this appears at first to indicate that we really cannot make a good slope estimate. This is not so at all. The point is that because of the variations which we know to exist, *we cannot make a perfectly accurate estimate of the true slope.* However, we can *make an estimate* and, together with it, *make an estimate of its accuracy.*

We are able to do this because some of the basic theorems of probability permit us to calculate the standard deviation of the slope estimate from a knowledge of the standard deviation of the basic random variable—per capita sales revenue. We probably would have a good idea of the latter standard deviation's numerical value because we have observed sales revenue values generated in all market areas over some period of time. The details of determining the standard deviation of the slope estimate are given in a technical note at the end of the chapter. Here, let us suppose that we have calculated it to be $0.50 per capita.

Now let us suppose that our sales revenue goal has the value 3.3. If the experiment is conducted and the resulting slope estimate is 3.25, we would be in somewhat of a quandary. We are trying to determine whether the slope is greater than or less than 3.3, and we realize that our slope estimate is subject to error. Since the slope estimate is not quite 3.3, the apparent answer is that we are beyond the optimal point on the sales response curve. We realize, however, that if another group of areas had been selected, the slope estimate might well have been 3.35, which would have resulted in an entirely different conclusion.

The resolution of this dilemma is simply to recognize our ignorance and to make the best possible use of the information we have. To do so, we simply define (*before* conducting the experiment) three ranges of the slope estimate. These ranges indicate the outcomes which we will attribute to the experiment. If the slope estimate is very small, say nearly zero, we are fairly sure that the true slope is less than 3.3. The reason for this is that we know that the standard deviation of the slope estimate is 0.50, and we know that the characteristic normal distribution of random variables which are means (discussed in the technical note at the end of the chapter) is such that it is extremely unlikely that an estimate of zero would result if the true slope were greater than 3.3. Moreover, if the slope estimate is 7.0, we are rather certain that the true slope is greater than 3.3 (using the same argument). For slope estimates near 3.3, we are unsure as to whether the true slope is greater or less than 3.3.

Therefore, the three ranges which we define for the slope estimate are in terms of (1) a range which almost certainly has a true slope greater than 3.3, (2) a range which almost certainly has a true slope less than 3.3, and (3) a range for which we shall express ignorance as to the exact numerical value (but within which we can still draw some conclusions which are meaningful to decision making). These ranges are best defined in terms of the response goal (3.3) plus and minus some number of standard deviations of the slope estimate. For instance, ranges might be those shown in Table 9-3.

This table indicates that if the slope estimate is more than two standard deviations from the goal, we shall conclude that the true slope is other than 3.3. If the slope estimate lies within the igno-

TABLE 9-3   POSSIBLE RESULTS OF EXPERIMENT

Range of slope estimate	*Conclusion*
Below 2.3	Slope probably less than 3.3
2.3 to 4.3	Ignorance range (?)
Above 4.3	Slope probably greater than 3.3

rance range, we can only conclude that the slope is likely to be in the general vicinity of the optimal slope (for, indeed, recall that 3.3 is the *optimal* slope). Our ignorance range is rather wide, and yet the managerial action implied by each possible result is rather clear. If the slope estimate is below 2.3, we should not increase advertising and should possibly consider reducing it. (We might conduct another experiment aimed at estimating the potential for a *reduction* from the a_1 level.) If the slope is within the ignorance range, we should probably retain the *status quo,* since we have no very good reason for doing anything else. If the slope estimate is above 4.3, we should consider an increase in advertising expenditure, for this indicates that we are likely to be operating on the steep portion of the response curve.

Note again that *all of this can (and should) be done prior to conducting the experiment.* One of the obvious reasons is that this is the only way we can be sure that the numerical slope estimate has not affected our determination of the ranges in Table 9-3. After all, if we know the numerical value of the slope estimate, there is enough arbitrariness in the procedure to permit us to set the result as anything we want it to be (assuming that the actual experimental slope estimate is not extremely close to 3.3). If we set the ranges up before the fact on a logical basis, we can be sure that our personal biases are not dictating the experimental result.

Of course, we are not necessarily restricted to two levels of spending in such an experiment. An experiment at three or more levels may be used for gaining insight into the curvature of the response curve. In particular, for a new product it is often desirable to determine the advertising response *threshold*—the level of advertising below which changes in the advertising expenditure make little or no difference. Of course, no one is certain that such a level exists for all products, but it almost certainly does for some.

Du Pont, one of the leaders in the experimental determination of advertising's effect on sales, conducted such a three-level experiment in 13 cities for Teflon cookware coating. For the control group there was no Teflon advertising. For the other two groups either five or ten commercials per week were run. No detectable increase in market share was observed for the "medium" level (five commercials per week), but market share doubled (on the average) for the markets with ten commercials per week. Du Pont subsequently doubled its advertising budget in a major national campaign. Thus, the experiment, with its apparent determination of a threshold level for advertising, clearly affected Du Pont's decision making processes and may have prevented the company from "wasting" money by spending below the threshold level.

Test marketing and experimentation. We have previously pointed out the similarities of experimentation and test marketing. A test marketing program is, in its entirety, an experiment which is aimed at estimating the effect of the product on consumers. In Chapter 6 we made note of the secondary learning objectives of test marketing, e.g., to test various prices, advertising campaigns, messages, and the like. There, however, the primary focus was on the product-oriented decisions. Should marketing be expanded? If so, where? Only in discussing the testing and control aspects of test marketing did we touch on the effect of advertising, and then the orientation was toward a goal other than direct measurement of advertising effectiveness in the same sense in which we are discussing it here. In that context we were concerned with more or less informal procedures for detecting gross changes in marketing performance. Here, the approach is more formal, since we are concerned with advance planning for advertising effectiveness measurement.

The selection of test areas was discussed in Chapter 6, using representativeness as a primary criterion. There, no problems of matching were encountered since the treatment to be applied in each area was identical. The treatment was the *package* consisting of the product itself and its associated controllables. The problems involved in randomly selecting market areas were extensively discussed in Chapter 6.

If test market areas are to be used for estimating the effect of media advertising, the first thing that must be recognized is that the effect which will be measured is unlikely to apply to anything but an introductory period. It is unlikely that the effect measured in a test market will bear any relationship to that which might be experienced in a stable market (and, indeed, in the same test markets after they become stable markets). Thus, the usefulness of such information may be rather limited. It is important, nonetheless, because introductory periods are so important to a product's future. Moreover, certain gross conclusions which can be made from advertising effectiveness measurements in test marketing may be generally applicable to the product. For instance, an indication that one medium is superior to another in the test markets may well have general applicability.

The way in which the ideas of representativeness, randomness, and matching can be integrated into the practicalities of a test marketing program should now be apparent. One can select the basic test markets on a *representative* basis or on some other basis dictated by the organization's makeup and objectives. *Then,* the test markets can be *matched* according to whatever outside influence is regarded as significant, and the matched pairs can be assigned to the control and experimental groups *randomly.* Although this process is not so simple to accomplish as it is to describe, it is the best way to integrate the practicalities of the real world and the theoretical necessities for obtaining rigorous results.

Other more complex experimental designs are available which can be used for advertising experimentation, both in conjunction with test marketing programs and for advertising effectiveness measurement for established products. Most of these have proved useful in the wider context of *promotional* effectiveness. Hence, we shall defer consideration of them to the next chapter, which deals with other forms of sales promotion.

Experimentation using consumer panels. It is also feasible to conduct advertising experiments in which the measurement of results is accomplished through a consumer panel. The most extensive use of a panel for this purpose is in a project called the Milwaukee Advertising Laboratory. Since this project deals with most of the salient points of the subject, we shall describe it in some length.

The MAL consists of two groups of households in the Milwaukee area that record their purchases of 41 specified product categories in a weekly diary. There are 750 matched households in each group, the matching having been accomplished on the basis of geography and demography. *Both newspaper and television advertising to these groups is strictly controlled.* Both Milwaukee newspapers can deliver *different* newspaper ads to the two groups as desired, and muting devices have been installed on the television sets in each panel home. The muting device blanks out television commercials when activated by the transmitters of the local TV stations. Thus, both the newspaper and the television messages to which each group is exposed can be controlled, and the groups' purchase records are available through the diaries.

Such a system makes possible a wide range of media experimentation. One can test the effectiveness of the *timing* of advertising, the relative effectiveness of media, the effectiveness of various sorts of messages, etc. The possibilities are virtually limitless. Of course, the kind of product whose advertising is to be tested must have a purchase rate which makes it feasible to use consumer diary purchase records. For instance, detergents, toothpaste, and cereals would appear to be ideally suited to such measurement, but life insurance and refrigerators would not.[26]

The primary feature which sets MAL off from other panel testing programs is the availability of a matched control group. It has been possible to establish a control to some degree in printed media "split runs"—the printing of different ads in two "editions" of a newspaper or magazine issue—but never previously has a carefully matched control group been available for television advertising effectiveness testing.

Although this experimental method is not yet accepted by all in the field of advertising, it clearly has tremendous implications. Since it utilizes less than two-tenths of 1 percent of the total audience in any "blackout" situation, one can perform advertising experimentation while simultaneously conducting a normal marketing program in the area. The need for very high or very low expenditures, as might be required to obtain the necessary information from another kind of experiment, is virtually eliminated. Of course, the establishment of such an experimental program requires agreement among numerous sources, the dissent of any one of which may seriously impair its effectiveness.[27]

MESSAGE EFFECTIVENESS

Most measurement directed toward the determination of advertising message effectiveness—sometimes called "advertising research" or "copy research"—takes place at the communications level of the organization's goals. In such instances the primary concern is whether certain themes and layouts are more effective than others *with respect to memorability, interest, attractiveness, understandability, etc.* The implicit assumption is that more effective communication (defined in these terms) contributes to the goals of the organization. As we have already pointed out, while this seems to be a plausible assumption, there is no existing proof of its validity. In fact, there are those in advertising who believe that there is no connection between the communications and sales objectives of advertising. Jack Haskins, an Indiana University researcher, has been quoted as saying that "learning and recall and retention measures seem, at best, irrelevant to the ultimate effects desired, the changing of attitudes and behavior. This conclusion is based on a review of seven advertising studies and twenty-one published

---

[26] See the Kroeger reference for further details on MAL.
[27] At the time of this writing, one Milwaukee television station is not participating in MAL.

non-advertising studies."[28] This is clearly an extreme position. Most advertisers have a favorable reaction to the measurement of communications variables. Yet, the controversy is real and clear.

The objective of advertising has been depicted as moving consumers through successive states. Before a product is advertised, the consumer is *unaware*. He then is brought through successive states of *awareness, comprehension, and conviction* to the state of *action* in which he makes a purchase. The measurement of message effectiveness is usually directed toward the detection of the middle three of these states—awareness, comprehension, and conviction. The techniques by which such measurements are accomplished differ considerably. Effectiveness tests are conducted before the advertisements are presented to the public, using opinion and attitude ratings, projective techniques (techniques of clinical psychology aimed at providing insight into the motivational patterns of consumers), and content analyses. Subsequent to the appearance of the advertisement, measurement is conducted through memory tests, tests involving attitude changes, etc.

From the viewpoint of quantitative analysis, the two most striking features of message effectiveness measurement are its almost complete *lack of direct association with the sales goals of the organization* and the *measurement difficulties which often preclude numerical measurement of any kind*. There is a good deal of quantitative literature concerned with *ways to use various measurement techniques*. For instance, survey sampling is based on the scientific methods of statistics and is considered a primary device for measuring message effectiveness. Yet, the quantitative analysis which has been performed in this area is largely oriented toward methods for obtaining information (in terms of bias and reliability) rather than toward the *use to which the information will be put*. Since our emphasis here is on decisions and methods for analyzing decisions, these techniques are not germane.

It should be emphasized that we are not decrying the significance of message effectiveness or of the techniques which are currently in use in that area. There is no question that quantitative analysts have paid too little attention to the advertising message. In concerning themselves with the dollars expended on advertising, they have too frequently forgotten what the advertising said. Moreover, one is led to believe that they have not chosen to be aware of the possibility that message content can, in some instances, be of such overwhelming importance that questions of budgeting and media allocation are relatively trivial. Nonetheless, any investigation into this area here would open a Pandora's box which would never

---

[28] *Sales Management,* Feb. 19, 1965, p. 66.

allow us to return to the mainstream of marketing management and decision making. We shall simply refer the interested reader to the Lucas and Britt book in the references (for an extensive discussion of the methods of message effectiveness measurement) and to the Green and Tull text (for treatment of the statistical design and interpretation aspects of message effectiveness measurement).

Competitive Advertising Strategies

Each of the models of advertising decision situations which we have discussed has omitted consideration of one of the basic elements of any decision situation—*competitive actions*. In this section and in the succeeding chapter we shall emphasize the role of competition in promotional decisions. As we pointed out in Chapter 3, whether one wishes to view the outcome array as being of the strategy–state of nature, strategy–competitive action, or strategy–generalized state of nature variety depends on the particular decision situation at hand. In some cases the effect of competition may well be neglected, and in others it may be of paramount importance. Here, we wish to develop enough of the methods for the situation in which competition is of significance so that the reader will be able to choose his conceptual basis wisely and to recognize the advantages and limitations of his analytic method. Moreover, the change in viewpoint from the consideration of the firm in virtual isolation, which has been a feature of most previous models, to the consideration of competitive actions as one of the two major elements which impinge on the decision situation is usually enlivening and useful. The structuring of problems in the competitive framework serves to give new insight into the scope of business decision situations.

THE ADVERTISING GAME

The problems of both advertising budgeting and allocation may be approached by a methodology called *game theory*. This theory, first developed by Von Neumann and Morgenstern (see references), deals with economic situations in which the competitive aspect is paramount. Hence, our section title—the advertising game.[29]

[29] There is some relationship between the kind of game we are discussing here and parlor games such as chess, checkers, and poker. However, another kind of game which is spoken of in the literature of management is entirely unrelated. This game is a *simulation* of a business environment which usually has some sort of learning or training as its objective.

The Basic Game Model. To illustrate a simple advertising game, let us consider that the sales which our organization and its single competitor achieve in each market area depend on our advertising allocation *and* the advertising allocation of our competitor. Further, let us assume for simplicity that we can predict the sales *due to advertising* which will result from various allocation combinations, and that both we and our competitor have only two possible strategies in each of two markets: we both may allocate two-thirds of our respective budgets to either area 1 or area 2 (and therefore one-third to the other). Some of these assumptions are quite unrealistic, of course. Nonetheless, they are descriptive of some real-world situations (say, one in which a budget was sufficient only to buy three indecomposable advertising units and a failure to advertise in a market was not acceptable to the firm), and they provide insight into the nature of the competitive problem.

Suppose the outcome array *for our firm* is that shown as Table 9-4. Two strategies are available to us—"allocate two-thirds to

TABLE 9-4   OUTCOME ARRAY IN TERMS OF OUR SALES GAINS

Strategy	C_1 (*Allocate* $\frac{2}{3}$ *to area* 1)	C_2 (*Allocate* $\frac{2}{3}$ *to area* 2)
S_1: allocate $\frac{2}{3}$ to area 1	\$5	\$2
S_2: allocate $\frac{2}{3}$ to area 2	\$3	$-\$2$

area 1" and "allocate two-thirds to area 2"—and the same two are available to our competitor. The outcome descriptors in Table 9-4 may be thought of as the total sales (due to advertising) accruing to *us* in each of the outcome situations. If both we and our competitor allocate two-thirds of our respective budgets to area 1, the sales *gained* by us through advertising *in both markets* is \$5 million, and if we both allocate two-thirds to area 2, our organization will *lose* \$2 million in sales.

But to whom do we lose the sales? To our competitor, of course. Therefore, our competitor's outcome array is one in which each entry is just the negative of the corresponding entry in our array. A situation such as this, in which one competitor's gains are the other's losses, is called *zero-sum*; i.e., the sum of gains is always zero. This is the situation which exactly applies if you and I were to bet a dollar on the outcome of a football game, since if I win, my gain is \$1 and your gain is $-\$1$, for a net of zero. However, it is not necessarily the case that this will apply equally well to advertising. It is certainly possible that some net gains can be the result of an advertising game, i.e., that a particular combination of strategies can result in an outcome in which *both competitors gain* something. This might be the case if advertising is effective in creating demand for a new product category. However, it is certainly possible that a product may have a stable market in which few net gains or losses are made by any seller, the only significant changes being switches from brand to brand which result in gains for some and losses for others.

If such is the case, and if we are willing to use "sales gain" as a proxy for utility, we may view the outcome array of Table 9-4 as depicting a decision problem under uncertainty. We select the uncertainty case because it is unlikely that we will have any meaningful information concerning the likelihood that our competitor will select any particular strategy.

The decision criteria for uncertainty which were discussed in Chapter 3 might be called on in this case. However, although we may apply the same basic reasoning which was used there, the zero-sum competitive characteristic of this decision situation sets it apart as a special case of decision making under uncertainty. Here, rather than being faced with an indifferent nature, or a competitor whose desires are not clearly known, we know that a *rational being* (although perhaps it would be more pleasant if we did not characterize him as such) *is consciously attempting to hurt us as much as possible in this situation.* Why is he doing this? Of course, the answer is that whatever we win, he loses, and vice versa. We do not have here any cozy situation of "perfect" competition (in the sense discussed in the pricing chapter) but rather of cutthroat competition.[30]

Since we are aware that our competitor will do whatever he can to hurt us in this decision situation, we may do well to extend our analysis of the problem a bit beyond a simple application of the Laplace, Hurwicz alpha, or some other uncertainty criterion. To show why we should not apply one of these criteria hastily, suppose that the outcome array of Table 9-5 were applicable and that we were going to use the Laplace criterion.

Under the Laplace criterion we assign equal likelihood to C_1 and C_2 and calculate the expectations

$$E(S_1) = \tfrac{1}{2}(\$14) + \tfrac{1}{2}(-\$6) = \$4$$
$$E(S_2) = \tfrac{1}{2}(\$10) + \tfrac{1}{2}(-\$4) = \$3$$

Since $E(S_1)$ is greater, the criterion tells us to select strategy S_1.

[30] The economist describes a market environment in which each firm must simply accept the prevailing equilibrium price as "perfect" competition. The degree to which this terminology is in conflict with the layman's view of competition and the cutthroat competition discussed here is apparent. (Luce and Raiffa call this situation "strict" competition.)

However, a quick look at the outcome array from the point of view of our competitor reveals that this is not a reasonable choice. Our opponent gains by selecting C_2, whatever strategy we choose, and loses under C_1, whatever strategy we choose. (Recall that the outcomes are in terms of *our* gains. Hence, a positive number is a loss to our opponent.) Being both rational and too clever to allow a formal criterion to lead him astray, our competitor will almost certainly select C_2, for to do otherwise would result in a certain loss, and to select C_2 will result in a certain gain.

Knowing that our competitor is likely to perform such an analysis, and believing that he will conclude that he should select C_2, we should obviously decide to choose S_2, rather than S_1 as the Laplace criterion told us. We would select S_2 because by doing so, we lose only $4 million rather than the $6 million which we would lose under S_1. Rather than thinking that our criterion has led us astray, however, we should consider that it was never meant to cope with the zero-sum game situation. In a cutthroat game

TABLE 9-5   OUTCOME ARRAY

Strategy	C_1	C_2
S_1	$14	−$6
S_2	$10	−$4

situation such as we have illustrated, one should analyze *both* the alternatives (strategies) available to the decision maker and those available to the competitor. In effect, we analyze *two* decision situations—ours and our competitor's—and try to outwit him so that a reasonably good outcome is achieved.

Let us apply the maximin gain (minimax loss) *reasoning*—but not the criterion—which was discussed in Chapter 3 to the decision problem illustrated by Table 9-4. In doing so, we shall view both our problem (S_1 versus S_2) and our competitor's problem (C_1 versus C_2). First, in our decision problem, we note that the worst outcome for both strategies is that in the right column—a gain of $2 for S_1 and a loss of the same amount for S_2. Using a maximum approach, we should choose S_1 *to maximize our minimum gain.* The maximin gain will be the $2 gain associated with the upper right outcome.

Now, viewing our opponent's problem in the same fashion, we look at Table 9-4 and think in minimax terms (since the numbers in that table are *losses* to him). He would therefore isolate the worst outcome for each strategy (the highest number in each column) and select the best (smallest) of these. The highest values in each column are $5 and $2 respectively and the lowest of these is the latter. Therefore, our competitor's minimax strategy is C_2 with its associated minimax outcome of $2.

But a strange and wonderful thing has taken place! Both we and our competitor have arrived at the same outcome through our analysis. (Really, of course, it is our analysis of what his analysis may be.) And it is indeed strange and wonderful, for consider what happens if we begin at this point and *inform our competitor of the fact that it is the maximin strategy which we shall choose.* In other words, we phone our competitor and tell him that we have decided to choose S_1. Surely it would be foolhardy to give him such competitive advantage. But is it really? Looking at Table 9-4, we see that *this information will have no effect on his action.* By his minimax choice of C_2, our competitor is already assuring himself of the least loss under our strategy S_1, and he would make no change if he knew with certainty that we would choose S_1. Similarly, if he tells us that he is going to choose C_2, it has no effect, for our choice of S_1 has already dictated that we obtain the maximum gain which is possible under his choice of C_2.

Hence, the overall result of our analysis is that:

1. We can assure ourselves that we will gain at least $2 by choosing S_1 (regardless of our competitor's choice).
2. Our competitor can assure himself that he will lose no more than $2 by choosing C_2 (regardless of our choice).
3. If either chooses any other strategy, no such assurances can be obtained. Therefore, being prudent businessmen, both of us should accept this analysis and take the consequence—a $2 gain by us, and therefore, a $2 loss for him.

But what if our competitor (dastard that he is!) won't play by these carefully constructed rules? The answer is that if he doesn't, he has everything to lose and we have everything to gain (as long as we do abide by the rules). Table 9-4 shows that if he plays dirty and chooses C_1 instead of C_2, he loses (and we gain) $5 instead of $2. Similarly, if we neglect the rules and choose S_2 rather than S_1 (and our competitor follows the rules and chooses C_2), we will lose $2 rather than gain $2. Therefore, since both of us are rational, we will both abide by the minimax solution: we will choose S_1 and our opponent will choose C_2.

As one might suspect, this fascinating situation does not always arise in such games. When it does, its appearance depends on a special characteristic of the outcome array. However, when it does occur, the best strategy, according to the minimax argument, is clear. Of course, this assumes that *all* of the information which is available is depicted in the outcome array and that our competitor has exactly the same information and knowledge as we do. This is not always the case, because the entries in the outcome array represent predictions of things which are yet to occur. In any particular problem situation, it is possible that we and our competitor may have different estimates or different models of the problem situation.

The situations in which the previous analysis and results hold are called *saddle point* outcome arrays. Such arrays are characterized by an outcome element which is *simultaneously* the lowest value in its row and the highest value in its column. Note that the upper right entry in Table 9-4 had this property. This element is called the saddle point of the array. In such cases *it is best for both competitors to select their respective minimax strategies and to settle for the saddle point outcome.*

Mixed Strategies. If the outcome array does not display a saddle point, some additional analysis is necessary. In such a case, we must introduce the concept of a *mixed strategy*. A mixed strategy is *a rule for selecting one of the available (pure) strategies according to a probability distribution.* In other words, a mixed strategy

TABLE 9-6   OUTCOME ARRAY FOR ADVERTISING ALLOCATION

Strategy	C_1	C_2
S_1	$3	$1
S_2	$2	$4

in the game shown in Table 9-4 would be any two numbers, each between 0 and 1, which sum to unity. Hence, each of the pairs $(\frac{1}{2}, \frac{1}{2})$, $(\frac{1}{4}, \frac{3}{4})$, $(0.60, 0.40)$, and $(\frac{7}{8}, \frac{1}{8})$ would be feasible mixed strategies if they specified the respective probability of choosing S_1 or S_2.

But why would anyone ever use a mixed strategy? In doing so, the decision making authority would be ceded to a random device—a blatantly ridiculous thing for the marketing manager even to consider! However, it may not be so ridiculous at all. Let us consider the previous game situation with a new outcome array as shown in Table 9-6. In this array there is no saddle point, so the previous analysis is of little help. Our maximin strategy is S_2 and the maximin outcome is a gain of $2 million. Our opponent's minimax strategy is C_1 and his minimax outcome is the *loss* of \$3 million indicated at the upper left. Hence, if both we and our opponent blindly accept the maximin (minimax) reasoning, we will select S_2 and he will choose C_1. But let us first analyze his decision problem *as we believe he may analyze it.*

Our competitor's security level[31] is maximized by a choice of C_1. Also C_1 is a good action for him to choose against S_2 (since the result from C_2 is worse). Hence, we might believe that our opponent is likely to select C_1. If that is so, we should choose S_1 and get \$3 million rather than \$2 million gain. However, if our opponent *analyzes our analysis of him,* he will figure that we will choose S_1. Then, he will actually choose C_2 (because it is better for him under our choice of S_1). That being the case, we should choose S_2 rather than S_1. But our competitor will then think that . . . , etc. One can see that this kind of argument is *cyclic* and *endless.* Its basic nature—"if he thinks that we think that he thinks that we think . . ."—can easily lead us to a psychiatrist rather than a solution to our advertising problem.

The key to resolving this apparent dilemma is that *we must withhold from our competitor knowledge of the strategy which we will use.* If at any time he knows (or can figure out) what we will do, he can take advantage of it. Of course, in this simple situation we cannot but win. But he can take advantage of any knowledge to ensure that we get no more than the minimum possible for our efforts. Since the same statements may be applied to us by him, our opponent also desires to withhold knowledge of his strategy from us.

One way to conceal the strategy which will be chosen is to choose it randomly. Although this is a decision making technique which would at first appear to be anathema to the scientist, it is obviously an effective way of preventing our opponent from outguessing us, and, as we shall show, it also serves to raise our (expected) security level. Additionally, since we shall choose the likelihoods which make up the random device which will be used to select a strategy, we have not really ceded our authority to chance.

Although the secrecy argument is the primary one for mixed strategies, there is another school of thought which treats the psychology of the situation. Suppose, the argument goes, that we set up a random device, such as a carnival spinner, to select our "pure" strategy. On any particular spin the security level of the indicated pure strategy may be poor indeed. So we have two choices: to be obliged to follow the spinner's result or to be able to override

---

[31] Recall that security level is the lowest outcome (or utility) of which one can be certain regardless of the strategy chosen by our competition or nature.

it and select a better strategy. But, of course, if we do allow ourselves to override the random device, we are really back to the "if he thinks that I won't follow the random device, then I think that . . . ," kind of cycle again. Psychologically we should prefer to follow whatever the spinner tells us to do (remembering that *we* have selected the probabilities) because this course does not permit us to fall prey to our own human frailties. As Luce and Raiffa put it:

It is not unlike the person who wants to go on a diet. He announces his intention, or accepts a wager that he will not break his diet, so that later he will *not* be free to change his mind and to optimize his actions according to his state at *that* time—e.g., to eat an ice cream sundae.[32]

Now let us consider the choice of the probabilities in the mixed strategy, for once we decide to use a mixed strategy, these probabilities become our decision variables. Suppose that our competitor selects C_1 and we choose between S_1 and S_2 by flipping a coin. We then receive $3 million or $2 million, each with probability $\frac{1}{2}$. Since both of these outcomes are at least as great as our maximin security level of $2 million, this is clearly preferable. If C_2 is chosen by our competitor, our expected gain is $$\frac{5}{2}$, which is better than our security level. Of course, since these are dollars and not utilities, the arguments of Chapter 3 tell us that we cannot base decisions on such calculations. But if we assume that the dollar gain is a valid proxy for utility, we can reasonably do so.

If we decide to choose *a* mixed strategy (not necessarily the equiprobable one corresponding to a coin flip, but some *best* mixed strategy), we may represent our feasible mixed strategies as

$$(p_1, 1 - p_1)$$

where p_1 is the probability of selecting S_1 and $1 - p_1$ is the probability of selecting S_2. Our *expected sales gain, if our competitor chooses C_1*, is

$$3p_1 + 2(1 - p_1)$$

or

$$p_1 + 2 \tag{9-31}$$

If our opponent chooses C_2, our expected sales gain is

$$1p_1 + 4(1 - p_1)$$

or

$$4 - 3p_1 \tag{9-32}$$

[32] R. D. Luce and H. Raiffa, *Games and Decisions,* John Wiley & Sons, Inc., New York, 1957, p. 75.

The (expected) security level of the mixed strategy is simply the lesser of these quantities. If the value of p_1 is $\frac{1}{4}$, for example, expression (9-31) has the value $\frac{9}{4}$ and expression (9-32) is $1\frac{3}{4}$. Hence, the *expected* security level is $\frac{9}{4}$. If the maximin *reasoning* is applied *to these expectations,* the result is that we should *choose p_1 so that this expected security level is as large as possible;* i.e., we should *maximize the minimum expected sales gain.*

If we let u represent this (unknown) value which is the maximum of the minimum expected sales gains for all values of p_1, we know that

$$p_1 + 2 \geq u \tag{9-33}$$

and

$$4 - 3p_1 \geq u \tag{9-34}$$

We know that these inequalities hold because the left side of each inequality is an expectation of sales gain for one of our competitor's possible actions and *u is the smaller value of the two, for a given p_1. We wish to choose p_1 so that this smaller value is as large as possible.*

If we graph the equalities corresponding to expressions (9-33) and (9-34) together with the lines $p_1 = 0$ and $p_1 = 1$ to account for the inequality ($0 \leq p_1 \leq 1$), since p_1 is a probability, in the same fashion as we plotted the inequalities in the specialized media selection model, we get the situation shown in Figure 9-10.

Now, we wish to select p_1 to maximize the minimum expected sales gain. For any value of p_1, the expected sales gain if our competitor chooses C_1 is represented by all points on the line which slopes upward to the right in Figure 9-10. This is the plot of the equality corresponding to expression (9-33). Hence, it is labeled "C_1." Similarly, because the right-downward sloping line represents the equality corresponding to (9-34), it is labeled "C_2"; i.e., it is the expected sales gain if our competitor chooses C_2.

The smaller of these two expectations, for any value of p_1 between 0 and 1, is denoted by the circles. It is easy to see that this minimum value is represented by the expectation for C_1 for one range of p_1 and by the expectation for C_2 for another range of p_1. For example, if p_1 is $\frac{1}{4}$, our expectation under our competitor's choice of C_1 is the point on the line labeled "C_1" which is above $p_1 = \frac{1}{4}$. This point is labeled "A" in the figure. Our expectation under our competitor's strategy C_2 is the corresponding point on the line labeled "C_2," B. Therefore, the minimum of the two is the expected sales gain under C_1, A. For $p_1 = \frac{3}{4}$, exactly the opposite occurs, so that the minimum expected sales gain corresponds to our opponent's choice of C_2 (at the point D rather than E).

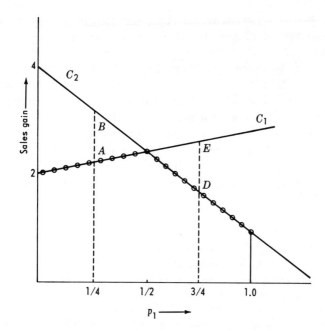

FIGURE 9-10   GRAPHICAL ILLUSTRATION OF MIXED STRATEGIES

We wish to choose p_1 so that this minimum value will be as large as possible. Looking at Figure 9-10, we see that this occurs at the peak—the point of intersection of the two lines. The value of p_1 here is $\frac{1}{2}$ and the corresponding sales gain is $2\frac{1}{2}$ million; hence u is $\frac{5}{2}$. These numerical values may be found by simply equating the left sides of expressions (9-33) and (9-34) and solving for p_1 and then u.

This says that we can ensure that the smallest value of the *expected* sales gain will be at least $2\frac{1}{2}$ million by selecting a mixed strategy with $p_1 = \frac{1}{2}$. In other words, we are to flip a coin in selecting between S_1 and S_2.

Of course, we chose this example to work out with simple numbers. There is nothing sacred about the probability $\frac{1}{2}$. If one changes the numbers in the outcome array, the best probability might well be any number between 0 and 1. Secondly, we should recognize that this form of graphical analysis is limited to the situation in which we have only two pure strategies.[33] However, the same *concepts* may be applied to any number of strategies,

---

[33] The reader who knows game theory will recognize that we can solve graphically if *either* we or our opponent has only two strategies. See Luce and Raiffa.

and a best mixed strategy may be developed by using techniques of game · theory. (Luce and Raiffa is a basic reference on these points.)

Also, our opponent can apply exactly the same sort of analysis as we do. In fact, we may analyze his analysis just as we did in the simpler case. That we discover his best mixed strategy (or he discovers ours) does not matter, since the mixed strategy determines that we shall choose a pure strategy randomly. Therefore, even if we and our competitor know which mixed strategy the other will use, it does either little good. (We will leave the analysis of our competitor's decision problem for the exercises at the end of the chapter.)

Technical Notes

ESTIMATING THE PARAMETERS OF THE BASIC DYNAMIC ADVERTISING MODEL[34]

To use the basic dynamic model of expression (9-8) in advertising budgeting, or in the fashion described in expression (9-29) for determining dynamic sales response, it is necessary to estimate the parameters λ, M, and r—the sales decay constant, the sales saturation level, and the sales response constant. The determination of the sales decay constant is simplest.

To estimate λ, the "experiment" which needs to be conducted is the absence of advertising. Of course, if no advertising had been in progress, this need not be an experiment, at all, but rather, simply an after-the-fact observational analysis. In fact, it is even feasible to hypothesize an experimental sales decay occurring at the current level of advertising, if that level is quite low. The adjustments in interpretation which would need to be made in the model are apparent; the current advertising level is simply equated with zero advertising. In either case the sales decay could be investigated and a least-squares line fitted on semilogarithmic paper to obtain an estimate of λ. To illustrate this, consider the semilog plot of Figure 9-11, in which time is plotted on the horizontal axis and the natural logarithm of sales revenue R on the vertical axis. The least-squares fit to these data is the line shown. Its general form will be

$$y = ax + b$$

[34] This technical note requires only algebra and the basic concepts of regression analysis.

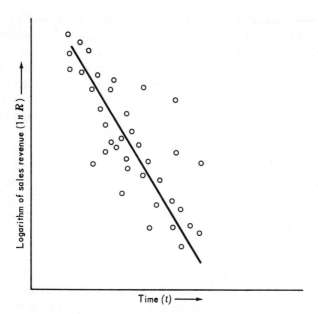

FIGURE 9-11 LEAST-SQUARES FIT TO OBSERVED SALES DECAY PATTERN

with a the slope and b the vertical intercept. In these terms, this is

$$\ln R = at + b$$

The vertical intercept is simply the logarithm of the sales revenue at the zero—$R(0)$. Hence,

$$\ln R = at + \ln R(0)$$

or

$$\ln \frac{R}{R(0)} = at$$

Taking antilogs, we have

$$\frac{R}{R(0)} = e^{at}$$

or

$$R = R(0)e^{at}$$

which is identical to expression (9-6) with

$$a = -\lambda$$

Therefore, *the slope of the least-squares line is a valid estimate of the negative of the sales decay constant* λ.

The sales saturation level M may be roughly estimated by simple experiments in which advertising is increased and continued for some time period. The sales achieved will undoubtedly not continue to increase, but rather to level off, giving a rough idea of the saturation level.[35] Alternatively, some characteristic of the potential consumers for the product may permit an estimate based on the proportional composition of the market; e.g., the proportion of tinted blonds approximates M for a new blond hair tint.

The sales response constant r may be estimated in the fashion illustrated in the discussion of the basic dynamic model. In the case of a new product it is simply the sales generated per advertising dollar when sales revenue is zero. For an existing product with known sales saturation level, the sales generated per advertising dollar at the sales revenue level R is just

$$\frac{r(M - R)}{M}$$

If an experiment is conducted at various levels of R, the resulting sales generated per advertising dollar can be equated to this quantity and various values of r calculated. The average value might then be taken as the estimate of the sales response constant.

ON THE DETERMINATION OF SALES RESPONSE[36]

The form of the sales response as a function of time which is implied by the basic dynamic model may be determined to be expression (9-29) by beginning with expression (9-8). Let

$$b = -\left(\frac{rA}{M} + \lambda\right)$$

and

$$a = rA$$

Expression (9-8) is then

$$\frac{dR(t)}{dt} = a + bR(t)$$

or

$$\frac{dR(t)}{dt} - bR(t) = a$$

which is a linear differential equation of the first order. The integrating factor

$$e^{-bt}$$

---

[35] See the Vidale and Wolfe reference for a discussion of such an experiment.
[36] The level of mathematics required for this technical note is basic differential equations.

can be used to obtain

$$\frac{dR(t)}{dt} e^{-bt} - bR(t)e^{-bt} = ae^{-bt}$$

The left side is the derivative of

$$e^{-bt}R(t)$$

Hence,

$$e^{-bt}R(t) = \int ae^{-bt}dt$$

or

$$e^{-bt}R(t) = -\frac{a}{b} e^{-bt} + C$$

which is

$$R(t) = -\frac{a}{b} + Ce^{bt} \qquad\qquad (9\text{-}35)$$

At $t = 0$, $R(0)$ is

$$-\frac{a}{b} + C$$

Hence,

$$C = R(0) + \frac{a}{b}$$

Substituting C into expression (9-35), we get

$$R(t) = \frac{a}{b} e^{bt} - \frac{a}{b} + R(0)e^{bt}$$

which, converted back to r, A, M, and λ is exactly expression (9-29).

ESTIMATING THE SLOPE OF SALES RESPONSE CURVES [37]

The estimate of the slope of the sales response curve which can be made from a simple two-level experiment, as given by expression (9-30), is

$$b = \frac{\bar{r}_2 - \bar{r}_1}{a_2 - a_1}$$

The mean $\bar{r}_1$ is calculated as

$$r_1 = \frac{1}{n_1} \sum_{i=1}^{n_1} r_i$$

[37] This technical note requires a basic statistical knowledge equivalent to that required for the Dixon and Massey text in the references.

if there are n_1 areas in the control group and the r_i are random variables corresponding to the various trials in the choice of the control group of areas. Hence, if the sample is chosen randomly and each r_i has the variance of σ^2, the variance of $\bar{r}_1$ is

$$V(\bar{r}_1) = \left(\frac{1}{n_1}\right)^2 \sigma^2 + \left(\frac{1}{n_1}\right)^2 \sigma^2 + \cdots + \left(\frac{1}{n_1}\right)^2 \sigma^2 = n_1 \left(\frac{1}{n_1}\right)^2 \sigma^2 = \left(\frac{1}{n_1}\right) \sigma^2$$

Similarly, the variance of $\bar{r}_2$ is $(1/n_2)\sigma^2$ (assuming the same variance σ^2 for those sampling variables).

If there are n_2 areas in the experimental group, the variance of the difference $\bar{r}_2 - \bar{r}_1$ is simply

$$V(\bar{r}_2 - \bar{r}_1) = \frac{1}{n_1}\sigma^2 + \frac{1}{n_2}\sigma^2 = \sigma^2 \left(\frac{1}{n_1} + \frac{1}{n_2}\right)$$

and the variance of the slope estimate $\hat{b}$ is therefore

$$V(\hat{b}) = V\left(\frac{\bar{r}_2 - \bar{r}_1}{a_2 - a_1}\right) = \left(\frac{1}{a_2 - a_1}\right)^2 V(\bar{r}_2 - \bar{r}_1)$$

because the denominator is a constant. This is

$$V(\hat{b}) = \frac{\sigma^2}{(a_2 - a_1)^2} \left(\frac{1}{n_1} + \frac{1}{n_2}\right)$$

and the corresponding standard deviation is the square root of this quantity.

Having made the determination from known values of n_1, n_2, a_2, and a_1, and a knowledge of the variance of the basic variable σ^2, one needs only to select some multiple of the standard deviation of the slope estimate which is satisfactory to establish the slope estimate ranges (as shown in Table 9-3). The central limit theorem (see Dixon and Massey) tells us that the distribution of means for large sample sizes is approximately normal, and so we may use this distribution in deciding which multiple to use.

Summary

Decisions related to advertising, along with product decisions, are unquestionably the most significant aspects of the development of an overall consumer product marketing strategy. One of the unique features of advertising decisions is the nebulous nature of the basic advertising objectives of the organization.

If the sole objective of advertising is taken to be the motivation of sales and the making of profits, the problems of advertising budgeting and the allocation of advertising expenditures to various media, areas, etc., may be approached in a straightforward fashion.

Measuring the effect of advertising expenditures is the area of advertising analysis which offers the most difficulty. Universally valid measures or measurement techniques are available for neither individual advertising messages nor advertising media. The result is that the decisions of advertising must be approached in terms of the basic problem structure and the analytic principles which are applicable.

Advertising decisions in a competitive environment, like pricing decisions in a similar environment, must be approached in a special way. So, too, there are essential differences and similarities between advertising decisions and other promotional decisions. In the next chapter we shall illustrate these points of similarity and difference.

REFERENCES

Allen, R. G. D.: *Mathematical Analysis for Economists,* Macmillan & Co., Ltd., London, 1962.

Bass, F. M., and R. T. Lonsdale: "An Exploration of Linear Programming in Media Selection," *Journal of Marketing Research,* May, 1966, pp. 179–188.

Benjamin, B., W. P. Jolly, and J. Maitland: "Operational Research and Advertising: Theories of Response," *Operational Research Quarterly,* vol. 11, no. 4 (December, 1960).

Boyd, H. W., Jr., and R. Westfall: *Marketing Research,* rev. ed., Richard D. Irwin, Inc., Homewood, Ill., 1964.

Brown, D. B., and M. R. Warshaw: "Media Selection by Linear Programming," *Journal of Marketing Research,* vol. 2, no. 1 (February, 1965).

Cochran, W. G., and G. M. Cox: *Experimental Designs,* 2d ed., John Wiley & Sons, Inc., New York, 1957.

Crisp, R. D.: *Marketing Research,* McGraw Hill Book Company, New York, 1957.

Dean, J.: *Managerial Economics,* Prentice-Hall, Inc., Englewood Cliffs, N.J., 1951.

Dixon, W. J., and F. J. Massey, Jr.: *Introduction to Statistical Analysis,* 2d ed., McGraw-Hill Book Company, New York, 1957.

Engel, J. F., and M. R. Warshaw: "Allocating Advertising Dollars by Linear Programming," *Journal of Advertising Research,* vol. 4, no. 3 (September, 1964).

Friedman, L.: "Game Theory Models in the Allocation of Advertising Expenditures," *Operations Research,* vol. 6, no. 5 (September–October, 1958), pp. 699–709.

Garvin, W. W.: *Introduction to Linear Programming,* McGraw-Hill Book Company, New York, 1960.

Green, P. E., and D. S. Tull: *Research for Marketing Decisions,* Prentice-Hall, Inc., Englewood Cliffs, N.J., 1966.

Hildebrand, F. B.: *Methods of Applied Mathematics,* Prentice-Hall, Inc., Englewood Cliffs, N.J., 1952.

King, W. R.: "Non-Technical Constraints in Production Decision Problems," final report of research project sponsored by American Production and Inventory Control Society, 1965.

Kroeger, A. R.: "What's Going On in Milwaukee," *Television Magazine,* November, 1965.

Lucas, D. B., and S. H. Britt: *Measuring Advertising Effectiveness,* McGraw-Hill Book Company, New York, 1963.

Luce, R. D., and H. Raiffa: *Games and Decisions,* John Wiley & Sons, Inc., New York, 1957.

"SM Trend Report: Measuring Advertising Effectiveness," *Sales Management,* Feb. 19, 1965.

Vidale, M. L., and H. B. Wolfe: "An Operations Research Study of Sales Response to Advertising," *Operations Research,* vol. 5, no. 3 (June, 1957), pp. 370–381.

Von Neumann, J., and O. Morgenstern: *Theory of Games and Economic Behavior,* 3d ed., Science Editions, John Wiley & Sons, Inc., New York, 1953.

Williams J. D.: *The Compleat Strategyst,* rev. ed., McGraw-Hill Book Company, New York, 1965.

EXERCISES

1. Put yourself in the position of a small manufacturer who is about to embark, for the first time, on an advertising campaign addressed to the consumer market. What would your feeling be with regard to the objectives of advertising? If you heard advertising executives discussing *only* "exposures" and "awareness," what would you think?

2. Most advertising agencies are compensated through a commission system which gives them 15 percent of the dollars which an advertiser spends for radio, TV, newspaper, and magazine ads. What do you think of this system in terms of the basic advertising objectives? A few agencies have gone to a cost-plus fee system for some of their accounts. Does this seem better or worse to you?

3. What special situation exists concerning the objectives of advertising for new products? How might these objectives serve to change your view concerning ad agency compensation as indicated in Exercise 2?

4. What are the implications of interactive effects of various promotional devices (including media advertising) in terms of the "fear" psychology of advertising?

5. Comment on the fallacy of determining an advertising budget by relating spending to sales.

6. Repeat exercise 5 for competitive advertising budgeting and the "maximum spending" approach.

7. Suppose the sales revenue rate R is related to the advertising expenditure rate by $R = 1 + \log A$. If the product in question has

a unit variable relevant cost of $0.50 and a unit sales price of $1.50, what is the best advertising expenditure? In what terms must A be measured to give meaning to your answer? Over what range of A is this sales response function meaningful?

8. Give an argument as to why some arbitrary advertising expenditure which is larger than the optimal value determined in exercise 7 is not optimal. Use a graph of the sales response function as a basis for your argument. Do the same for an arbitrary advertising expenditure which is less than the optimal.

9. The figure below is a plot of the logarithm of monthly sales over time for an unadvertised product.
   What is the phenomenon demonstrated by this curve? What are its causes?

10. A test was carried out in which four areas were subjected to the same level of advertising (per period) in the following fashion:

   Area 1: received advertising in period 1 only
   Area 2: received advertising in periods 1 and 2
   Area 3: received advertising in period 2 only
   Area 4: no advertising

   The results obtained in terms of *cumulative sales* are shown below.

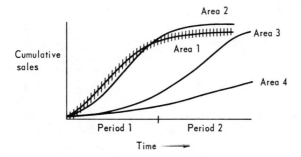

   a. What in particular does the comparison of the results of area 1 and area 3 indicate?
   b. What does the comparison of the results of areas 1 and 3 versus area 2 indicate?
   c. What does the comparison of the results of areas 1, 2, and 3 versus area 4 indicate?
   d. What is the overall advertising phenomenon which these data illustrate? (Particularly the results of areas 1, 2, and 3.)

11. Interpret the basic dynamic model of expression (9-8) in layman's terms.

12. Suppose that a company undertook advertising at a rate of $10,000 per month and this caused the sales rate to increase from $50,000 to $60,000 per month. If the sales saturation level is considered to be $100,000 per month, what sales response constant is implied? What sales rate would be expected to result from a further $10,000 increase?

13. Suppose the sales decay constant and sales response constant both have the value 2. What advertising expenditure is necessary to retain a $50,000 monthly rate of sales if the sales revenue ceiling is $100,000? What do you think of the position of a firm which is in this situation?

14. For each of the following sales response functions, (a) make a rough sketch of the function, (b) find the advertising expenditure rate which maximizes profit, and (c) find the sales rate at the optimal expenditure rate.

$$R = a \log bA$$
$$R = a + b \left(\frac{ca}{1 + cA}\right)$$
$$R = a + bA$$

a, b, and c are constants; R is sales revenue rate; and A is advertising expenditure rate.

15. A company must determine the advertising expenditure for two products whose individual sales responses are depicted as

$$R_1 = 2 \log b_1 A_1$$
$$R_2 = 4 \log b_2 A_2$$

(in millions of dollars per year). The incremental profit as a fraction of selling price (before advertising) which is applicable to each product is 0.2 and 0.3. Find the best advertising expenditure rates—A_1 and A_2—if the total advertising budget is $800,000.

16. From the point of view of the decision maker, what is the most significant difference between "technical" and "nontechnical" spending constraints?

17. What does the feasibility space of Figure 9-4 represent?

18. What are the differences and similarities between the idea of "effective exposures" in the "specialized media allocation model" and the implicit assumption concerning the effect of advertising in the "basic dynamic model"?

19. What was the form of the overall sales response function in exercise 15? What does this imply concerning the two products in question?

20. Why is it so important to spell out in advance each of the possible results of an advertising experiment, and what conclusion will be drawn (or action taken) in the case of each?

21. Explain the need for a control group in conducting an advertising experiment.

22. Apply the minimax reasoning to the abstract advertising problems depicted in the outcome arrays shown below.

	C_1	C_2
S_1	5	3
S_2	4	1

	C_1	C_2	C_3	C_4
S_1	2	5	3	3
S_2	1	-1	4	2

23. Explain the logic of the apparently ridiculous idea involving the use of a random device to select a (pure) strategy in a play of a game.
24. If the opposing player in a game situation were actually a large group of people whose aggregate actions were summarized in the set of C's, how would the minimax analysis need to be changed, if at all? Why?

10

Promotional Decisions

In this chapter we focus attention on the promotional decisions of the organization. In Chapter 9 the emphasis was on promotion of the product through advertising. Because advertising is big, controversial, and obviously important, it is best to treat it initially as though it were conducted in a vacuum. Of course, it is just one of the many forms of promotion which are used by most businesses, and since the effect of advertising is intrinsically intertwined with the effect of salesmen's calls, point-of-purchase displays, product demonstrations, and other nonrecurrent promotional devices, it can be viewed independently only at a sacrifice, the loss being the interdependent effects.

The basic objectives of all forms of promotion are the same—to produce sales. The communications subobjectives are also similar—to transform the potential customer from a state of unawareness to one of action. The methods which may be chosen to produce this transformation are diverse. For one product, personal selling may be directed at retailers, as in the case of packaged grocery products, and for another, salesmen may contact the potential customer directly, e.g., the "Avon lady." Even within a product class, promotional techniques may be widely varied, ranging from prize drawings ("You may already have won a trip to Hawaii") to free samples or "10 cents off" coupons.

There are also many variations of each basic promotional technique. The use of coupons offering price reductions presents a good illustration. Such coupons are common promotional devices for drug and grocery products. Usually they are mailed or delivered to the consumer and can be redeemed by him at the retail store when he purchases the specified product. The retailer is compensated for the revenue loss by the manufacturer. The objective of such promotions may be either market penetration (as with a coupon attached to a well-known product which offers another of the company's products at a reduced price) or increased consumer usage. In the latter case, a coupon offering a second unit of a product at a reduced price will serve to increase consumer's home inventories and thus permit a greater usage rate. The problem with such promotions is that it is difficult to control "misredemptions." Since no individual records are kept, the customer may redeem the coupon from a grocer without actually fulfilling the terms of the offer (buying the product). To control this problem, Rexall has introduced a "gift coupon" operation for other manufacturers, in which the customer must match a coupon which he receives through the mail and one contained inside the product's package. The matched pairs are then redeemed through Rexall for money certificates or trading stamps.[1] In developing this particular promotional device, Rexall has presumably found a way of using a variation of a time-tested promotional technique to advantage in achieving the various objectives of the seller and, at the same time, circumventing one of the major problems usually associated with coupons. In doing so, they have demonstrated that the only bounds on the forms which product promotion may take are the limits of creativity and ingenuity of the selling organization.

In fact, most of us are so bombarded with "technical" claims made concerning the worth of various products through media advertising that we suspect (and there is evidence to support the claim) that there is little real effect on the consumer. In such cases, nonadvertising promotional devices may carry the day for a product. A rather amusing report of a sale which was consummated through a simple promotional device offers a good illustration of this sort of situation:

My local supermarket—Gristede's—gets my toothpaste business. It's easier to buy there than to make a special trip to the drugstore. Still the purchase usually throws me for a loss. Here's why.

I'm a TV watcher who does not completely ignore the commercials. But, like most people, I find it hard to pinpoint the relative merits of hexachlorophene, GL-70, Gardol, and Irium. Which, for heaven's sake, will do most for my teeth?

Toothpastes with Durenamel or Fluoristan trouble me less. Fluoristan suggests that the paste may contain some kind of "decay preventing" fluoride, while Durenamel conveys a sense of making my tooth enamel

[1] See "Rexall Forms Program to Provide Coupons to Product Makers," *Wall Street Journal,* Mar. 15, 1966, p. 4.

harder—quite a trick and I do not believe it. Even less troublesome is a toothpaste that contains milk of magnesia. I know specifically what *that is*—though I wonder why the manufacturer has mixed it with toothpaste.

Price is a factor. I usually ask at the counter whether or not they have any special deals on toothpaste. Sometimes, there's a "five cents off" offer or a special "two tubes for 89 cents."

Not long ago, I was on a trip to Massachusetts and I needed a tube of toothpaste. Stopping at a general store in Mill River, I found the owner had completely eliminated my agony of choice. How? Which toothpaste did I buy? The one (Gleem) that came with a free box of crayons. They have nothing to do with oral hygiene but my two-year-old daughter will have plenty of fun with them.[2]

Undoubtedly almost every consumer, manager, and analyst would agree that product promotion is one of the most significant aspects of modern marketing. Each of us, in our role as a consumer, is bombarded every day with a bewildering array of advertisements and other promotions. Yet, the true extent of product promotion is perhaps best exemplified by those more subtle promotions which are not necessarily designed to induce purchases *now*. These longer-range promotions are often devised to *inform* in the short term, with the action (purchase) to be motivated later. For example, although few communities have undertaken extensive anti-water-pollution programs, the manufacturers of treatment equipment extensively promote the general idea and their particular wares through the country. The same is true in the field of urban transportation. In both these cases the modern marketing technique of *demand creation* is being employed, and the best way to create demand is through effective communications.

Also, second-order product promotion is not uncommon. For instance, a major steel company has a traveling exhibit displaying advanced designs of autos, railroad equipment, and other items, none of which are produced by the steel company. Of course, these items are made of steel. Hence, the steel company is promoting the sale of the products of its customers—the manufacturers of autos and railroad equipment. In doing so, it is also promoting the sale of its basic steel products on a level which is once-removed from its customers.

Promotional Budgeting and Allocation

The concepts of analysis for determining the promotional budget are akin to those for advertising budgeting. Whether we are discussing media expenditures, sales manpower, or free samples, the

[2] "Sales Management Goes Shopping: To Pick a Toothpaste," *Sales Management,* Feb. 19, 1965, pp. 91–92. Quoted with permission.

basic response characteristic is of paramount importance. That characteristic—declining marginal returns—indicates that one should not continue promotional resource expenditures indefinitely, but rather only up to some point on the response curve. Therefore, there is a "best" promotional budget which can be determined.

In this light the basic model of promotional budgeting is identical to the basic model of advertising budgeting. One needs only to interpret the expenditure in Figure 9-1 as being the expenditure level for whatever promotional device is being considered and the sales rate as being generated by that promotional expenditure. Thus, the same basic concepts and methods are applicable to all forms of promotional budgeting, whatever may be the particular promotional device.

It should be noted that these comments on the similarity of promotional budgeting problems for all forms of promotion *do not imply that the response curve is the same for all forms.* It is quite likely that the response curves for media advertising and for promotional coupons are quite different, for instance. All that is assumed is that both curves have the general declining marginal return characteristic which is common in resource expenditure situations. Since the solution to the basic promotional budgeting model is in terms of the slope (first derivative) of the response curve, any curve with the declining marginal return property which describes the appropriate response phenomena may be used.

The problems of allocating expenditures to various promotional activities, whether subject to an overall budget constraint or unconstrained, also have a basic similarity. Therefore, the basic static model of advertising allocation and its associated concepts may well be extended to include nonadvertising promotional media such as free samples, coupons offering price reductions, etc. The changes which need to be made in interpreting the symbols in the basic models are readily apparent. The only additional difficulties are those of *measurement.*

In effect we are saying that one may apply the same general ideas and methods of analysis to decisions involving any type of promotional endeavor. Rather than repeating these ideas and models with different symbolism here, we leave it to the reader to refresh himself concerning them by returning to Chapter 9. The use of these ideas and models in nonadvertising promotional contexts will be illustrated in the exercises.

Having made the point of the basic similarity of promotional decisions, we shall focus attention in subsequent sections of this chapter on two specific promotional facets—personal selling and performance measurement. Personal selling is one form of promotion which is sufficiently pervasive and unique to require individual attention. Moreover, the measurement problems associated with personal selling decisions serve to clarify the general measurement problems of promotion. The next major section will treat these problems and managerial decisions of personal selling. In the subsequent major section we shall discuss general promotional measurement problems and seek to clarify the essential similarities and differences of various forms of promotion. The last major section will discuss a major analytical tool which has been applied to a wide range of promotional decision problems.

Personal Selling

The significance of personal selling to virtually all business is illustrated in a personal way to anyone who travels. If one schedules an airline flight for any Friday afternoon, boards the airplane, sits down, and engages in conversation with his seatmate, he is likely to find that his new acquaintance's occupation is in some way related to sales. Frequently, he is either a salesman returning home from a week's round of calls, a sales manager making a quarterly visit to the western office, or an area manager heading to the home office for a two-week refresher course. In any event, unscientific "experiments" conducted by the author and his associates have produced a large number of such results.[3]

A more objective measure of the importance of personal selling is reflected in the estimate[4] that 55 percent of expenditures for major promotional forms are devoted to sales worker compensation (excluding travel expenses, training costs, and sales administration) whereas only about 36 percent of such expenditures are for advertising and about 9 percent for other forms of promotion.[5] Therefore, although the expenditures made in promoting products through advertising are more visible, they are not so great as those made on personal selling. Additionally, when one recognizes that over 10 percent of the nation's labor force is engaged in personal selling

---

[3] One begins to suspect that the airlines are the travel arms of sales departments. In fact, if one realizes that only a relatively small proportion of the American public have *ever* flown, the dependence of the airlines on repetitive travelers, such as salesmen, is vividly illustrated.

[4] T. H. Spratlen, "An Appraisal of Theory and Practice in the Analysis of Sales Effort," unpublished doctoral dissertation, Ohio State University, 1962. Based on data from Internal Revenue Service, *Printers' Ink,* and U.S. Bureau of the Census. Cited in Matthews et al., p. 337 (see references in Chap. 6).

[5] The relatively small percentage expenditure for forms of promotion other than advertising and personal selling partially explains our deemphasis of their details in this treatment.

activities of some nature, their significance becomes even more apparent.

Many volumes have been written concerning salesmanship and the psychological factors which are the basis of personal selling. It is beyond the scope of our treatment to deal with these here. However, just as we pointed out that the content of the advertising message might in some instances be of such overriding significance as to make decisions involving total dollar expenditures pale in importance, so must we consider the potential importance of the "content" of the direct sales contact (versus the sales expenditure in terms of number of salesmen, etc.). Another illustration from the magazine *Sales Management* demonstrates both the importance of content and the interaction of various promotional forms.

> Recently, when I needed the brake drums turned on my 1963 Chevrolet, I took it to eight auto dealer repair shops in Dallas.
>
> All had some excuse for not taking on a job this small: "The man who does this is out sick today," or "We turn them but we don't remove them; you have to get somebody else to do that," and even simply "Our machine is broken."
>
> Finally, I took it to a small Ford dealer in a suburban town. He agreed to do it—it would take about 30 minutes. By the way, he wondered, had I had a chance yet to test-drive the new Mustang?
>
> When I came back from my test drive, he told me that my '63 Chevy still had a high value and made me an excellent offer for it as a trade-in on the new Mustang.
>
> When he reminded me gently that my car had more than 50,000 miles on it I decided to buy. My new-car warranty will cover all the repairs nobody else wants to make, and I got courteous and efficient service.
>
> All the other dealers missed a sale by being blind to the opportunity offered by this $3 repair job.
>
> Needless to say, one of the local dealers asked me why I had traded with the "other man" instead of him.
>
> It took 20 minutes, but I told him.[6]

Here the ingenuity of the salesman in utilizing effectively one aspect of his service in promoting another, and in utilizing various promotional forms—free demonstrations, repair services used in effect as a "loss leader," and high trade-in allowances—obviously interacted effectively with the personal salesmanship of the individual in producing a sale.

In this discussion we shall concern ourselves primarily with the decisions involved in the management of a sales force, rather than with the personal qualities and approaches which contribute to

[6] "Sales Management Goes Shopping: Eight Missed a Sale," *Sales Management,* Feb. 19, 1965, p. 91. Quoted with permission.

effective selling, however important they may be. The reason for this is that we know so little concerning such psychological aspects and, in particular, we can say little which has not already been effectively said elsewhere.[7] A primary goal will be to analyze the problems of measuring sales force performance. In a later section the measurement of the interactive effects of personal selling and other promotional forms will be taken up.

A basic measurement technique which is a necessary part of any treatment of sales promotion, personal selling, and sales force decisions is the *sales analysis*.

SALES ANALYSIS

Sales analysis is the process of understanding one's sales effort and results in a fashion which is diagnostic and serves to provide information useful for decision making purposes. Just as the physician takes measurements of heart rate and blood pressure before he diagnoses the ills of a patient, so must the sales manager determine certain factual information concerning sales before he can make meaningful sales force decisions.

Of course, many of the necessary measurements are products of the accounting system; yet one frequently finds that the form in which accounting records provide the basic data is not the best one for decision making purposes. For instance, one large insurance company[8] has separate accounting systems for life insurance and fire insurance. Therefore, although many agents sell both, it is virtually impossible to obtain consistent information concerning the total performance of individual agents. One finds that the large number of agents precludes the maintenance of detailed historical records on each. Hence, each segment of the enterprise maintains detailed records only on large-volume producers (in their area); i.e., the life department has records of large life producers and the fire insurance department has records of large fire producers. If one wishes to view total performance of agents, one can obtain detailed data only on those agents who produce·large sales volume in both categories simultaneously.

Sales analysis is not a glamorous part of marketing. The basic approach involves the compilation of hard cold facts about product sales. In effect, the result of a sales analysis may be thought of as a "status report" on a firm's sales. Additionally, to place the

---

[7] See, for example, the Still and Cundiff or the Canfield text in the references.
[8] Insurance companies provide excellent examples of many points related to personal selling because the vast proportion of insurance sales are made on a direct personal sales basis.

current sales status in perspective, some historical sales information may often be incorporated in a sales analysis.

The aspects of sales analysis which are sales-oriented derive from a need to know and understand the sales which are currently being garnered by a firm. This is a basic starting point for any managerial decision regarding personal selling (and for many more).

A logical procedure for the analysis of sales-oriented sales information might begin with a compilation of data on *total sales revenue* (by product) *over some historical period*. The long-term trend of sales revenue serves to place the current situation in proper perspective. Another information display which fulfills a similar purpose is a historical presentation of *market share* over the same period. This serves to relate past sales results to the results experienced by competitors during the same time periods. Any consideration of one absolute sales measure (e.g., sales revenue) without reference to a relative measure (e.g., market share) may be very misleading.

Additionally, detailed data on one or more aspects of sales by product line, item, market area, sales district, individual salesman, customer classification, and individual customers usually are incorporated into a sales analysis. Of course, in particular situations some of these levels may be inapplicable. In others, it may be impossible to obtain the relevant data. In the latter case, one of the primary values of a sales analysis may be in focusing attention on those areas in which information is necessary before logical analysis of sales managerial decisions can be made. In any case, the basic idea is clear—to gather sales data at whatever level may be necessary to enhance decisions. If, for example, one must decide how much sales effort is to be allocated to various retailers, it is necessary to know sales results at the retail level. If the appropriate question is the staffing of a district sales office, the sales results produced at that level form a basic unit of necessary information.

The difficulties in obtaining such sales data, particularly at the retail level, are usually great. Retail sales audits on a continuing basis are usually very expensive and time-consuming. One alternative for some products is a service, such as that provided by a subsidiary of Time, Inc., which involves the collection of sales data on food products in selected market areas. A manufacturer who uses this service obtains retail sales data for his and competitors' products on a continuing basis. These data are provided by retail chains, compiled by the service by means of computers, and furnished to the manufacturer for a fee.

Other expenditure-oriented information provides an effective complement to the sales-oriented data. For instance, a percentage breakdown of market share and sales revenue or profit contribution by sales territory should include information on the proportion of advertising, personal selling, and other promotional expenditures in each territory. It may also be desirable to study the detailed components of each of these expenditures as related to the sales produced. A detailed study of how salesmen spend their time—often referred to as a "time and duty" analysis—may be performed as a guide for management action.

The direct results of such a sales analysis may be of great importance. The basic data are available to perform *comparisons* and *evaluations* in the manner discussed in a slightly different context in Chapter 6. Management action can then be taken to correct gross inequities. Such a result was recently witnessed by the author when a marketing manager was for the first time presented with data on regional office sales staff and sales revenue. Since the staff size variation from office to office was great, further investigation appeared to be warranted. Part of the variation was explained by geography; e.g., sales personnel in Utah were faced with problems, such as long intercustomer distances, different from those in New York. However, other large variations indicated that there was a great deal of variability in sales efficiency. Actions were then taken to ensure that the field personnel were informed of the measures which were being applied and to solicit their reactions to such measures. Several areas reported existing circumstances which militated against efficiency and yet could be cleared up rather simply by corporate-level action. The net direct result of the sales analysis was an increased understanding of the organization, on the part of both field and corporate personnel and, hopefully, more effective future operations. In addition to the direct results of a sales analysis, the procedure provides the necessary data base from which sales managerial decision problems may be approached.

OBJECTIVES OF PERSONAL SELLING

The basic profit objectives of the organization impute profit objectives to each sales unit and individual salesman in the organization. If the profit contributed by each salesman is maximum, the organization's overall profit will be maximum.

The profit objective, obvious as it may sound, may be in conflict with the sales objective enunciated earlier—to sell. The critical point which may cause the conflict is *what products are sold*. In many multiproduct organizations, a single salesman is responsible for a line or perhaps for a wide range of products. It goes without saying that most salesmen concentrate on those products or items which are easiest to sell, bring them the greatest personal re-muneration, and with which they are most familiar. Further, it is apparent that those items or products are not necessarily the ones which earn high profits for the organization. Hence, to sell in large volume does not necessarily result in proportionately large profits. Some of the sales management decision areas which can be used to resolve effectively any existing conflict of sales objec-tives will be discussed later.

Here, we wish to focus attention on the major objective—to sell—and its relationship with the subobjectives of personal selling. The subobjectives may be classified as *retentive* or *conversional*. The salesman seeks to sell by retaining customers he already has or by converting others into customers (or both). This distinction, which may at first appear to be, like many similar ones, of purely pedagogic significance, is actually one which pervades the day-to-day operations of every salesman.

To be successful, each salesman must have some plan. The plan should designate, if only in a broad way, *whom* he will contact and *how much* time he will devote to each. Of course, this is not to say that every good salesman schedules his calls and time precisely in advance, or even that many do. Most, however, have a general plan which will not permit them to while away the day in conversation over an alcoholic beverage unless there is the defi-nite possibility (and probability) of a reward in the form of a sale. For such a plan, the first basic decision to be made concerns the retentive versus conversional sales subobjectives. Should the salesman devote time to an established customer hoping to get additional orders and to maintain his satisfaction, or should he prospect for new customers by visiting those who have not yet bought from him?

The significance of this conflict is further exemplified by the realization that salesmen are more and more ceasing to be order takers and becoming service-oriented. The salesman who visits a retailer, wholesaler, or industrial consumer is frequently called upon to render assistance to the customer. From the telephone company's "communications consultant," who analyzes communica-tions needs, to the paper manufacturer's representative, who may spend days in attempting to discover the cause of the difficulty being experienced by a printer's presses, the salesman is a technical and service consultant to his customers. In some industries—the chemical industry, for example—the need for technical sales is so great that few nonengineers are hired for sales positions. In fact, in some oligopolistic market environments, the primary com-petition between firms is conducted on the basis of the service provided by each firm's sales representatives.

In such instances the retentive versus conversional sales objective conflict is paramount, for although one may not produce immediate additional sales by spending time with current customers, the long-term goodwill which is created may be invaluable. On the other hand, prime prospects for additional sales may be neglected if only retentive goals are pursued.

These comments imply that the individual salesman plans for his personal time allocations. In fact, it may be the prerogative of the sales manager to do this. In most cases, he should at least develop guidelines for making such decisions. The guidelines can then be passed along to the salesman in the form of suggestions, instructions, or as a subtle part of his training. In later sections we shall investigate the determination of such guidelines and their operational implementation.

A significant point which needs to be emphasized here is the absence of effective control and incentives common in many sales organizations. Many salesmen receive all or a part of their remuneration in the form of sales commissions. It is well known that commissions make the sales field highly lucrative for the individual who can sell effectively. Yet, the monetary success of individual salesmen often presents a significant problem to the sales manager, for what incentive can he offer to the salesman of 20 years who has finally "made it" and seems content to live off his repeat sales commissions? In effect, the idea of personal utilities for money is exemplified by this kind of salesman. The additional $10,000 which he might make by actively converting prospects is partly reduced by high tax rates. But even more important is the fact that his marginal utility is low for these commissions. What will it buy him that he does not have already? Of course, this is a problem with respect to each of an organization's personnel, but the additional compensation as a direct and immediate result of additional effort which is a part of the sales commission structure is somewhat peculiar to sales. This direct compensation makes the motivational problem more apparent and more pronounced than it is among executive personnel.

The lack of control and incentives in many sales organizations means that the sales manager finds it difficult to implement the guidelines for pursuing the conflicting sales subobjectives. He may well know that a significant proportion of time devoted to prospecting is likely to be profitable, but he may find it difficult to motivate the successful salesman to do it. One must recognize that the sales personality is not one which easily complies with instructions from above. Moreover, good salesmen are in great demand, and so the manager must take great care to retain his personnel. Probably the best approach in such a case is one made on the basis of pride; sucessful salesmen are usually proud of their accomplishments and, by the nature of their environment, are usually competitive in nature. Most companies realize this and establish awards for desired levels of sales performance. Often such awards are relatively inexpensive; yet if they are true indicators of accomplishment and if their presentation is accompanied with sufficient fanfare and publicity, they can serve as incentives beyond the value of mere money.

One school of thought concerning the basic conflict between retentive and conversional sales objectives holds that the best overall results may be achieved by having different individuals seek the two objectives independently, i.e., one kind of salesman who exclusively seeks to maintain present customers and another kind devoted to converting noncustomers into customers. The high apparent costs associated with this tradition-breaking approach might well be more than compensated for by better utilization of the different talents required for effective pursuit of the two basic sales subobjectives.[9]

SALES FORCE DECISIONS

In previous sections we have discussed the basic informational requirements of sales force decisions and the objectives of personal selling, together with some comments on the implementation difficulties associated with sales force decisions. Here, we shall concern ourselves with sales force decision problems from the standpoint of the executive who must plan, organize, and control the selling function.

Retentive Allocation of Sales Effort. One of the primary allocation problems of sales management has to do with the *time* spent by salesmen in calling on existing customers. There is a good deal of evidence that the sales revenue response to sales effort is generally described by a curve like that shown in Figure 10-1.[10] This curve describes the sales revenue gained from a single customer as a function of the *total time* spent with him. The form of the curve is typical: a small time being too little to be adequately perceived by the customer, an effective range, and a range of saturation in which the effort is more than can be properly appreciated by the customer. Such a response curve is similar to that hypothesized for advertising and other promotional forms.

Since each salesman has only limited time available to spend

---

[9] See the Kahn and Shuchman paper in the references.
[10] See Waid, Clark, and Ackoff or Brown, Hulswit, and Kettelle, for example.

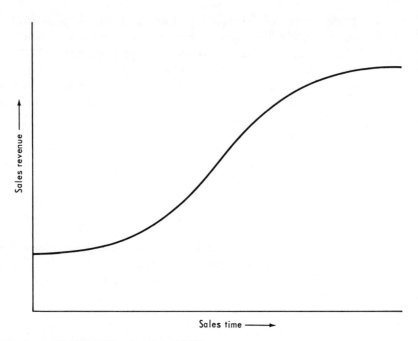

FIGURE 10-1   SALES RESPONSE TO SALES EFFORT

in servicing existing customers, his time allocation to each is of obvious importance. Moreover, this importance is amplified by the preponderance of empirical evidence revealed by sales analyses of many organizations, which indicates that a *large proportion of total sales revenue is accounted for by a relatively small proportion of the total number of customers.*[11] Of course, it is possible that this evidence reflects how salesmen *do* spend their time, rather than indicating how they *should* spend it, but one suspects that there are important ramifications of the empirical evidence in the normative analysis of sales effort allocations.

Of course, the response curve indicated in Figure 10-1 is applicable to a single customer, and one is not likely to be aware of the curve for each customer. To estimate such a curve for an individual customer with any degree of reliability would be impractical since it would require a number of data points (sales and salesmen's calls) which might well span a very long period of time. Even if it were possible to obtain such curves, a large number of customers would require estimating and analyzing an impossibly large number of curves.

It may well be possible to *categorize* customers according to similarity of response patterns. If such a categorization could be

[11] See, for example, the Waid, Clark, and Ackoff paper.

made, data on those customers in each category could be used to estimate a single response curve which would then be used to represent each customer in the category. This would reduce the difficulty involved in estimation and, provided that the sales response patterns for the various groups were sufficiently different, would permit making an approximately optimal allocation of sales effort.

The difficulty in this procedure lies in the classification of customers. For some products a natural classification on the basis of the nature of the customer's operation may evolve. All customers who distribute the product to retailers might be in one category, and all retailers who are dealt with directly in another. Or, if only retailers are involved, all hardware stores might naturally fall into one category and all discount stores into another. The customer's size, competitive status, and other such factors may also be useful in making such categorizations.

We shall not give any detailed discussion of the determination of optimal sales effort levels in this case since this problem may be approached either in the same fashion as the media allocation problem of the previous chapter or as a special case of the model to be developed for retentive versus conversional effort in the next section.

It is of greater interest here to treat the case in which no response curves can be reliably estimated for various customer groups. In one such situation[12] many combinations of customer characteristics were used as a basis for groupings, but essentially random patterns resulted, i.e., scatter diagrams of sales revenue versus number of calls for various customer groups showed no existing pattern which might suggest an underlying response curve. This led the analysts to believe that the organization might be operating on the upper *saturation* portion of the response curve. In this region, changes in sales effort would not be expected to produce consistent changes in response. To illustrate this, consider the underlying response curve of Figure 10-1 and some actual observations which might be made, which would scatter about the basic curve in some pattern of random fluctuation. Such a situation is shown in Figure 10-2, in which the dots represent observed points which scatter about the underlying response curve.

If the organization is actually operating in the saturation region, only those points included in the block at the upper right will be observed. This subset of the scatter of observed points is shown as Figure 10-3. Note that the distribution is essentially random. Hence, even though the points result from an underlying response

[12] See Waid, Clark, and Ackoff.

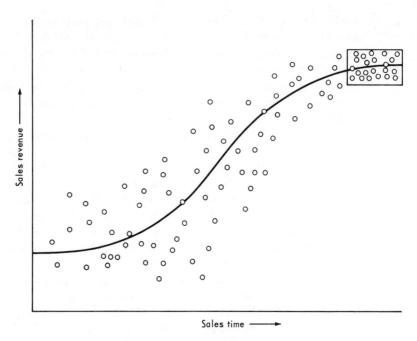

FIGURE 10-2   SCATTER DIAGRAM AND UNDERLYING RESPONSE CURVE

pattern, they appear to be randomly distributed. The reason, of course, is that we are observing only a very narrow range of values, and in that particular range the response curve is nearly horizontal. If one tried to fit an S-shaped response curve of the type shown in Figure 10-1 to the data points of Figure 10-3, one would be doomed to failure, for although the S-shaped curve is really applicable, it does not apply over the region of the data which are available. Since the investigator obviously does not know the underlying curve—indeed, this is what he is trying to estimate—he will not be cognizant of this difficulty.

In the particular case described, after some further justification of the hypothesis that operations were being conducted on the upper portion of the curve, the conservative action of reducing the sales effort per customer to the *lowest level, which had actually*

FIGURE 10-3   PORTION OF SCATTER DIAGRAM FROM FIGURE 10-2

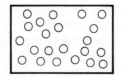

been utilized over some past period of years was implemented. Thus, if a particular customer had been called on 40, 65, and 50 times in each of the past three years, the plan for the coming year would allocate 40 calls to the customer. The implicit assumption, of course, is that all levels represented points which were "high enough" not to be below the best time allocation on the response curve. Therefore, the effort devoted to each customer could be reduced *at least* to the lowest observed level without doing any harm. In fact, the results reported from these actions in this circumstance were favorable.[13]

The selection of a managerial action in this way, without a detailed knowledge of the underlying relationships, exemplifies one of the primary advantages of quantitative analysis. Although in this case the exact nature of the response curve was not known, a general knowledge of its shape, together with the theory of optimal allocation of effort (as discussed in the advertising context in the previous chapter), permitted the choice of a managerial action, which, if not optimal, was decidedly superior to the one which was in use. Hence, even though one is not always able to apply directly the optimization models of quantitative analysis, the models may well form the basis for an informed guess as to the best strategy.

Retentive versus Conversional Allocation of Sales Effort. The determination of how much time a salesman should devote to the retention of existing customers or to the conversion of new customers must be based on the *likelihood of conversion* and the *likelihood of retention.* Presumably, each hour spent with a prospect should (up to a point) increase the likelihood of making a sale, whereas each hour devoted to a customer should increase the likelihood of retaining him as a customer. Since the total available sales time is restricted, to spend time with a prospect (and increase the conversion probability) is to neglect an existing customer (and decrease the probability of retention).

If we assume for the moment that all customers and potential customers are equiprofitable—that each existing and converted customer will produce the same future profit—the retention versus conversion problem may be approached by using the curves given in Figure 10-4. The upper curve gives the *probability of retaining an existing customer* at various levels of sales effort—hours per month. The lower curve gives the *probability of converting a noncustomer into a customer* at the same sales effort levels. These

[13] Analysis of the data for several years was made with consistent results for all years before this action was taken. See Waid, Clark, and Ackoff for details.

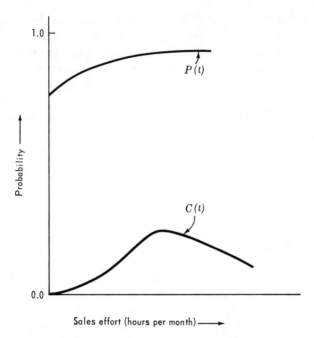

FIGURE 10-4   PROBABILITY OF RETENTION AND CONVERSION

probabilities are symbolized as $P(t)$ and $C(t)$ respectively. Of course, the curves represent general shapes, which seem to be validated by admittedly meager data,[14] rather than precisely defined functions. Note that the top curve indicates a rather good likelihood of retaining an existing customer even if no sales effort is allocated to him. The lower curve has the familiar S shape with a drop-off at higher levels, indicating that too much time spent with a prospect will probably annoy him and thus decrease the likelihood of his being converted.

To determine the time which *should* be devoted to conversional efforts, we may seek to maximize the ratio $C(t)/t$, i.e., to maximize the average conversion probability "obtained" per unit of time invested. This ratio is maximized at the point of tangency of the line drawn from the origin, as shown in the lower curve in Figure 10-5. This may be easily demonstrated by pointing out that the slope of any line drawn from the origin to a point on such a curve represents the average change in the quantity plotted on the vertical axis per unit of the quantity plotted on the horizontal axis. Since we desire to maximize such a quantity—$C(t)/t$—the line having the largest numerical value for its slope will determine

[14] See Brown, Hulswit, and Kettelle, for example.

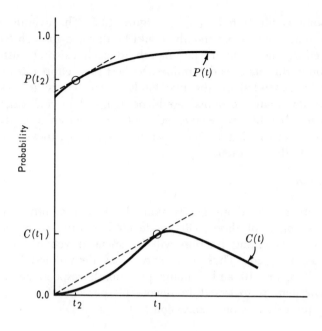

Sales effort (hours per month) ⟶

FIGURE 10-5 DETERMINATION OF OPTIMAL CONVERSIONAL AND RETENTIVE SALES EFFORT

the point on the curve at which $C(t)/t$ is maximized. One can easily see in Figure 10-5 that all other lines extending from the origin to a point on the curve must lie below the one shown. Hence, the illustrated line must be the one with greatest slope, i.e., with maximum $C(t)/t$. The point t_1 in Figure 10-5 is therefore the amount of time which should be devoted to each prospective customer. The corresponding conversional probability is $C(t_1)$.

To determine the time to be devoted to retaining each present customer, we need only to recognize the basic allocation principle as developed in the previous chapter—that to achieve the best overall result, the marginal "return" for all possible activities (in this case, conversion and retention) should be equal; i.e., the slope of the response curves should be equal. The slope of the conversional curve at the point t_1 is simply the slope of the tangent line, $C(t_1)/t_1$. Hence, the time to be devoted to retaining a customer is that point where the slope of the retention probability curve has this value, i.e.,

$$\frac{dP(t)}{dt} = \frac{C(t_1)}{t_1} \qquad\qquad (10\text{-}1)$$

This point is illustrated as t_2 in Figure 10-5. The overall result, then, is that t_1 hours per month should be devoted to each conversional effort and t_2 hours per month to each existing customer.

Although this analysis determines the best sales effort rates which should be expended on the two kinds of customers, it does not tell one how many accounts should be assigned to each salesman. If the number of prospective accounts per salesman is n_c and the number of existing accounts per salesman is n_R, these numbers must satisfy the equation

$$t_1 n_c + t_2 n_R = 170k \tag{10-2}$$

because there are about 170 working hours per month and only some fraction k of those will be devoted to directly productive work. The remaining fraction will be spent traveling, attending sales meetings, etc. Thus, if t_1 and t_2, as determined by using Figure 10-5, are 10 and 3 hours per month respectively and if each salesman is expected to devote 60 percent of his time to directly productive labor, expression (10-2) is

$$10n_c + 3n_R = 102 \tag{10-3}$$

which must be (approximately) satisfied by some determination of n_c and n_R, the number of each type of account to be assigned to each salesman.

In practice it would not seem to be prudent to abandon existing customers in order to devote more time to conversional effort. Therefore, initial conversional time should probably simply be the time left over after all retentive efforts have been expended (at the rate of t_2 hours per month for each existing customer). If the existing customer group becomes small for any particular salesman at any time, since he is devoting only t_2 hours per month to each, his potential customer group must automatically be expanded. The additional customers who are successfully converted thereby replenish the existing customer group for the next month. If the existing customer group becomes predominant for some salesman at some time, he should continue to devote t_2 hours per month to each, and it can be expected that lapses (customer losses) will occur in the future. The equilibrium level, which may be used as a guideline for determining the mix of existing and potential customers for a salesman, will be at the point where expected conversions and expected losses per month are equal. The "best" conversional probability is $C(t_1)$ (see Figure 10-5). Hence the expected number of conversions per month is

$$C(t_1)n_c$$

Similarly, the expected number of losses is

$$[1 - P(t_2)]n_R$$

At equilibrium

$$C(t_1)n_c = [1 - P(t_2)]n_R \qquad (10\text{-}4)$$

Expressions (10-2) and (10-4) may be simultaneously solved for n_c and n_R. For example, if t_1 is 10 and t_2 is 3, as used in expression (10-3), the corresponding probabilities, as determined from Figure 10-5, might be $C(t_1) = 0.20$ and $P(t_2) = 0.90$. Then, expression (10-4) is

$$0.20n_c = 0.10n_R$$

Taken in conjunction with expression (10-3), this gives $n_R = 12.75$ and $n_c = 6.38$. The guideline, then, is to assign to each salesman thirteen existing customers and six potential customers.

Personal Sales Effort Allocation. Assigning the *number* of potential and existing customers to a salesman is only one aspect of the allocation of sales effort. The other major element involves the *selection* of the particular customers who are assigned. In considering this aspect of sales effort allocation, we are relaxing the restrictive assumption of equally profitable customers which was made in the last section. Here, we recognize that the profitability of different customers may be widely variable and that consideration of such variability can lead to overall greater profit.

Allocation of salesmen on basis of customer characteristics. In some instances it may be most meaningful to consider the differences in profitability which are inherent in different kinds of customers. Suppose, for example, that the time to be spent with potential and existing customers and the mix of the two types of customers is fixed. In such a case we may be able to categorize customers according to some essential characteristics such as the type of business or geographic location. The groupings of customers may be on any basis deemed significant to the problem of allocating sales effort. For example, if we sell to electronics firms, it may be important for our salesman to understand their business and problems. After a salesman acquires such knowledge, his time is probably best spent on this type of customer rather than on organizations in the construction industry. Or, if a number of customers are located in a particular geographic area, it is probably best to allocate some fixed number of salesmen to that area and not to have salesmen going from area to area. The latter practice

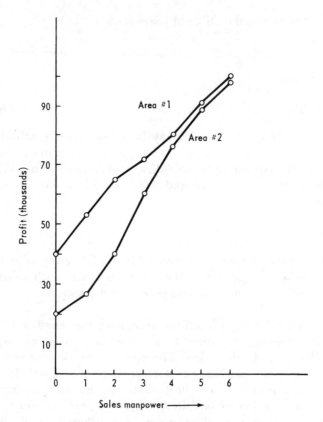

FIGURE 10-6   RESPONSE CURVES FOR TWO AREAS

would require them to expend a disproportionate amount of their
time in travel, with resulting lower sales and higher cost.

If the profitability of each of a number of various customer
groups displays a different pattern of response to sales effort, it
is essential to consider this variability in assigning salesmen to customers. As a simple illustration, suppose that we have estimated
the response curves shown in Figure 10-6 for two geographic areas
and wish to determine the *number* of salesmen to be assigned
to each area. These curves are similar to the idealized response
curves shown previously, except for the particular measures chosen
to represent sales effort input and performance output. Both
curves generally have, at the upper extremes, the declining
marginal return property which indicates that the first salesman
to be assigned to an area calls on the most productive customers
and each successive salesman calls on customers who are ever-decreasing in their profit contribution. At some point, all existing
customers will be covered and some salesmen must necessarily

devote all of their time to conversional effort directed at noncustomers with ever-decreasing profit potential.[15]

If a total of six salesmen is available for assignment to the two areas, to determine the best allocation, we need only to consider the various possible assignment combinations and the total profit which will be achieved under each. Table 10-1 illustrates such data. The profits in the third and fourth rows are taken from Figure 10-6, and the last row is simply the sum of the respective profits achieved in the two areas.

TABLE 10-1 PROFIT FOR VARIOUS SALES MANPOWER ALLOCATION OF SALESMEN TO TWO AREAS

Number assigned to area 1	6	5	4	3	2	1	0
Number assigned to area 2	0	1	2	3	4	5	6
Profit in area 1	100	90	80	72	65	53	40
Profit in area 2	20	27	40	60	77	89	98
Total profit	120	117	120	132	142	142	138

Table 10-1 indicates that the highest profit results from assigning one or two salesmen to area 1 and four or five to area 2. Even allowing for possible errors in our estimates of the profits which can be obtained in the two areas, it is clear that the profit levels for allocations of four, five, or six salesmen to area 2 are significantly higher than those for one, two, or three. Therefore, we should be certain that we always allocate more salesmen to area 2. Since we probably would not wish either to withdraw entirely from area 1 or to be represented there in a minimal fashion which is not even an adequate level of showing the flag, we would probably choose to assign four salesmen to area 2 and two to area 1.

In arriving at such a conclusion, the reader should note that we have effectively combined objective and subjective analysis. This is a simple illustration of the integration of scientific analysis and the judgment of the decision maker to which we devoted great attention in earlier chapters. The concern given above to the aspects of the real-world problems which are "not included in the numbers" (e.g., our concern for showing the flag) and our judgments concerning the effects of errors in the measurements

[15] The lower portion of the curve for area 2 does not possess the decreasing marginal return property. We are viewing these curves as being empirically derived, and there are two possible explanations for this. Its S shape might reflect a phenomenon which is actually being experienced in the area. On the other hand, it may be that the random variations inherent in the observations have caused a particular relationship of successive points which is not indicative of the underlying structure.

may often lead us to the selection of a strategy which is not the strategy indicated as best by the mathematical model. In selecting a different strategy, we are superimposing judgment and experience on objective analysis. We are not afraid to do so, because we know that the model is an approximation of the real world. *In most such cases the solution to the model serves as a guide to the strategy we choose. Moreover, the strategy actually selected will usually be very similar to that which the model indicates to be best.* In this illustration it is, in fact, one of the two equally good best strategies.

If we were subsequently to lose a salesman who could not be immediately replaced, we would perform the same analysis again, using Table 10-2. This table incorporates the same data as the previous one, but it is set up for a total of five, rather than six, salesmen. This table indicates that the best allocation of the five salesmen is four to area 2 and one to area 1. An allocation of all five to area 2 would not be much less desirable but would result in the total neglect of area 1, probably an undesirable long-term proposition.

TABLE 10-2   PROFIT FOR SALES MANPOWER ALLOCATION OF FIVE SALESMEN TO TWO AREAS

Number assigned to area 1	5	4	3	2	1	0
Number assigned to area 2	0	1	2	3	4	5
Profit in area 1	90	80	72	65	53	40
Profit in area 2	20	27	40	60	77	89
Total profit	110	107	112	125	130	129

Allocation of salesmen on basis of customer characteristics and salesman characteristics.   In many cases, both the variation in the profitability of customer groups *and* the variation in ability of the sales personnel are significant enough to be incorporated into sales effort allocation decisions. Often these two aspects, the sales *potential* of a customer group and the relative abilities of salesmen, are treated independently when they would better be treated as interdependent factors affecting the outcome of the decision problem.

Consider the trivially simple situation shown in Table 10-3, for example. The table shows the expected profit which will result from the assignment of each salesman to each territory. These data indicate that salesman *A* is more able than *B* (because he will produce greater profit in either territory) and that territory II is better than territory I (because greater profit will be gained there regardless of the salesman assigned).

**TABLE 10-3 EXPECTED PROFIT FOR VARIOUS
SALESMAN-TERRITORY COMBINATIONS (THOUSANDS)**

Salesman	Territory	
	I	II
A	$10	$20
B	$ 9	$13

The decision maker has only two alternative strategies:

S_1: *A* to I and *B* to II
S_2: *A* to II and *B* to I

If he chooses S_1, he is choosing the profits on the main diagonal of Table 10-3, for a total profit of $23,000. If he chooses S_2, he will obtain a total profit of $29,000. Thus, he should choose S_2. The result, which can be generalized, is that the *best salesman should be assigned to the territory with the greatest potential, the second-best salesman to the territory with second-best potential, etc.*

Knowing this result, the sales manager can make territory assignments without constructing a table such as that shown. Of course, this assumes that there is a salesman who is uniformly best (for all territories), one who is uniformly second-best, etc. It also makes the same assumption concerning territories. In many situations this will not be the case. For instance, the best salesman in one territory may be best because of his particular affinity with the types of businesses which predominate there. In another territory he may not be better than many others, because different kinds of business customers are prevalent there. In making assignments in such situations, no general rule of thumb such as "assign the best to the best" is applicable.[16]

The situation in Table 10-4 demonstrates the case in which there is no uniformly best salesman or territory. There, four salesmen are to be assigned to four territories, one to each. No one salesman is superior to all others in all areas, and no area is uniformly better than all others. The profit attained is the result of *both* the area's potential and the salesman's ability, and an able salesman may do better in a poor area than a poor salesman in a good area. In this case the best assignment is not at all obvious.

[16] The intuitive appeal of the rule "assign the best salesman to the best territory" makes it appear to be a logical rule of thumb. In reality, however, it is a *decision rule* which can be mathematically developed. See, for example, Sasieni et al., p. 194.

TABLE 10-4   EXPECTED PROFITS (THOUSANDS OF DOLLARS)

Salesman	Territory			
	I	II	III	IV
A	1	8	4	1
B	5	7	6	5
C	3	5	4	2
D	3	1	6	3

The profit-maximizing assignment may be determined either by enumerating the total profit associated with each of the 24 possible assignment strategies or by using the simple algorithm given in the next section. Here, we wish only to illustrate the concepts involved and the difficulty in determining the best strategy—even in the simple case involving only four salesmen and hence, 24 available strategies. The illustration is made more vivid when we recognize that if we had to assign 20 salesmen to 20 territories, the total number of possible strategies is the fantastic number 2,432,902,008,176,640,000. In this case, we would clearly not enumerate all of the strategies. Instead, it would be necessary to resort to an efficient computational algorithm such as that to be given.

The best assignment in the problem of Table 10-4 is

A: II
B: IV
C: I
D: III

with a resulting total profit of $8 + 5 + 3 + 6 = 22$ (thousand dollars). The reader who has attempted to determine this for himself will find that the algorithm in the next section is much less tedious than enumeration, even for this very small problem.

Note that this solution *cannot* be interpreted in any terms which are similar to those of the rule of thumb given previously. In this case there is neither an overall best salesman nor a best territory, but rather merely salesman-territory combinations which produce various degrees of profit.

Technical Note on an Algorithm for the Salesman Assignment Problem.[17]   The assignment problem just treated—that of assigning a group of salesmen to territories in an optimal fashion consid-

[17] This technical note involves only algebra. It is set off as a technical note only because it entails a solution algorithm which is not necessary at first reading and which may not be of interest to all readers.

ering both the differences in the salesmen's ability and the differences in each area's potential—involves a large number of strategies which are available to the decision maker. If n salesmen are to be assigned to an equal number of territories, one to each, there are $n! = n(n-1)(n-2) \cdots (2)1$ different strategies. This is easily seen by recognizing that an assignment of one salesman to one territory is equivalent to the selection of one element from the $n \times n$ assignment array (such as that shown for $n = 4$ in Table 10-4). Therefore, the first salesman to be assigned can be assigned to any one of n territories. Since each territory requires only one salesman, the second salesman has only $n-1$ possibilities, given that one salesman has already been assigned. Similarly, the third salesman has only $(n-2)$ possibilities and so on, until the last salesman has only a single possibility. The total number of ways in which these n assignments can be made is therefore

$$n(n-1)(n-2) \cdots (2)1 = n!$$

since the first can be assigned in any of n ways, the second in any of $n-1$ ways, etc.

Since $n!$ is a rather large number even for n of a modest size, some efficient device is needed for sifting through the available strategies in this "simple" decision problem under certainty. We shall demonstrate an algorithm, using several examples. The first point that we need to note is that our criterion of optimality, the criterion for choice of the best strategy, is *maximization of profit*. We shall find that it is simpler for us to think in minimization terms[18] so that we need to recognize that *the maximization problem can be transformed to a minimization one by using the negative of the profit function*, i.e., the negatives of each of the profits in Table 10-4, for instance. To demonstrate this, consider the profit function in such a decision problem to be that denoted by P in Figure 10-7. The negative of this function is shown as $-P$. The maximum value of the profit function is achieved at a value of the decision variable which has been denoted as x_0. Note that the *minimum* of the negative P function, which is simply the mirror image of the P function, occurs at the same value (x_0). Thus, *if we find the value of the decision variable which corresponds to the minimum of the negative of P, the maximum value of P will occur at the same point*. This gives us a simple device for converting a maximization criterion into a minimization one if for some reason it is mathematically convenient for us to do so. We do this in the case of the assignment model by multiplying the dollar quantity corresponding to each salesman-territory combina-

---

[18] The reason for this will quickly be made apparent.

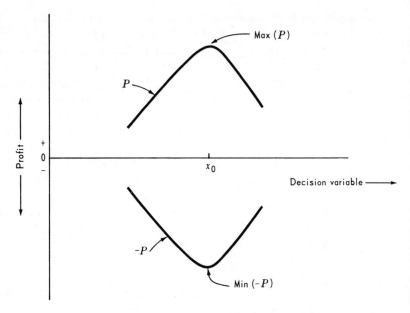

FIGURE 10-7   MAXIMIZATION OF PROFIT

tion by -1. We emphasize again that *we do this for no reason other than the mathematical convenience which it produces for us.*

Using the problem of Table 10-4 to demonstrate, we simply multiply each element by -1 and think in terms of minimization.

TABLE 10-5   ASSIGNMENT TABLE USING NEGATIVE OF PROFITS

Salesman	Territory			
	I	II	III	IV
A	-1	-8	-4	-1
B	-5	-7	-6	-5
C	-3	-5	-4	-2
D	-3	-1	-6	-3

Thus, we get Table 10-5. To solve the problem, we wish to choose elements in Table 10-5, one and only one in each row and column, such that their *sum* is as *small* as possible. We choose exactly one element in each row because this indicates that each salesman can be assigned to only one territory, and exactly one element in each column since this indicates that each territory is assigned to only a single salesman. We concentrate on the *sum* because we are using *total* profit (from all territories) as a proxy for utility. We make the sum as *small* as possible because we have transformed from the maximization of profit to the minimization of negative profit for mathematical convenience.

Suppose that we choose the *smallest* element in the first row of the table and subtract it from every other element in that row, i.e., choose —8 in the first row and subtract it from each element to get Table 10-6. Any assignment in this table involves a sum which is 8 greater than it previously was (because we have *subtracted* a —8 from *each* element in the first row and exactly *one* element in the first row is included in every assignment). For

TABLE 10-6   FIRST REVISED ASSIGNMENT TABLE

7	0	4	7
—5	—7	—6	—5
—3	—5	—4	—2
—3	—1	—6	—3

instance, the assignment consisting of the main diagonal elements can be seen to have a total sum of —15 (—1 —7 —4 —3 = —15) in Table 10-5, and —7 (7 —7 —4 —3 = —7) in Table 10-6. However, *the assignment which minimizes the sum in Table 10-5 also minimizes the sum in Table 10-6.* Of course, the value of that minimum sum will be different (by 8) in the two tables. In fact, it is generally true that *we may add any constant to every element of any row or column in this array* and find that the *assignment which minimizes the sum in one table will also minimize the sum in the revised table.*

TABLE 10-7   SECOND REVISED ASSIGNMENT TABLE

7	0	4	7
2	0	1	2
2	0	1	3
3	5	0	3

Hence, let us proceed to subtract the minimum (least algebraic) entry in each row from *every element in that row*. This gives Table 10-7. If we then subtract the smallest element in each *column* from every element in the column, we get Table 10-8.

Now, consider what we have done. We have changed the assignment table to one without negative values, and we have done this in a fashion which assures us that the assignment which minimizes the sum of the elements in Table 10-8 is the same one which minimizes the sum in the original problem. Finding that assignment now depends only on the obvious fact that *since there are no negative elements in Table* 10-8, *the minimum sum cannot be less than zero.* Hence, if we can find an assignment which has a sum which is identically zero, it must be one which corresponds to the minimum sum.

To find such a sum, we simply look for zeros which appear alone in their respective rows or columns. For instance, each of the last two columns has only a single zero. If there is an assign-

TABLE 10-8   THIRD REVISED ASSIGNMENT TABLE

5	0	4	5
0	0	1	0
0	0	1	1
1	5	0	1

TABLE 10-9   ELEMENTS IN OPTIMAL ASSIGNMENT

	I	II	III	IV
A		✓		
B				✓
C	✓			
D			✓	

ment having a zero sum, it must include these elements because there must be exactly one element chosen in each and every row and column. Similarly, there is only one zero in the first row. Having identified these, the assignment indicated in Table 10-9 may be seen to be one having a zero sum. The reader should note that this corresponds to the solution given in the previous section. In the original assignment problem array of Table 10-4, this corresponds to a maximum total profit of 22 (thousand dollars), $(8 + 5 + 3 + 6)$.

In some cases, then, this simple and straightforward procedure leads us to a best assignment. We simply multiply all profit entries in the assignment table by -1 and then revise the table by subtracting the smallest element in each row and column from every element in that row or column. This leads us to a revised table which has a number of zero entries. We then use a trial-and-error approach to determine an assignment which includes only zero elements in this table. If we can do so, the corresponding assignment is the best one and the actual maximum profit can be determined by summing the corresponding elements *in the original assignment table* (e.g., Table 10-4).

However, it is not necessarily true that the procedure of subtracting the lowest element in each row and column from every element in the row or column will lead to a table in which a zero-sum assignment can be determined. If it is not possible to do so, a further simple extension of the algorithm will suffice. Consider, for example, that we have reduced a 5×5 salesman-territory profit array by taking the negative of each element and following the subtraction procedure, and that the result is that shown in Table 10-10.[19] If we attempt to develop a zero-sum assignment here, we

TABLE 10-10   REVISED 5 $\times$ 5 ASSIGNMENT TABLE

5	[0]	8	10	11
0	6	15	0	3
8	5	[0]	0	0
[0]	6	4	2	7
3	5	6	[0]	8

choose the indicated elements (since each is the only zero in either its respective row or column) only to find that our next natural choice—the single zero appearing in the last column—will not be allowable (because it appears in the third row and we have already chosen a zero entry in the third row; i.e., we have already assigned the third salesman and cannot therefore assign him again). Any other sequence of selecting these five points would lead us to the same dilemma. Therefore, it is not possible to make a zero-sum assignment by using this table.

However, we still desire to find the minimum-sum assignment. To do this, we *draw the minimum number of horizontal and vertical lines through the rows and columns of the table which are adequate to cover all of the zeros.*[20] In this case the lines are

[19] Taken from Sasieni et al., p. 191

[20] It is not necessary to have the least possible number of such lines, but the procedure is more efficient if we use as few as we can to cover all zeros easily.

those shown in Table 10-11. Now *we subtract the smallest element not covered by a line from all uncovered elements and add this quantity to all elements lying at the intersection of two lines.* In this case the smallest uncovered element is 3, which leads us to Table 10-12.

TABLE 10-11   REVISED TABLE WITH LINES DRAWN THROUGH ZEROS

~~5~~	~~0~~	~~8~~	~~10~~	~~11~~
0	6	15	0	3
~~8~~	~~5~~	~~0~~	~~0~~	~~0~~
0	6	4	2	7
3	5	6	0	8

TABLE 10-12   ASSIGNMENT TABLE SHOWING BEST ASSIGNMENT

8	[0]	8	13	11
0	3	12	0	[0]
11	5	[0]	3	0
[0]	3	1	2	4
3	2	3	[0]	5

This creates a table having zeros in some new positions. Using this table, we now try to find a zero-sum assignment. We do this, of course, just as we did before, by first searching out those zero elements which are unique in a row or column, etc. A zero-sum assignment is shown in Table 10-12. It represents a profit-maximizing assignment in the profit array of the original problem. If we still were unable to find a zero-sum assignment, we would draw new lines in *this* table and repeat the procedure until we arrived at a table in which a zero-sum assignment can be found.

To summarize the algorithm, we:

1. First multiply every element in the profit table by −1.
2. Subtract the minimum element in each row and column from every element in the respective row or column.
3. Seek a zero-sum assignment in the resulting table. If one is found, it is the profit-maximizing assignment in the original problem.

4. If one is not found, draw lines to cover all zeros.
5. Subtract the smallest element not covered by a line from all uncovered elements and add this quantity to all elements at the intersection of two lines.
6. Repeat steps 3 to 6 until a zero-sum assignment is found.

Salesman Recruitment and Selection. The problem of obtaining the best possible people to perform an organization's personal selling tasks is of great importance to the sales manager. The ultimate success of the enterprise depends on sales, and sales depend, to some large degree, on the ability of the salesman. Since the cost of recruiting and hiring a salesman may well be in the $15,000 area, one cannot adopt a policy of hiring anyone and allowing him to try his luck. Moreover, the (unknown) costs in customer ill will which might result from such a policy would probably be prohibitively high.

The three major aspects of recruitment and selection from the viewpoint of managerial decisions are:

1. The determination of individual characteristics which can be used to *predict* sales performance.
2. The development of procedures for finding and interviewing potential sales candidates.
3. The *selection* of salesmen from among the applicants.

Determining predictive characteristics. It is a generally accepted maxim in the sales field that there are general characteristics of individuals which can be used to foretell sales success. Moreover, there are obvious requirements that any organization places on individuals it hires; they cannot be over 60 years old in some cases, for example. A summary of all personal characteristics which are believed to reflect on sales success is a basic necessity for rational recruitment and selection decisions.

The fashion in which such characteristics are compiled is often largely subjective. For instance, a company might believe, relying on past experience, that it is better to hire people who have previously been successful in sales jobs—and to pay the higher salary which they will probably demand—than to hire totally inexperienced salesmen. Therefore, the sales experience characteristic would be one of the criteria chosen. Sometimes more objectively based characteristics indicated by surveys in which the records of groups of good and poor salesmen are analyzed with a view to finding essential differences between the groups. In one such study of life insurance agents, for example, it was found that sales candidates who eventually turned out to be productive agents were family men, not financially independent ("hungry," as one observer put it), who already owned life insurance on themselves. A disproportionate percentage of the successful group also belonged to a number of civic and social organizations and had sales experience.

Recruitment. Having such basic information on the characteristics of desirable sales candidates, the sales manager can approach the second problem—designing a program to ferret out likely candidates. In many respects the sales manager's recruitment task is similar to one aspect of media advertising. In determining the predictive characteristics, he has defined a potential "audience" whom he wishes to sell on the idea of accepting sales positions. His problem is to *reach* the individuals in his audience. (In fact, he may actually do this by media advertising.) This is the same view we adopted in the specialized media allocation model of Chapter 9.

If it has been determined that young, inexperienced, but well-educated candidates are desirable, the manager may decide to design a program of calls on college campuses. There, he can interview graduating students who are interested in a sales career. The colleges which he visits would be selected on the basis of the programs offered; e.g., a chemical firm would visit schools teaching chemistry and chemical engineering if it wished to hire technical salesmen.

Advertising in the classified section of the newspaper may also be desirable, especially if no sales experience is required (since many salesmen tend to change jobs on the basis of personal contact rather than in response to ads). Or some personal recruitment by salesmen in the field may be in order. The recruitment program possibilities are virtually limitless, but since the problems involved seem to defy adequate quantification, we shall not treat the details of recruitment further.

Selection. The salesman selection problem is one which can be quantified and analyzed. Basically, the selection problem involves the identification of those candidates who are to be offered sales positions from among those who are contacted in the recruitment effort. Various tools are available to aid in this task. Among these are usually the applicants' responses to questions on application forms, personal references, credit and business references, interviews, and various tests which may be administered.

These tools are equivalent to second-order predictive devices which are applied after the initial delineation of predictive characteristics and the accomplishment of the recruitment program. In essence, the general characteristics are applied first and the size of the group of potential candidates is brought to manageable proportions; e.g., old men, persons who do not speak English, those without sales experience, etc., may be eliminated. After candidates who meet the initial criteria are found, additional predictors are developed for sales success.

It is possible to make predictions of sales success on totally subjective bases. For example, the vice-president who interviews candidates and makes selections on the basis of "looking in his eyes to see if he can cut the mustard" is using a highly subjective device for predicting sales success. Or a formally structured interview in which the applicant is rated on mental alertness, speech, appearance, etc., may be used to predict his eventual performance. Scores on mental ability tests, interest tests, personality tests, or sales aptitude tests may also be used.

In any event, whatever the tools used by a particular firm, they serve to define a set of predictors of sales success. These predictors may then be used to indicate the degree of success to be expected from each of the available candidates. This may be accomplished in several ways. In some instances a single predictor may be considered adequate to predict future sales success. Thus, a critical sales aptitude test score may be so defined that applicants who do not achieve this score are not considered further and those who do are hired. Operationally, such a critical score would usually be combined with other predictors, so that the achievement of at least the critical score does not ensure the candidate's being hired, but rather being considered further. The recommendations resulting from interviews and references are most often used in conjunction with a test score in this fashion to determine those who will be appointed.

Frequently various numerical measures such as test scores are available as predictors. It is desirable to combine these scores with other numerical ratings which may be available[21] in order to develop a single prediction of success which incorporates the predictive value of all available predictors. Various statistical techniques have been used for these purposes. One method which may be used is to approach the predictive problem in a manner similar to that which we used for predicting sales results in potential market areas in Chapter 6.

The basic nature of the two predictive problems is the same. In the one case data are available on market area characteristics which can be used to predict marketing success. In the other, the predictors are test scores and other numerical information which appear to be valid and reliable predictions of the success of salesmen. In the case of market areas, data may be available on the predictors and eventual sales results in a number of test market

---

[21] For instance, personnel recruiters often make numerical ratings based on the results of interviews with applicants. See the Nunnally reference for further discussion.

areas. In the case of salesmen, data on past experience with tests administered to potential salesmen and the eventual success of those appointed may be available from sales department records. The use of these data in making predictions may be similar for the two different predictive problems.

The reader should recall that one of the approaches taken to the market area prediction problem in Chapter 6 was to use a qualitative measure of performance for market areas. This also seems to be meaningful in the case of predicting the success of salesmen, since the idea of imposing a quantitative performance measure such as "sales revenue generated" or "commissions earned" would tend to discriminate against those salesmen who are assigned overwhelmingly conversional activities, those assigned to territories which have low potential, etc. This is the approach which we shall adopt here. There are two primary reasons for this. First, it appears to be a useful approach. Secondly, it is possible to integrate this predictive approach effectively with the salesman assignment model discussed previously in a fashion which permits a comprehensive treatment of performance prediction *and* salesman assignment.[22]

To do this, let us consider a dichotomous job performance measure which we shall term success (S) or failure (F); i.e., every sales appointee is either a success or a failure (just as every market area in Chapter 6 was assumed to be either good or bad). If such a performance measure can be applied to salesmen with whom the organization has past experience, it should be possible to estimate the probabilities (based on past relative frequencies):

q_S = probability that a randomly selected candidate will be a successful salesman

q_F = probability that a randomly selected candidate will be unsuccessful as a salesman

where

$$q_S + q_F = 1$$

In other words, q_S and q_F are simply (estimates of) the probability of sales success or failure for individuals who meet the broad criteria which are initially used to predict potential sales success (e.g., age, marital status, experience, etc.).

If the same quantitative measures (test scores, ratings, etc.) are being applied to the current selection problem, the probability distribution of the measures may be estimated for salesmen who

---

[22] Most of the remainder of this section is based on a portion of the King reference.

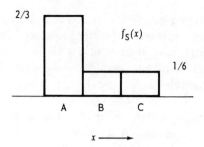

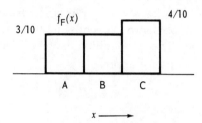

FIGURE 10-8   TEST SCORE PROBABILITY DISTRIBUTIONS FOR SUCCESSFUL AND UNSUCCESSFUL SALESMEN

have been evaluated as successes or failures in the past. We shall denote these distributions as $f_S(x)$ and $f_F(x)$ respectively. Here, x is assumed to be a single predictor variable. Figure 10-8 illustrates an overly simple situation which might exist with these probabilities. There, it is assumed that a single test, on which the applicant may score either A, B, or C, is to be used as a predictor of success. The sketch shows that

$$f_S(A) = \tfrac{2}{3}$$
$$f_S(B) = \tfrac{1}{6}$$
$$f_S(C) = \tfrac{1}{6}$$

i.e., that two-thirds of those salesmen who have been successes in the past have scored A on the test and one-sixth have scored B and C respectively. Additionally,

$$f_F(A) = \tfrac{3}{10}$$
$$f_F(B) = \tfrac{3}{10}$$
$$f_F(C) = \tfrac{4}{10}$$

i.e., 30 percent of past unsuccessful salesmen have scored an A, 30 percent a B, and 40 percent a C on the test. Note that although the test appears to be a good predictor, it is not a perfect one. For instance, although more successful people score well than do not, some significant proportion of successful people have attained each of the possible scores. Similarly, unsuccessful salesmen have achieved each score, although their tendency is toward lower scores.

Given these basic estimates, which are made from past records of salesmen who have been evaluated as successes and failures, the decision maker is interested in administering the test to a group of sales applicants and using the test score achieved to predict the success level which will be achieved by each. In particular we shall choose to make this prediction in terms of the *probability that the individual will be a sales success*. We shall denote this probability as $p(x)$, since it obviously depends on the particular test score which is attained. The probability may be calculated to be[23]

$$p(x) = \frac{q_S f_S(x)}{q_S f_S(x) + q_F f_F(x)} \tag{10-5}$$

Suppose that only 20 percent of candidates who meet the general criteria are estimated to be capable of sales success (this might be the case if inexperienced candidates were considered, for example). Then,

$$q_S = 0.2$$
$$q_F = 0.8$$

and the probability that an applicant who scores an A on the test will be successful, using expression (10-5), is

$$p(A) = \frac{0.2 f_S(A)}{0.2 f_S(A) + 0.8 f_F(A)} = \frac{0.2(\frac{2}{3})}{0.2(\frac{2}{3}) + 0.8(0.3)}$$

or about 0.36.

To judge the value of the test in selecting salesmen, we need only to recognize that if we were to appoint salesmen on the basis of the general criteria alone, only 20 percent of them would be successful (on the average) since q_S is 0.20. If we use the test and appoint only those who score an A, about 36 percent will be successful. This is a rather significant increase which undoubtedly would lead us to use the test.

Let us look further by calculating the probability that someone who scores a B on the test would be successful in a sales job.

---

[23] The reader who is not versed in basic probability is asked to accept this expression on faith for the moment. It is based on a straightforward application of basic probability concepts which are known as Bayes' theorem.

$$p(B) = \frac{0.2\,f_S(B)}{0.2\,f(B) + 0.8\,f(B)} = \frac{0.2(\frac{1}{6})}{0.2(\frac{1}{6}) + 0.8(0.3)}$$

or about 0.12. Therefore, only 12 percent of those scoring B on the test will be predicted to be successful. Moreover,

$$p(C) = \frac{0.2\,f_S(C)}{0.2\,f_S(C) + 0.8\,f_F(C)} = \frac{0.2(\frac{1}{6})}{0.2(\frac{1}{6}) + 0.8(0.4)}$$

or about 0.09. Hence, only 9 percent of those scoring C will be predicted to be successful.

If this is the only test available for prediction, a rational choice would be to appoint only those candidates who score an A on the test. Although this will still produce a rather large percentage of failures, this result is inherent in the job which is in question and the general criteria which were used for screening purposes. The use of the test and this decision rule for appointment is predicted to result in a significant increase in the proportion of successful appointees—from 20 to 36 percent. As we pointed out earlier, the decision maker would probably consider it in combination with the nonquantitative predictors such as interview results and references and select each salesman on a personal basis.

One of the primary advantages of this approach is that it is equally applicable to multiple predictors. Consider, for simplicity, that two tests are available and that only two scores are possible on each—A or B on test I and 100 or 50 on test II. Since none of the analysis which we have done depends on the number of possible outcomes, this limited situation will be adequate to describe the methodology. Table 10-13 indicates the possible results which might have been found in the case of past successful salesmen. These data indicate that 40 percent of past successful

TABLE 10-13   TEST RESULTS FOR SUCCESSFUL SALESMEN

Test I	*Test* II	
	100	50
A	0.4	0.2
B	0.3	0.1

salesmen have scored *both* an A and 100 on the two tests, 30 percent have scored a B on the first and a 100 on the second, etc. In essence, Table 10-13 is a two-dimensional probability distribution which may be symbolized as $f_S(x,y)$, where x is the score on the first test and y is the score on the second. In this terminology Table 10-13 reduces to

$f_S(A,100) = 0.3$
$f_S(A,50) = 0.2$
$f_S(B,100) = 0.3$
$f_S(B,50) = 0.1$

Table 10-14 gives similar information for past salesmen who have been failures.

TABLE 10-14   TEST RESULTS FOR UNSUCCESSFUL SALESMEN

Test I	Test II	
	100	50
A	0.1	0.2
B	0.2	0.5

Thinking of this information as the probability distribution for failures, we have

$f_F(A,100) = 0.1$
$f_F(A,50) = 0.2$
$f_F(B,100) = 0.2$
$f_F(B,50) = 0.5$

The overall impression of these two tables is that successful salesmen tend to do well on both tests and unsuccessful ones do poorly on both. Of course, neither is a perfect predictor, nor are the two taken in conjunction.

If we again assume that $q_S = 0.2$ and $q_F = 0.8$, we may calculate, using an expression completely analogous to (10-5),

$$p(A,100) = \frac{0.2\,f_S(A,100)}{0.2\,f_S(A,100) + 0.8\,f_F(A,100)}$$

which is

$$p(A,100) = \frac{0.2(0.3)}{0.2(0.3) + 0.8(0.1)}$$

or about 0.43. Thus, 43 percent of those who attain the best score on both tests are predicted to be successful salesmen. Compared with the figure of 20 percent for those candidates who survive the initial screening, this is a significant proportion and would appear to justify the use of the two tests in selecting salesmen. In the next section we will illustrate a unified treatment of salesman selection and assignment decisions which incorporates these basic ideas.

Selection and assignment.[24] The predictive aspects of salesman selection and the allocation features of salesmen assignment may be effectively combined into a single comprehensive approach. To do this, we need only to consider that the same factors which led us to argue that the assignment of salesmen to customer groups should depend on both customer differences and differences in salesmen are equally applicable to the prediction of sales performance for a salesman candidate. In other words, when we predict the future performance of applicants, we should consider that their performance may be dependent on the particular sales job to which they are assigned. For instance, one individual may perform well if assigned to a group of electronic firms as accounts, but poorly if assigned to call on millinery shops. Or a door-to-door salesman who has poor speech habits may do much better if assigned to a lower-class neighborhood than if he were assigned to a fashionable area. Other more subtle differences are likely also. In predicting success, we should take such differences into account.

To demonstrate how we might do this, let us suppose that there are three distinct customer groups to which we wish to assign salesmen. For the moment, we shall also assume that we have three individuals to be hired and assigned. We do this because the fact that we have no experience with these individuals presents the same difficulty as in the "pure" selection problems; i.e., we must predict their performance levels on the basis of test scores, ratings, etc. Later we shall consider the question of having more individuals available than there are positions to fill.

Let us suppose that we can make meaningful estimates of probabilities which are analogous to q_S and q_F *for each position.* For instance, we might know that there is a greater likelihood of being successful on a retentive basis than on a conversional one. Therefore, if the organization decided to adopt the idea which we discussed previously—that salesman should be assigned *either* to retaining existing customers or to converting noncustomers, but not to both—the position corresponding to totally conversional effort would entail a lesser likelihood of success for many candidates. However, some candidates might well be better suited to the conversional task than to the retentive one; thus, their particular probability of success would probably be higher for the conversional position than for the retentive one.

However, the overall differences in the nature of the positions would undoubtedly mean that candidates who had only met the

[24] This section is based on a portion of the King reference.

original criteria and had not yet been interviewed and tested would probably have greatly different likelihoods of success in the various positions because of the nature of the positions themselves. Considering three customer groups (rather than just two), let us suppose that the probabilities are those shown in Table 10-15. These probabilities indicate that the third position is probably a difficult one and the second is an easy one, since the third has a low success probability and the second a high one. Of course, if it is not meaningful to estimate these probabilities in a particular situation, we might just estimate q_S and q_F as given previously. In that case, all entries in the first column of Table 10-15 would be q_S and all entries in the second column would be q_F.

TABLE 10-15   PROBABILITIES THAT A RANDOMLY SELECTED INDIVIDUAL WILL BE A SUCCESS OR FAILURE IN THREE POSITIONS

Position	Success	Failure
1	$\frac{1}{4}$	$\frac{3}{4}$
2	$\frac{9}{10}$	$\frac{1}{10}$
3	$\frac{1}{10}$	$\frac{9}{10}$

Now, let us suppose that we have tests which can be used to predict individual success in the three kinds of positions. In most cases, a number of tests would be used. For example, a sales aptitude test in conjunction with a mechanical aptitude test might be valuable in predicting sales success in a position requiring the salesman to call on machinery manufacturers. For simplicity, we shall illustrate the use of a single test score. The same simple change which was used in transforming the single-predictor variable treatment to a multivariate approach in the preceding section is also applicable here.

If only three possible scores can be made on the test, our information concerning past test scores and performance may be that given in Table 10-16. These data indicate the probability that individuals who are successes or failures in each position will achieve the various test scores (based on past experience). For example, two-thirds of those who were successes in positions of the first type scored A, one-sixth scored B, and one-sixth scored C. On positions of the third type, nine-tenths of successful salesmen scored A and only one-twentieth scored B and C respectively.

Suppose that the three available candidates score A, B, and C respectively on the test. The assignment question is then which position to assign to each. To determine this, we can calcu-

TABLE 10-16 PROBABILITY DISTRIBUTIONS OF TEST SCORES

Position	Performance	Test score		
		A	B	C
1	Success	$\frac{2}{3}$	$\frac{1}{6}$	$\frac{1}{6}$
	Failure	$\frac{3}{10}$	$\frac{3}{10}$	$\frac{4}{10}$
2	Success	$\frac{1}{3}$	$\frac{1}{3}$	$\frac{1}{3}$
	Failure	$\frac{5}{10}$	$\frac{3}{10}$	$\frac{2}{10}$
3	Success	$\frac{9}{10}$	$\frac{1}{20}$	$\frac{1}{20}$
	Failure	$\frac{2}{10}$	$\frac{3}{10}$	$\frac{5}{10}$

late the probability that each individual will be successful in each position. For instance, the probability that the individual with a score of B will be a success in the second position is

$$p_2(B) = \frac{\frac{9}{10}(\frac{1}{3})}{\frac{9}{10}(\frac{1}{3}) + \frac{1}{10}(\frac{3}{10})}$$

or 0.91. This calculation is completely analogous to that of expression (10-5) *as applied to the second position*. The same kind of calculation can be made for each job, using the test scores achieved by the three candidates and the probabilities of Tables 10-15 and 10-16. The reader may verify that the resulting probability array is that shown as Table 10-17. This table shows the probability

TABLE 10-17 PROBABILITIES OF SUCCESSFUL PERFORMANCE IN THREE POSITIONS

Candidate (test score)	Position		
	1	2	3
1(A)	0.43	0.86	0.33
2(B)	0.16	0.91	0.02
3(C)	0.12	0.94	0.01

that each candidate (with his test score shown in parentheses) will achieve success in each of the three positions. Note that all three candidates have a high likelihood of success in the second position, which is presumably easy, but only one can be assigned there.

To determine the best assignment on the basis of this information, we need some indication of the *difference in the worth of successful versus unsuccessful performance in each job*. Such a measure is an organizational utility in the sense discussed in Chapter 3. A utility measure might be estimated by using the Von Neumann—Morgenstern standard gamble technique (see Chapter 3) or these "worths" might be developed by using techniques of job evaluation (see the Jucius reference). For this example we shall assume that we can estimate the *average profit difference resulting from successful versus unsuccessful performance in each of the positions.* If these differences are 2, 1, and 3 (thousand dollars) respectively for the three positions, we may determine a standard assignment table, in terms of *expected profit,* by multiplying the three columns of Table 10-17 by 2, 1, and 3 respectively. The resulting expected profit table is shown as Table 10-18.

TABLE 10-18   EXPECTED PROFIT RESULTING FROM VARIOUS ASSIGNMENTS (THOUSANDS)

Candidate	Position		
	1	2	3
1	0.86	0.86	(0.99)
2	(0.32)	0.91	0.06
3	0.24	(0.94)	0.03

If we choose *maximization of expected* profit as our criterion, we may solve this assignment problem by using the algorithm which was previously given. The solution is the one indicated by the circled elements in Table 10-18. Thus, on the basis of our predictions of the sales success of the three candidates, we should assign them to the second, third, and first positions respectively.

Having integrated the predictive aspect of salesman selection with the problem of salesman assignment, we need now only to demonstrate the use of this approach in a more typical situation in which *there are more available candidates than positions to be filled.* We shall give an example in which there is only one excess candidate, but the method is equally applicable to any number.

Suppose that the previous example is changed by deleting the third position. Hence, three candidates are available for two positions and the relevant expected profit array is just Table 10-18 with the last column omitted. The standard approach to solving an assignment problem of this type depends on having an equal number of candidates and jobs. To satisfy our need for a third position in this case, we can *simply insert a fictitious third position in which everyone is predicted to achieve an expected profit of zero.* Having done this, we may then proceed to solve the fictitious assignment problem represented by Table 10-19.

TABLE 10-19   EXPECTED PROFITS IN FICTITIOUS ASSIGNMENT PROBLEM

Candidate	Position		
	1	2	3
1	(0.86)	0.86	0.0
2	0.32	0.91	(0.0)
3	0.24	(0.94)	0.0

The formal solution to this assignment problem, determined by using the algorithm previously given is shown by the circled elements in Table 10-19. The interpretation of this assignment is simple. The first and third candidates are hired and assigned to the first and second positions respectively. Since the second candidate is assigned to the fictitious position by the model's solution, he is not selected for appointment. The third position in this fictitious problem is profitless. It corresponds to a failure to be selected for appointment as a salesman. Since each candidate will produce the same profit if he is assigned to it, no unfair advantage is given anyone in solving the assignment problem in this manner. There will be exactly one assignment to the third position in every possible strategy and this assignment will contribute nothing to the total expected profit levels associated with any of the strategies. Thus, assignments are made on the basis of the best combination of expected profit levels in the first two (real) positions; the candidate who is left over is assigned to the third (nonexistent) job. In this way the predictive aspect of salesman selection, which was previously incorporated in the salesman assignment decision, can be amplified in its usefulness, for the predictions of sales success are being used to solve simultaneously both the selection and the assignment problems of sales force management.

Technical Note on the Use of Bayes' Theorem in Salesman Selection.[25]   Suppose that an event A can occur only after one or the

[25] This technical note requires an understanding of *conditional* probability. Any basic probability text will contain a more extensive discussion of the details of Bayes' theorem.

other of two events, S or F, has occurred. (S and F are mutually exclusive and exhaustive.) Suppose further that the probability of occurrence of S and F is known and that the conditional probability of A given either S or F is known. In other words, the following probabilities are known:[26]

$P(S)$
$P(F)$
$P(A|F)$
$P(A|S)$

If we wish to know one of the conditional probabilities such as

$P(S|A)$

we may proceed by recognizing, from the basic definition of conditional probability, that

$$P(S|A) = \frac{P(AS)}{P(A)} \qquad (10\text{-}6)$$

The intersection event in the numerator (to be read as "the probability of occurrence of both A and S") may be written as

$$P(AS) = P(S)P(A|S) \qquad (10\text{-}7)$$

and, since the events S and F are mutually exclusive and exhaustive, the denominator may be written as

$$P(A) = P(AS) + P(AF) \qquad (10\text{-}8)$$

Inserting (10-7) and (10-8) into (10-6), we have

$$P(S|A) = \frac{P(S)P(A|S)}{P(AS) + P(AF)} \qquad (10\text{-}9)$$

Each of the probabilities in the denominator may be rewritten in the conditional form to obtain

$$P(S|A) = \frac{P(S)P(A|S)}{P(S)P(A|S) + P(F)P(A|F)} \qquad (10\text{-}10)$$

This expression, which is essentially Bayes' rule, gives the conditional probability sought in terms of known probabilities.

In the context of salesman performance prediction, the events S and F are respectively that a salesman is either a success or a failure, and A may be taken to be a particular test grade. Thus,

[26] Probabilities are indicated by $P(\quad)$. Probabilities such as $P(A|S)$ are conditional probabilities, which are to be read "the probability of A given the occurrence of S."

we know $P(S)$ and $P(F)$—the probability that a candidate who has survived the initial screening will be a success or failure respectively. We also know $P(A|S)$ and $P(A|F)$, i.e., the probability of successful and unsuccessful salesmen achieving the test score in question. Therefore, expression (10-5) is identical to (10-10), with obvious differences in notation.

Salesman Compensation. Another important area of sales force decisions is the remuneration of salesmen. One can recruit and select those salesmen who have the greatest potential for success, but if the financial motivation, which drives each of us to some degree, is not great, few are likely to achieve the success predicted for them. Of course, there are nonmonetary motivational devices which may prove to be very effective. Among these are fringe benefits, prizes, awards, the granting of special privileges, status symbols, etc. In fact, in recent years such simple motivations and rewards as wall-to-wall office carpeting and special parking spaces are becoming increasingly prevalent at the executive level. In particular, in the case of a salesman who has previously achieved financial success, and therefore has small marginal utility for additional compensation, as we have noted earlier, nonfinancial incentives may be of greatest value. However, in the majority of cases and in the context of the overall organizational profit result, the compensation of salesmen ranks as the most straightforward, simple, and effective device for motivation.

The design of a good compensation plan is the major decision problem in the compensation area. Of course, there is no such overall best plan. The best plan in any particular circumstance depends on three primary factors—*sales functions, profitability,* and *competition.*

The functions which the salesman is expected to perform must be of paramount importance in determining his compensation. If he is compensated for some functions and not for others, he will naturally tend to concentrate on those which involve the greatest payoff (to him). What the organization wishes him to do is to concentrate on those which have greatest profit payoff to the organization. Thus, the second important factor is *profitability.*

The result of these two factors may be summed up in two simple principles. First, *a salesman should be compensated for all organizational functions which he is expected to perform.* Secondly, *the amount of compensation for each function should be determined on the basis of its relative profitability to the organization.*

Thus, if a salesman is expected to perform technical services for the customer, to erect point-of-purchase displays, or to do anything else which is desired by the organization but does not have results reflected in the immediate sales revenue dollars which are credited to him, he should be paid for his services. The level of such payments should be appropriately related to the relative long-term profitability of each activity. Unfortunately this is extremely difficult to measure quantitatively for many activities. The difficulties occur because many sales activities are not meant to produce revenues directly. Thus, the services performed by a salesman may result in good will, which eventually produces added sales. But how is this to be measured? Similarly, time spent in erecting point-of-purchase displays is chargeable to promotional costs; yet, as we have discussed earlier, the measurement of promotional effects is neither simple nor inexpensive. Because of these difficulties, the two basic principles in dealing with sales functions and profitability must usually serve only as general guidelines in the development of a compensation plan.

We should emphasize that the application of these basic principles is not an easy task, even if the measurement problems are not considered. For example, the filing of sales reports is a typical sales function which contributes to long-term sales success of both the organization and the individual salesman. But the effects of these reports are so far removed from the salesman's day-to-day activities that they often seem to him to be useless nuisances which hinder his concentration on selling. The trade-offs between the various activities which the salesman is expected to perform should be constantly borne in mind by the manager. For instance, to include a bonus for correct and timely submission of sales reports in a compensation plan might result in better reporting. But would it result in overall benefits to the organization? Such trade-offs between the value of direct selling and the value of nonproductive activities are of great significance to the organization in the design of the compensation plan.

No generally valid principles, other than those already presented in the context of allocating expenditures to various advertising media, are easy to determine for this allocation problem. The basic problem here is the same allocation problem discussed for advertising media, i.e., the salesman's allocation of time and effort to various activities. The allocation problem is made more complex here because the manager is once removed from the actual allocation decisions. The manager must design a compensation plan which motivates the salesman to allocate his effort in the most effective fashion. Coupling this with the measurement difficulties, we can easily see why no quick and easy answers are available. Perhaps it is sufficient to note that the manager must be constantly aware of this problem so that, in his zeal for administration, he does not load excessive nonproductive activities onto the salesmen. This is best illustrated by the amusing series of communications within a fictitious sales department (source unknown):

First Letter from Salesman to Sales Manager:
Dear Boss: I seen this outfit which they an't never bot a dime's worth of nothing from us and I sole them a couple thousand dollars worth of guds. I am now goin to Chawgo.

Second Letter from Salesman to Sales Manager:
Dear Boss: I cum here an sole them ten grand.

Sales Manager's Memo to Sales Department (with above two letters appended):
We ben spending to much time hear tryin to spel insted of tryin to sel. Let's watch those sails. I want everybody should reed these letters from Gooch who is on the rode doin a grate job fur us and you should go out and do like he done.

The third factor which affects the design of a compensation plan, competition, serves as a constraint on both the amount of compensation which can be paid and the design of the plan. It is a truism to say that good salesmen will not long be retained if one's compensation plan compares unfavorably with that of competition or if the total compensation which individuals can achieve is significantly less than they could obtain elsewhere. Valid information on the compensation plans of competitors and compensation levels of individual competitive salesmen is difficult to obtain. The most direct source is generally available to the salesmen, rather than the sales manager, because the salesman is in informal contact with representatives of competing firms as he pursues his work in the field. Thus, the manager may be forced into a position of bargaining with salesmen who have better information than he has.

The most common types of salesman compensation plans are *straight commissions* and *salary plus commissions.* The most significant managerial decision problems involve the commission aspect, for commissions serve as the monetary incentive to produce greater sales. The two important aspects of most commission schedules are *the commission rate to be paid for each item and the relationship of the rate to volume.*

We have already noted that the commission rate should be related to profitability. If it is not, the salesman will not be induced to perform in the fashion which is best for the organization. If commission rates are profit-related, the salesman's individual objectives (remuneration) and the organization's objectives become identical, or at least the actions which serve to achieve one also achieve the other. Thus, the commission rate may be different for different products or different customers. If the rate is different for the various products which a single salesman sells, the profit-oriented rate will tend to deliver the salesman from the temptation to concentrate sales effort on those items which are *easy* to sell rather than those which are profitable.

The commission rate is often related to volume by the use of a sliding scale; e.g., the salesman is paid 5 percent of the first $200,000 in revenue and one *additional* percent of each additional $100,000. The economies of scale which are achieved in large orders by a single customer, or in a single area or time period are passed on to the salesman.

Often a *marginal approach* may be adopted for commission schedules. For instance, average profit may be ignored and a salesman may be paid a higher commission for conversional results than for sales from existing customers. Or, higher rates may be paid on items which are overstocked. In such instances, the *extra* revenue in excess of costs is being sought in a short-run approach to salesman compensation.

We shall illustrate a *compensation formula* which is used for area sales supervisors by one company. These supervisors are not salesmen but rather, local sales managers; yet, the particular plan illustrates most of the salient points concerning all sales compensation plans. These supervisors are paid according to the formula

$$\$5,000 + 0.05a + 0.20b + 0.50c$$

where a = total commissions earned by salesmen who are supervised

b = commissions earned by *new* salesmen who are supervised

c = unspent portion of administrative budget

This compensation plan entails the payment of a base salary, 5 percent of total commissions which are earned by supervised salesmen, an extra payment of 20 percent for those commissions earned by new (less than two years) salesmen, and a payment (penalty) dealing with budget expenditures. Thus, the supervisor is heavily motivated to aid new salesmen in becoming successful, for he receives 25 cents for every dollar that they earn (5 percent plus 20 percent). He may spend money from his administrative budget to provide them with clerical services, training, etc., but if he exceeds the budget, he absorbs one-half of the overage. This sort of plan forces the supervisor to consider explicitly the trade-offs which exist and the allocation problem which is created for him by the plan. If the compensation rates are keyed to long-term organizational profitability, the compensation formula should serve to align the objectives of the firm and the supervisor, so that actions which benefit one will also benefit the other.

Control and Compensation. Control of the sales force is one of the primary operational objectives of sales management. The ability of management to implement each of the decisions which are made is directly related to the degree of control which it exerts over the sales force. The most direct control device is the compensation plan, for if a sales manager cannot order a salesman to erect displays, or to push particular items, he can *compensate the salesman in a fashion which makes it his own best interest to do as management wishes*. This is the basic element of sales force control.

The importance of control is amplified in situations in which a resort to authority is not only undesirable but impossible. If the salesman is an employee, and all else fails, he can be ordered to take whatever actions management wishes. But in some sales situations the salesman is an independent businessman. For instance, many insurance agents and brokers are not employees of the insurance company for whom they sell. Thus, the companies exercise little authority over them. All control must be of a motivational nature.

One aspect of sales force control which is of great importance is the *identification of areas of sales difficulty*. For instance, since the attrition rate in sales is generally high, the sales manager knows that some of his recently appointed salesmen will not be successful. It is to the organization's benefit if these failures can be identified early so that control actions can be taken. In some cases adequate training may not have been provided, and if the lack can be detected early, the salesman may be retrained. To fail in early detection is to allow the salesman to plod along with unsatisfactory results so long that he loses hope and desire and decides that the job is not for him. In other cases, it may be desirable to eliminate salesmen who perform badly because they may be causing irreparable harm to the company through their sales inadequacies.

In one situation in which this aspect of control is of particular significance, the salesmen are paid on straight commissions. Because new salesmen are not expected to earn commissions which are adequate to support their families immediately, a *drawing account* is set up. This account, in effect, allows the new salesman to borrow against future commissions, up to some specified amount. If the salesman does not eventually achieve a commission level at which he can repay these loans, the organization incurs large losses. Therefore, the problem of early identification of potential failures is extremely important, for early identification allows early

control action in the form of retraining to increase earning potential
or dismissal (to cut the organization's loss).

Superficially the early identification problem appears to be trivial.
One simply notes the commissions earned in the early months of
selling and takes action in those cases in which commissions are
too low. But what is too low? The answer, of course, must be
stated in terms which are relative to the performance of other
salesmen.

One approach to this identification problem is shown in Figure
10-9, which shows a lot of total commission earnings and the
number of months of selling activity for past salesmen. The upper
line represents the (smoothed) *average* total commissions earned
by new salesmen who have become successful in the past. The
lower line represents the level of total commissions above which
95 percent of past successful salesmen have earned. Each new
salesman's total commissions to date are plotted on the graph, and
a comparison is made with its relationship to the two lines.
A salesman whose earnings graph lies above the top line is doing
well—better than about 50 percent of past successful salesmen.
A salesman whose graph falls below the bottom line is a potential
candidate for difficulty. In the latter case, the reasons for this rela-

FIGURE 10-9   SALESMAN COMPENSATION GRAPH

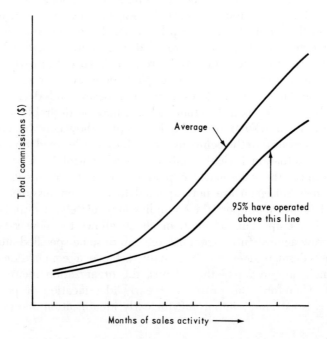

tively poor performance should be investigated. Other cases would probably warrant careful watching to ascertain whether the trend is up or down.

Using a graph such as this, the sales manager may *manage by exception.* In this case, the exceptions are the salesmen who demonstrate early poor performance. Concentration of management effort on them may permit the correction of small difficulties which are hampering their sales results, or it may allow the organization and the deficient salesman to part company before extensive damage is done either to the firm's sales or to the individual's morale and self-confidence.

The Traveling Salesman. Without arguing that the problem of the traveling salesman is a major sales force decision, we shall mention it in passing. The *problem* is interesting for several reasons. First, it belongs to a category of real problems which must be faced by salesmen and sales managers. Secondly, it is very similar to the basic assignment problem discussed earlier; yet, it is different enough to provide new insight into the structure of the model. To the analyst the traveling salesman problem has special significance because it is one of those few mathematical models which are simple in form but have proved intractable from the viewpoint of developing a general solution algorithm.

TABLE 10-20   TRAVELING SALESMAN MILEAGE TABLE

From	To			
	Home office	1	2	3
Home office	0	80	40	50
City 1	80	0	20	70
City 2	40	20	0	50
City 3	50	70	50	0

The basic situation is easily described. Suppose that a salesman is to visit each of a specified set of cities once and only once. He begins his trip from the home office and must return to the home office after having visited all of the cities. Suppose that he desires to plan an itinerary in a way to *minimize the total distance which he must travel.* The mileage data shown as Table 10-20 apply to a situation involving the home office and three cities which must be visited. The zero elements on the diagonal of the table indicate that there is no distance between a city and itself. Part of the table is redundant because the distance from city 1 to city 2 is equal to the distance from city 2 to city 1 and similarly for all other pairs.

A solution to this problem is much like the solution to the basic assignment model which we discussed in terms of salesman assignment. Considering each element in the table as describing a "from-to" combination, we would like to choose one element in each row and column such that their sum is a minimum. However, there are additional restrictions. First, we cannot choose any of the zero elements on the main diagonal, since they correspond to a trip from a city to itself. Additionally, we must go to each city in a single continuous trip. To understand this, consider the assignment indicated by the check marks in Table 10-21. This as-

TABLE 10-21   POSSIBLE SOLUTIONS

	Home office	1	2	3
Home office		✓	✗	
1	✓			✗
2		✗		✓
3	✗		✓	

signment is a feasible solution to the assignment problem which underlies this traveling salesman problem. Also, it satisfies the condition of no assignments on the main diagonal. However, the tour which it represents may be laid out in two independent parts.

HO → 1 → HO
2 → 3 → 2

A basic requirement of the traveling salesman problem is that no such independent "loops" exist in the solution; i.e., the salesman must begin at the home office and go once to each city, eventually returning to the home office. For instance, the assignment indicated by the crosses in Table 10-21 satisfies this condition (and all others). It may be diagrammed as

HO → 2 → 1 → 3 → HO

The laborious method of determining the best itinerary is to enumerate each of the feasible ones, sum up the mileages, and choose the one with the least mileage. This is done in Table 10-22. Each of the feasible solutions is called a tour.

Table 10-22 shows that each of the tours

HO → 2 → 1 → 3 → HO

and

HO → 3 → 1 → 2 → HO

will be 180 miles, and that no other tour will require less travel. Therefore, the salesman should use one of these as his itinerary.

The reader will recall that earlier we presented a simple algorithm for solving the assignment problem. The basic feature of such an algorithm is that it will lead, with certainty, to the best solution in a finite number of steps. No such algorithm is available for the traveling salesman variation of the assignment problem. As a result, the mathematics of the problem solution are of great interest to analysts. For operational solutions to such problems,

TABLE 10-22   TOTAL MILEAGE FOR VARIOUS TOURS

Tour	Total mileage
HO–2–1–3–HO	180
HO–1–2–3–HO	200
HO–3–1–2–HO	180
HO–2–3–1–HO	220
HO–1–3–2–HO	240
HO–3–2–1–HO	200

efficient approximation techniques have been developed. Any of these may be used to obtain the solution to a problem of modest size if a computer is available.[27]

Brand Switching Analysis

In this section we shall deal with a kind of marketing analysis which focuses on the time behavior of customers who make repeated purchases of a product class but may from time to time switch from one brand to another. *The usefulness of this sort of analysis in marketing is not restricted to promotional decisions;* yet, because the greatest usage has occurred in that area, we shall treat the methodology in the context of promotion. The primary model for brand switching analysis is the *Markov process.*

[27] See, for example, the Little et al. paper in the references.

SIMPLE MARKOV PROCESSES

The basic element of a Markov process, described in the marketing context, has to do with various *states* which may be ascribed to an individual customer at a point in time. In brand switching models the state is usually *the customer's preference for a particular brand*. In many marketing situations an individual's brand preference changes over time. Thus, he buys brand A for a while, then brand B, then A again, etc.; i.e., he *switches* from brand to brand.

The aggregate brand switching behavior displayed by large groups of customers from one time period to another may be described probabilistically. Thus, if we can determine empirically that 10 percent of the customers who prefer brand A in a given time period switch to brand B in the next, we may ascribe the probability 0.10 to this particular change in brand preference for this time period transition. Figure 10-10 diagrammatically illustrates such a situation for two brands, A and B. The arrows connecting the two brands indicate the four possible switches which might occur, i.e., A to B, B to A, A to A, and B to B. The number on each of the lines is the relevant probability. Thus, there is a probability of 0.8 that a customer who prefers A will continue to prefer A in the next time period and a 0.2 likelihood that he will switch preference to B. Similarly, there is a 0.6 likelihood that a customer who prefers B will retain his preference and a 0.4 probability that he will switch to A.

It is important to note that this simple probabilistic treatment of brand switching *does not imply that the behavior of an individual customer is being described at random*. The individual may switch brands consciously, with malice aforethought, and for good and valid reasons. Nonetheless, *the behavior of large groups may be best described in probabilistic terms*.

This probabilistic description may be summarized in a *transition matrix*—an array whose elements give the probabilities of the various changes which may occur. If p_{ij} is the probability that a

FIGURE 10-10 BRAND SWITCHING

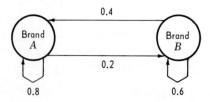

customer's preference will switch from brand i to brand j from one period to the next, the transition matrix is

From	To	
	A	B
A	p_{AA}	p_{AB}
B	p_{BA}	p_{BB}

The upper left element gives the probability of the switch from A to A, the upper right element is the probability of the switch from A to B, etc. In the two-brand situation described in Figure 10-10, the transition matrix is

From	To	
	A	B
A	0.8	0.2
B	0.4	0.6

The most significant characteristic of a transition matrix is that *the probability elements in each row must sum to unity.* In loose terms this is equivalent to saying that some transition must occur. If a customer begins with a preference for A in one period, in the next *he must have some preference*—either for A or for some other brand. Hence, the probability that he will make one of the mutually exclusive and exhaustive switches is 1.

Now suppose that each of the two brands has exactly 50 percent of the total market in some period, that the market is of a fixed size, and that the given transition matrix is applicable. In the second period brand A will retain 80 percent of its customers and also take away 40 percent of B's. Thus, the market share for brand A in period 2 will be

$$50\%(0.8) + 50\%(0.4) = 60\%$$

This indicates that 80 percent of the original 50 percent who prefer A will continue to do so, and 40 percent of the 50 percent who prefer B will switch to A.

Correspondingly, the market share for B in the second period must be 40 percent. It is made up of the 60 percent of the original customers who are retained and the 20 percent of A's customers who are obtained, i.e.,

$$50\%(0.6) + 50\%(0.2) = 40\%$$

If the same probabilities hold for brand switching in going from the second to the third period, A will again retain 80 percent of its share (which is now 60 percent of the total market) and will again attract 40 percent of B's customers (who now represent only 40 percent of the market). Thus, A's market share in period 3 will be

$$60\%(0.8) + 40\%(0.4) = 64\%$$

Similarly, B will retain 60 percent of its 40 percent share of the market in period 3 and will lure away 20 percent of A's 60 percent share. Hence,

$$40\%(0.6) + 60\%(0.2) = 36\%$$

will be B's market share in period 3.

If we continue to assume that the same transition matrix holds from one period to another, we can continue these calculations to obtain the market share data shown in Table 10-23. The most

TABLE 10-23   MARKET SHARES OF TWO BRANDS

Period	Brand A's market share	Brand B's market share
1	50%	50%
2	60%	40%
3	64%	36%
4	65.6%	34.4%
5	66.24%	33.76%
6	66.496%	33.504%

obvious feature of the data of Table 10-23 has to do with the apparent approach to a $\frac{2}{3}$–$\frac{1}{3}$ division of the market after only a few time periods.

If we introduce the concept of a *steady state*—the equilibrium state approached by the system as the number of transitions approaches infinity—we may describe the steady state as being a $66\frac{2}{3}$–$33\frac{1}{3}$ percent division of the market in this situation. This numerical division, which we have arrived at on the basis of intuition, can be shown mathematically to be valid.[28]

[28] See the Feller text for this proof.

Another interesting aspect of the steady state is that *it is independent of the initial market shares.* This has an intuitive basis if
we think of the steady state as being that attained after an infinite
number of transitions. Clearly the effect of the transition probabilities eventually becomes predominant over the initial conditions.
Table 10-24 illustrates the market share distribution which would

TABLE 10-24   MARKET SHARES OF TWO BRANDS

Period	Brand A's market share	Brand B's market share
1	80%	20%
2	72%	18%
3	64.8%	35.2%
4	65.92%	34.08%
5	66.368%	33.632%
6	66.5472%	33.4528%

be attained by using the same transition matrix, but beginning
with *A* having 80 percent of the market and *B* having only 20
percent. Note that the steady-state market shares which are approached after only a few time periods are the same as those approached in Table 10-23, even though the beginning market shares
are very different in the two cases.

MARKOV PROCESSES AND PROMOTION

If we suppose that brand *A* is our brand and that brand *B* represents either our single competitor or the aggregate of all competitors, we may think of a two-state process in which the states simply
represent customers and noncustomers. In other words, a consumer
either is or is not a customer of ours in any given time period.

The basic objective of product promotion in these circumstances
is to *effect changes in the transition probabilities.* In particular,
promotion seeks to increase the rate at which noncustomers become
customers and to decrease the rate of the converse switch, i.e.,
to increase the probabilities in the first column of the transition
matrix and to decrease the probabilities in the second column correspondingly. For example, suppose that the previously given transition matrix is applicable and that it is possible to alter this to

$$\begin{bmatrix} 0.9 & 0.1 \\ 0.5 & 0.5 \end{bmatrix}$$

by an expenditure of $700,000 on promotional activities. If the initial market share breakdown is 50–50, Table 10-23 shows the series of market shares which will be experienced in six time periods. With the same initial 50–50 breakdown and the new transition matrix (which may be "purchased" for $700,000), the equivalent distribution is shown in Table 10-25. This table shows that, beginning from the same 50–50 split of the market, after six time periods the resulting market share is about 83 percent, whereas in the old matrix it was less than 67 percent.[29] Thus, the promotion has resulted in a significant increase in steady-state market share.

TABLE 10-25   MARKET SHARE OF TWO BRANDS AFTER PROMOTION

Period	Our market share	Competitor's market share
1	50%	50%
2	70%	30%
3	78%	22%
4	81.2%	18.8%
5	82.48%	17.52%
6	82.992%	17.008%

The basic promotional decision problem is whether or not we should perform the promotion. The answer, of course, is determined by whether or not the cost of $700,000 is outweighed by the profit gained from the increased share. If we assume that no other changes occur in the distribution of market shares, the gain which we derive from the promotion will last forever; i.e., we expect to get closer and closer to a steady-state $\frac{5}{6}$–$\frac{1}{6}$ split of the market rather than closer and closer to a $\frac{2}{3}$–$\frac{1}{3}$ split. But since the promotional expenditure occurs *now* and the returns accrue over the indefinite future, we must consider the *time value of money*.

Let us assume for simplicity that the size of the total market is constant and that each percentage gain in market share during a given year is worth $20,000 in annual profits. If the time periods in Tables 10-23 and 10-25 are years, one can see that the market share gains in each of the six years are those shown in Table 10-26. In the first year the market share under a policy of no promotion and that under one of promotion are identical. In the second year a share of 60 percent would be attained if there were no

[29] It can be shown mathematically that the steady-state market share in this case is $\frac{5}{6}$, and in the case of the old matrix it is $\frac{2}{3}$. See the Feller text for this proof.

promotion (Table 10-23) and a share of 70 percent with promotion. The other figures in the third column of Table 10-26 are obtained similarly. The fourth column of the table indicates the annual profit equivalent of the market share gain, calculated at $20,000 per percentage point. The present value of this annual profit equivalent is then shown. The present value is calculated by using a 5 percent annual opportunity rate of return and assuming that the profit

TABLE 10-26   COMPARISON OF PROMOTION VERSUS NONPROMOTION STRATEGIES

Period	Market share if no promotion (Table 10-23)	Market share under promotion (Table 10-25)	Gain in market share for period	Annual profit equivalent of market share gain	Present value of annual profit equivalent
1	50%	50%	0%	$0	$0
2	60%	70%	10%	$200,000	$195,000
3	64%	78%	14%	$280,000	$254,000
4	65.6%	81.2%	15.6%	$312,000	$269,500
5	66.24%	82.48%	16.24%	$324,800	
6	66.496%	82.992%	16.496%		

accrues at the end of the previous year. Thus, the $280,000 annual profit equivalent is obtained after two years and its present value is[30]

$$\frac{\$280,000}{(1.05)^2}$$

or about $254,000.

Observing the last column of Table 10-26, we can see that the present value of the stream of annual profit *gains* for the first three years is about $714,000. Therefore, the current promotional outlay of $700,000 is more than offset by the present value of the profit gains in only the first three years, and the outlay should be undertaken. Of course, this is a toy example. Generally such pronounced changes in the transition matrix could not be "bought" so cheaply by promotional outlays.

We should also point out that it may well be possible that a promotional outlay would *not* be justified in such an instance. Even though this simple example involves endless profit gains, the present value of the endless stream is finite, and this present value may well be either less than or greater than the promotional outlay in any particular instance.

---

[30] Chapter 5 gives the details of the calculation of present values.

IMPLICIT ASSUMPTIONS IN BRAND SWITCHING ANALYSIS

There are a number of implicit assumptions in the use of a Markov model to describe consumer brand switching behavior which should be carefully considered by the marketing manager who may wish to use the model. First, the distribution of market share to the various brands at any time period is assumed to depend only on the distribution in the *previous* time period and the transition probabilities. If we make the analogous statement for an individual consumer, we can say that *his probability of purchasing a given brand in any time period is dependent only on his previous purchase.* It does not depend on his entire purchasing history. We may think of a purchase history as simply a sequence of letters indicating the brands bought by a consumer. The sequence *ABCCCACCB* indicates that the consumer tried three brands—*A*, *B*, and *C*—then repurchased *C* twice, returned to *A*, etc. One would imagine that his entire history has a bearing on his likelihood of choosing the various brands at his next purchase. Yet, the Markov model assumes that he "forgets" his past purchases and bases his purchasing behavior only on the most recent purchase, the satisfaction derived from it, and the attraction of other brands. Such behavior as is inherent in the model is probably not a perfectly valid description of the fashion in which most consumers behave. Yet, if the abstraction is not too great, the model may be "accurate enough" to provide predictions which are meaningful for decision making.

The second difficulty entailed in Markov models of brand switching has to do with the time periods which are involved. We have assumed that the total market is of a constant size. Therefore, our model indicates that every customer makes a purchase in each time period. Of course, consumers just don't behave in this simple a fashion. The interval of time between purchases for different consumers varies widely.[31] One method of circumventing this difficulty is to introduce an additional state, "no purchase." Thus, in each period each customer is either in one of the states indicating a purchase of a particular brand or in the "no purchase" state.

An additional difficulty is introduced by this new state, however. Recall that each customer's probability of making the transition from one state to another depends only on the previous state. Thus, a customer who has failed to make a purchase in a given period is relegated to the "no purchase" state, and even his most recent purchase is forgotten; i.e., his transition probabilities then depend

---

[31] We have previously recognized this behavior in Chap. 6, where we corrected repeat sales data to account for those customers who had not had an opportunity to repeat.

only on the "no purchase" state in which he was placed and not on the state describing his last actual purchase.

Of course, there are other minor difficulties with the Markovian model of brand switching behavior. For instance, it does not account for variations in the quantity purchased by a customer during a time period. Yet it has proved useful in promotional analysis.[32] Some of the additional difficulties will be discussed in the context of the succeeding major section dealing with the measurement of promotional effectiveness.

Measuring Promotional Effectiveness

As we have previously pointed out in the context of advertising, the objective analysis of managerial decisions is best served when comprehensive effectiveness measurements can be made. This is no less true of each of the decisions involving the other promotional programs of the organization. All that was said on this point in Chapter 9 is equally appropriate with respect to the other promotional devices which the organization may use. All offer the same basic measurement difficulties, although, of course, many particular devices have features which obviate some of the general measurement problems. In fact, all of the previous discussion related to media effectiveness, the forms of the response function, and experimental designs may equally well be applied to other forms of promotion. All that we need to do is interpret each promotional form as a "medium." Thus, we would treat coupons offering a price reduction as a medium and attempt to determine the sales response generated by this form of promotion.

In this section, then, our task is a limited but important one. There is no need to repeat those things which have already been said. Here, we shall extend the discussion of measuring promotional effectiveness beyond that of the previous chapter in two ways. First, we shall consider the important question of *interactions* in the effect of the various promotional forms. Secondly, we shall give attention to the unique measurement aspects of some of those promotional models which have been discussed in this chapter.

MEASURING PROMOTIONAL INTERACTIONS

In the previous chapter we outlined the logical basis for experimental determination of the effectiveness of promotional expendi-

---

[32] See, for example, the Harary and Lipstein, the Howard, and the Maffei references. A useful summary of this kind of analysis is presented in the Herniter and Howard reference.

tures in the context of a simple two-level advertising experiment. In many situations involving one primary promotional form, such as media advertising for packaged grocery products, it may well be valid to concentrate attention on the measurement of effectiveness in the single promotional area. In other situations it is probably not meaningful to consider the effect of one promotional form without giving simultaneous consideration to other forms. The most obvious illustration of such a situation is any product which is sold through personal contact with the consumer. The sale of life insurance is a circumstance in which the effect of at least two promotional forms—media advertising and personal sales effort—is significant both in the absolute and in their interactive effect.

The reader should recall exactly what is meant by *interactive effects. Promotional devices are said to interact if the magnitude of the effect of one depends on the absolute level of the other.* This is undoubtedly the case for the effects of personal selling and media advertising in life insurance. At the lower end of the sales effort scale, it is unquestionably so. If there is no sales effort— i.e., no agents available from whom insurance may be purchased— the insurance company will be unable to detect any sales response to media advertising. At other levels of expenditures in one of the promotional devices, the sales effect of the other may vary over wide ranges. Such a situation is grossly illustrated by the curves in Figure 10-11. The curves show the sales response to TV advertising for two levels of personal sales effort expenditure, low and high. At the low level of personal sales effort expenditure, the sales revenue response to advertising is given by the lower curve, and at the high level of personal sales expenditure, it is given by the upper curve. The difference between the responses to television advertising at the two levels of personal sales effort expenditure indicates that a significant interaction exists.

Of course, the basic problem in promotional decision making is that we do not actually know the curves depicted in Figure 10-11. Therefore, we must estimate them experimentally. A general class of experimental designs which may be used to assess the effect of various promotional devices and to gain some insight into their interactive effects is composed of experimental setups called *factorial designs.*

Factorial designs of promotional experiments usually involve a number of promotional devices (the factors), each at a number of different *levels* of the controllable variables. For instance, if we wished to investigate the effects of TV advertising and personal sales effort simultaneously, we would choose two or more relevant expenditure levels for these two factors. If two expenditure levels

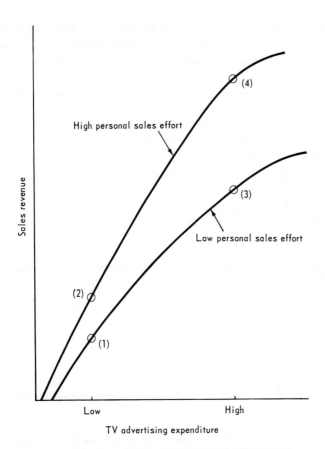

FIGURE 10-11   INTERACTION OF TWO PROMOTIONAL TECHNIQUES

were chosen for each, we would have a two-factor–two-level factorial experiment. Table 10-27 shows a natural arrangement for such a design. Expenditures for both TV advertising and personal sales are described by two levels, low and high, in this design. Four particular combinations of the levels are indicated by the

TABLE 10-27   TWO-FACTOR–TWO-LEVEL PROMOTIONAL DESIGN

TV advertising expenditure	*Personal sales expenditure*	
	Low	*High*
Low	(1)	(2)
High	(3)	(4)

four cells labeled (1) through (4). Each of these combinations is called a *treatment* (or treatment combination), a term emanating from early fertilizer experiments which were conducted by using factorial designs.

To conduct the experiment, we need only to select some areas randomly and assign them randomly to the various treatments. For instance, we could randomly select four areas and assign one to the low-low treatment, one to the low-high treatment, etc. Or, we might well select eight areas and assign two to each treatment.

Having done this, we would then measure the *response,* probably in terms of the sales revenue which is generated in each area (under each treatment combination). These data could then be entered in the appropriate cells of Table 10-27. Table 10-28 gives

TABLE 10-28   RESPONSES PRODUCED BY PROMOTIONAL EXPERIMENT (THOUSANDS)

TV advertising expenditure	*Personal sales expenditure*	
	Low	*High*
Low	$ 5	$ 7
High	$15	$23

hypothetical data which roughly correspond to the response curves in Figure 10-11. These data indicate that $5,000 in sales revenue resulted in the area which was treated with low levels of TV advertising and personal sales expenditure, $15,000 in sales revenue was produced by a high level of TV advertising in combination with a low level of personal sales expenditure, etc.

The *effect* of the television advertising and personal sales factors is the (average) change in response produced by a change in the level of the factors. For instance, the change in response produced by the change from a low to a high level of television advertising was $10,000 at the low level of personal sales effort and $16,000 at the high level of personal sales expenditures. The average change is therefore $13,000, i.e.,

$$\frac{\$(15 - 5) + \$(23 - 7)}{2} = \$13$$

This average effect is called the *main effect* of television advertising. The main effect of sales effort is similarly calculated to be the average change produced by the change from the low to the high level of this factor, i.e.,

$$\frac{\$(7 - 5) + \$(23 - 15)}{2} = \$5$$

or $5,000.

The *interactive effect* of television advertising and personal sales effort is defined as the *difference* in the effect of television advertising at the two levels of personal sales effort. Numerically it is calculated as one-half of this difference for purposes of analytical convenience, i.e.,

$$\frac{\$(23 - 7) - \$(15 - 5)}{2} = \$3$$

Note that this interaction is symmetric in the two factors. If we had calculated it in terms of the difference in the effect of personal sales effort at the two levels of television advertising, it would be

$$\frac{\$(23 - 15) - \$(7 - 5)}{2} = \$3$$

The advantages of a factorial design over designs which have a more intuitive basis can be made clear at this point. The novice experimenter might choose to vary one factor at a time, rather than to employ the factorial design, for instance. In doing so, he would obtain data points representing three of the four cells in Table 10-27. Table 10-29 shows such a situation. The cells there

TABLE 10-29   ONE-FACTOR-AT-A-TIME EXPERIMENTAL DESIGN

TV advertising expenditure	*Personal sales expenditure*	
	Low	*High*
Low	(1)	(2)
High	(3)	

are numbered in the same way as in Table 10-27. The one-factor-at-a-time approach gives information on the responses obtained in cells (1), (2), and (3). These three cells represent the *minimum* number of treatments which are necessary to obtain any relevant information on the effects of the two factors. Hence, this appears superficially to be a rather good design.

The effect of the advertising change is indicated by the difference

(3) — (1), and the effect of the personal sales expenditure change is indicated by the difference (2) — (1) in this simple design. Because we realize that errors are involved in these estimates, we would undoubtedly choose to repeat each treatment in at least one additional area to confirm our estimates of the effects. Of course, in reality we would recognize this beforehand and simply use at least six areas—two per treatment. We could then use the average of the duplicated responses to calculate the effects.

Note, however, that the four treatments in the factorial design gave us exactly the same sort of information. If we fill in the table with treatment (4), to obtain a factorial design exactly like that previously discussed, we estimate the effect of advertising using the average of the two differences (3) — (1) and (4) — (2). If no interactive effects exist, these two differences will themselves differ only because of experimental error, and their average gives an estimate of the effect of television advertising which is just as valid as that obtained from the duplicated treatments in the one-factor-at-a-time approach. Analogous comments hold for the estimation of the effect of personal sales effort. Thus, the *factorial design requires only four observations to obtain the same information which was obtained by six observations with the single-factor approach(if no interactions exist).*

If an interactive effect does exist, a factorial design is necessary to avoid misleading conclusions. For instance, consider the curves shown in Figure 10-11. The lower curve represents the sales revenue response to television advertising at the low level of personal sales effort, and the upper curve represents the same response at the high level of personal sales expenditure. The low and high levels of television advertising expenditure which are used in the experiment are indicated on the horizontal axis.

If the one-factor-at-a-time approach were used in this situation, the observations numbered (1), (2), and (3) would be obtained. Clearly this would give no indication of the worth of either of the promotional devices, since the increase in personal sales expenditure itself produced only the small effect (2) — (1) and the increase in television advertising expenditure produced the effect (3) — (1). There would be little reason to draw the inference that the high level of *both* of the factors would produce the large response indicated by (4) in Figure 10-11. Of course, the factorial design would detect this interactive situation.

When the interactive effect in such a situation is large, the main effects cease to have much meaning; i.e., the change produced in one factor averaged over the other factor is not a useful bit of information. If this is the case, we should only specify the effects in terms of the level at which they apply. For example, the effect of television advertising is indicated by $(3) - (1)$ *at the lower level of personal sales effort* and by $(4) - (2)$ *at the higher level of personal sales effort.*

The basic model which is used to analyze a factorial experiment of this kind is

$$R_{ijk} = M + A_i + S_j + D_{ij} + e_{ijk}$$

where R_{ijk} is the sales revenue response at the ith level of advertising (A), the jth level of personal sales effort (S), and for the kth trial (repetition) of the treatment. M is the true mean of all trials, A_i is the true mean of all trials at which advertising is at its ith level (measured from M), S_j is the true mean of all trials at which sales effort is at its jth level (measured from M), D_{ij} measures the interaction of advertising and personal sales, and e_{ijk} represents the experimental error.

TABLE 10-30 THREE-FACTOR–THREE-LEVEL FACTORIAL DESIGN

TV advertising expenditure	*Expenditure on point of purchase*	*Personal sales expenditure*		
		Low	*Medium*	*High*
Low	Low	(1)	(2)	(3)
	Medium	(4)	(5)	(6)
	High	(7)	(8)	(9)
Medium	Low	(10)	(11)	(12)
	Medium	(13)	(14)	(15)
	High	(16)	(17)	(18)
High	Low	(19)	(20)	(21)
	Medium	(22)	(23)	(24)
	High	(25)	(26)	(27)

The results of the experiment may be analyzed by using the statistical procedure called the *analysis of variance*. Basically this procedure takes the experimental data and estimates from them the most likely values of A, S, D, and the experimental error variance. Using statistical significance tests, we test the significance of each of these in terms of whether or not the observed responses can be accounted for by experimental error. We shall not go into the mechanics of this here. (See the Cochran and Cox reference.) Here, we shall only note that the significant effects of the factors and their interactions can be assessed in this way.

To conclude our discussion of factorial experiments in determining promotional effectiveness, we need to illustrate that we are not restricted either to two factors or to two levels in conducting a factorial experiment. For instance, if we also wished to determine the effect of point-of-purchase displays, and if three levels of all factors were used, we might use the factorial design indicated by Table 10-30. Treatment (14) in Table 10-30 represents the medium level of all three factors; treatment (26) indicates the high level of television advertising and point-of-purchase expenditure in conjunction with the medium level of personal sales effort. A total of 27 different treatments would be necessary in such a design. Aside from the required size, however, the interpretation and analysis are completely analogous to those discussed for the two-factor–two-level design discussed earlier. (See Cochran and Cox.)

TIMING PROBLEMS IN THE EXPERIMENTAL DETERMINATION OF PROMOTIONAL EFFECTIVENESS

Once a promotional experiment has been designed and the measure of response delineated, the major problem involved in analyzing the experiment is *timing*. Suppose, for example, that we think in terms of a single promotional device which has a positive effect on sales revenue. Our measurements of sales revenue should enable us to detect it. But they will do so only if the effect is significant *during the time period in which experimental measurements are taken*. Suppose that there is a time lag in the major response to promotion. If the experiment does not extend over a long enough time period, the experimental measurements may be concluded before the effect emerges.

One solution to this problem is to make the experiment *open-ended*. In an open-ended experiment the time period during which observations will be taken is not arbitrarily fixed in advance, but rather the conclusion date of the experiment is left to depend on the observed experimental responses. Of course, as in all phases of experimental design, *the procedure for determining the end of the experiment should be established in advance*. Thus, although we leave the decision as to when the experiment shall end until after the experiment is conducted, the *decision rule* for determining the ending point is established in advance. In this way our personal biases are removed from the experimental results.

A comparison of the possible results that could be expected with those that might result by chance is the basis on which the determination of the *stopping* rule must be made. However, the stopping rule should not consider only the last observation—say the last monthly sales revenue figure—but rather the entire sequence of monthly revenues generated after the promotional treatment. Let us illustrate these ideas in a simple example.

Suppose we are considering directing a specified increase in personal sales effort toward a given class of customers. Since in any given month a particular customer's revenue is quite likely to be above its normal level, we have the problem of separating the responses which are due to the increased promotion from those which are due to chance alone. Suppose that we decide to run the experiment for at least four months.

For any one customer the four successive revenue observations which we will have made after four months may each be either above or below normal. A simple way of categorizing the situations which might occur is to symbolize a month having an above-normal revenue from a customer with a 1 and a month in which a normal or less-than-normal revenue is obtained with a 0. Thus, the 16 combinations in the first column of Table 10-31 are those which could

TABLE 10-31   POSSIBLE EXPERIMENTAL SALES RESPONSES

Observed sales response patterns	Response month	Probability
1111	1	0.004
1110	0	0.012
1101	4	0.012
1011	3	0.012
0111	2	0.012
1100	0	0.035
1001	4	0.035
0011	3	0.035
1010	0	0.035
0110	0	0.035
0101	4	0.035
0001	4	0.105
0010	0	0.105
0100	0	0.105
1000	0	0.105
0000	0	0.316

possibly be observed for any single customer. For instance, the entry 1111 in the first column of the table indicates that above-normal sales revenue could be observed in each of the first four months after the initial application of promotional treatment. The second entry, 1110, shows that above-normal response is demonstrated in the first three months but not in the fourth month.

To impose a conservative judgment on the outcome of the experiment, it might be reasonable to conclude that only those response patterns which have an unbroken string of above-normal revenues at the end of four months are indicative of positive experimental results. Thus, the pattern 1111 indicates that a positive response occurred in the first month, the pattern 0111 indicates a positive response in the second month, and the pattern 1110 does not indicate any sustained positive response. The response month for each of the possible sales response patterns is indicated in the second column of Table 10-31. The zero entries indicate the patterns with no sustained response. In using this convention, we are being rather conservative since we probably are omitting some response patterns which are likely actually to represent positive promotional results. The pattern 1110 representing above-normal response in each of the months except the fourth is the best illustration of such a situation.

Naturally, during any given month, some customers will produce revenues which are above their normal level and some will produce below-normal revenues, whether or not additional sales effort is applied. To account for this, we need only to know, from preexperimental observations, *the percentage of customers who will experience above-normal revenues in any given month*. This estimate should be made for each calendar month so that the effect of seasonality will not be omitted. Suppose that this overall percentage in our illustration is 25 percent.

If 25 percent of our customers could be expected to have above-normal revenues in a given month, and if the successive months are probabilistically independent, the likelihood of each response pattern due to chance alone may easily be calculated. For instance, the probability for the pattern 0101 is simply

$$\tfrac{3}{4} \cdot \tfrac{1}{4} \cdot \tfrac{3}{4} \cdot \tfrac{1}{4} = \tfrac{9}{256}$$

or about 0.035. These probabilities are entered in the last column of Table 10-31.

Now suppose that our experiment involves 1,000 customers. The *expected* number of customers who would display responses occurring in each response month is shown in the second column of Table 10-32. These expectations are calculated simply by adding the probabilities corresponding to patterns indicating similar response months in Table 10-31 and multiplying by 1,000. For instance, the patterns 1011 and 0011 both indicate a response in the third month. Their total probability from Table 10-31 is 0.012 + 0.035, or 0.047. Thus, 47 of the 1,000 customers used in the experiment can be expected to display a response beginning in the third month.

The third column of Table 10-32 gives hypothetical data on the actual results of such an experiment. It is apparent that the actual response is better than the expected. But how much better? This question can be answered through the use of a statistical test called the *chi-square test.*

TABLE 10-32   COMPARISON OF EXPECTED AND ACTUAL RESPONSES

Response month	Expected number of customers displaying response	Actual number of customers displaying response
0	750	723
1	4	6
2	12	18
3	47	52
4	187	201

The chi-square test involves the calculation of a quantity which depends on the differences between the actual and expected number of observations for a number of categories. Here, we expected 750 customers to fall in the "no response" category and we actually observed 723; we expected 4 in the "response after one month" category and we actually observed 6, etc. The quantity to be calculated is

$$D^2 = \sum_{j=0}^{4} \frac{(O_j - E_j)^2}{E_j}$$

where O_j is the observed number in each of the categories (numbered 0 through 4 to correspond with the response months in Table 10-32) and E_j is the expected number in each category. Here, this quantity has the value

$$D^2 = \frac{(723 - 750)^2}{750} + \frac{(6 - 4)^2}{4} + \frac{(18 - 12)^2}{12} + \frac{(52 - 47)^2}{47}$$
$$+ \frac{(201 - 187)^2}{187}$$

which is about 6.55.

To use this quantity, we need to make use of the known (approximate) probability distribution of D^2. Since the probability distribution is known and the quantity D^2 sums up our information about the differences between the observations and expectations, we can use the particular D^2 value to ascertain the significance of the discrepancy. To do this we need *first* to recognize that we cannot make such a determination with certainty; i.e., we must assume some risk of making a mistaken judgment.

To describe this quantitatively, we choose a number b such that the probability that a D^2 value equal to or greater than b is some arbitrary small quantity, say 0.05, i.e.,

Prob $(D^2 \geq b) = 0.05$

We can determine this value of b directly from a table of the chi-square distribution.[33] Most basic statistics books will contain

TABLE 10-33   VALUES OF b FOR VARIOUS PROBABILITIES THAT D^2 IS GREATER THAN OR EQUAL TO b

	Prob $(D^2 \geq b)$			
	0.25	0.10	0.05	0.025
b	5.39	7.78	9.49	11.1

such a table. The relevant portion of the appropriate distribution is shown as Table 10-33.[34] From this table we find that the critical value of b for a probability of 0.05 is 9.49, i.e.,

Prob $(D^2 \geq 9.49) = 0.05$

The value of D^2 which we have calculated (6.55) is not so great as this critical value. Therefore, the differences between observed and expected values in this illustration are not great enough to produce a D^2 value which would be observed less than 5 percent of the time if the *probabilities which went into the expectation calculations were applicable.* (These are the probabilities given in Table 10-31.) In fact, if these probabilities are applicable, the observed results would be observed purely by chance more than 5 percent of the time. Thus, we do *not* have a clear indication as to whether the response patterns which we have observed ac-

[33] The limiting distribution of D^2 is the chi-square probability distribution having a number of degrees of freedom which is one less than the number of categories. The idea of degrees of freedom is not important to the understanding of the methodology of this section. See the Brunk reference for further details.

[34] This is a portion of the chi-square distribution for four degrees of freedom. This number of degrees of freedom is appropriate since there are five relevant categories. See Brunk.

tually reflect a positive effect of the additional promotional expenditure or whether they have occurred by pure chance.

If the calculated value of D^2 had been larger than 9.49, we could conclude that such a value would be observed less than 5 percent of the time if the probabilities of Table 10-31 hold. Therefore, it would be rather likely that the observed values actually could represent a significant departure from those patterns which would be realized under the normal functioning of the system. (The reader should recall that normal circumstances entail about 25 percent of the customers ordering at above-normal rates.) This conclusion would be compatible with having run the experiment long enough to detect a sales response.

Since the actual D^2 value which we have calculated (6.55) is not greater than the critical value, we cannot determine whether the observed sales responses are different from those which would be displayed under normal circumstances. Therefore, we would probably choose to allow the experiment to progress at least another month. At the end of the fifth month we could then perform the same analysis again. Of course, it is possible that such a procedure will permit the experiment to go on indefinitely. In practice, however, we usually have some idea of the meaningful time limits of experimentation, so that we can recognize a point beyond which we will not continue. Of course, the upper limit of experimental time should be determined in advance so that it forms a part of the stopping decision rule.

EXPERIMENTATION IN MARKOV BRAND SWITCHING MODELS

Markov models of brand switching behavior are typically formulated on the basis of unplanned observations made in the real world. The verification of the probabilities used in the Markovian transition matrix is, then, the first step which must be taken before the model is used for predictive purposes. This obviously requires that the purchase patterns of individual consumers be followed over time. To do this, a consumer panel is used.

Two general types of experimental verification are necessary for the use of Markovian models in promotion. First, the basic transition probabilities must be verified. Then, the effect of promotion on the probabilities must be estimated. We shall concentrate on the first of these aspects since it is the basic verification and requires only data which can be obtained from any of a number of existing consumer panels in a straightforward fashion. The second experimental determination would probably require a special panel such as that of the Milwaukee Advertising Laboratory (discussed in Chapter 9) for best results.

Suppose that the existing transition matrix which is estimated to be applicable in a two-brand Markov model is[35]

From	To	
	A	B
A	0.8	0.2
B	0.3	0.7

Further, suppose that 100 members of a consumer panel, all of whom originally were brand A users, are to be used in verifying the Markov model. The *expected number* of customers for each brand after various numbers of time periods is given in Table 10-34. (These expected numbers are derived in exactly the same fashion as were the market share data in Tables 10-24 and 10-25.)

TABLE 10-34   EXPECTED NUMBER OF CUSTOMERS FOR TWO BRANDS

Time Period	Expected number of brand A customers	Expected number of brand B customers
1	100	0
2	80	20
3	70	30
4	65	35
5	62.5	37.5

Now suppose that 75 of the 100 are actually observed to purchase brand A during the second period, whereas 80 were expected to do so. What implication does this have for the validity of the Markov transition matrix? The answer is that *there is a variability associated with the expected number of purchasers of a brand in any given period.* In essence, the number of purchasers of a given brand in any period is a random variable whose expectations we have calculated in Table 10-33. But, since it is a random variable, this number will take on only one of a number of possible values in any period, and it would be rather unlikely if it were to take on a value exactly equal to the expectation. The question is: How close is the actual value to the expectation? In other words,

[35] This illustration is adapted from the Howard paper in the references.

is the observed number 75 close enough to the expectation of 80 to be consistent with the assumed transition matrix?

The answer, of course, lies in the *degree of variability of the random variable.* If the actual value is within a range of variation around the expectation which is quite likely to occur by chance, then the value 75 might well be observed if the transition matrix is actually valid. On the other hand, if the variability of the random variable is such that it is quite unlikely that a value as far away from the expectation as is 75 would be observed by chance, then the validity of the Markov matrix is subject to question.

The appropriate measure of variability is the *standard deviation* of the random variable indicating the number of customers for each brand in any given time period. We shall not go into the details of the calculation of this standard deviation here but will simply refer the reader to the paper by Howard listed in the references. Concentration here is on the ideas and methodology which are applicable, given that the numerical standard deviation can be calculated.

Suppose that a numerical standard deviation of one is associated with the number of people who purchase brand *A* during the second period. The actual observed number of purchasers, 75, is thus five standard deviations away from the expectation of 80. It is therefore rather unlikely that this number would be observed if the Markov transition matrix were valid. On the other hand, if the standard deviation were 5, we would have little reason to doubt the transition probabilities, since the observed number is only one standard deviation from the expectation. Such comparisons can be made for various time periods so that the variability information which is contained in the standard deviation measure can be utilized to make an experimental assessment of the validity of the Markov matrix of transition probabilities which has been put forth as being descriptive of a particular brand switching environment.

Summary

The primary focus of nonadvertising promotional decisions is on personal selling. The decisions involved in the organization and management of a sales force contribute greatly to the sales results of the organization. Since sales force decisions directly involve individuals—the manager, the salesman, and the customer—and their relationships, they are necessarily complex.

A sales analysis is a diagnostic process of understanding a firm's current and past sales record, which serves as an input for sales force decisions. Such decisions as those involving the relative time salesmen should spend on retentive and conversional efforts, the amount of sales resources to be allocated to various customers, and the recruitment, selection, and compensation of salesmen cannot be objectively resolved without the basic data obtained through sales analysis.

Brand switching analysis is a method for utilizing the temporal buying behavior of customers to predict the future course of events. Because its primary utility has been in making promotional decisions, it is discussed in this chapter.

The important problems of measuring promotional effectiveness, which were first taken up in the advertising context in the previous chapter, are explored more extensively in this chapter. The idea of *interacting effects* is the major direction in which the concepts of measuring promotional effectiveness are extended. If interactive effects exist, great care must be taken to ensure that the measurements which are made are valid. The *factorial design* is a primary device for gaining insights into the magnitude of promotional interactions.

The discussion of the major problem structure areas of marketing concludes with the next chapter, "Distribution Decisions."

EXERCISES

1. What would you think might be the differences between a response curve for media advertising and one for a nonrecurrent promotional device such as a "10 cents off" coupon?

2. The sales revenue rate R is related to the promotional expenditure rate E by

$$R = 5 + 2 \frac{15}{1 + 3E}$$

What is the best rate at which to make expenditures for this promotional device?

3. Suppose that the basic dynamics model of advertising, as given in Chapter 9, is to be interpreted as describing some nonadvertising promotional situation. If a large promotional expenditure of A dollars is made at a point in time, rather than over some period of time, what is the sales rate achieved? What is the immediate sales increase resulting from the promotion? What is the *total* of additional sales generated by the promotion?

4. If short intense promotional campaigns were being considered for several products, could you use your answer to exercise 3 to indicate a basis for deciding which product should be promoted?

5. What are some of the pros and cons associated with the policy of having one group of salesmen who pursue only retentive objectives and another group who seek only conversional objectives?

6. If the observations shown in Figure 10-3 were made for some very

narrow range of sales time near the middle of the range shown in Figure 10-1, what conclusion is likely to have been drawn concerning the response of revenue to personal sales effort?

7. Show why the tangent line drawn in Figure 10-5 determines the highest value of the ratio $C(t)/t$.

8. Suppose that the conversional sales response curve and the retentive sales response curve are those shown below. How many hours per month should each salesman allocate to the two types of effort?

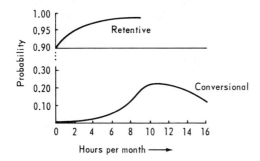

9. Using the results obtained in exercise 8, determine the number of each type of account to be assigned to each salesman.

10. Suppose that it is necessary to use the data of Figure 10-6 to determine the best allocation of seven salesmen to two areas. How might this be done?

11. (Adapted from the Brown, Hulswit, and Kettelle reference) A method of assigning salesmen to customers might simply be on the basis of the customer's potential revenue contribution. Suppose that the conversion rate of the optimal conversional sales time expenditure is 20 percent per month and the lapse rate at the optimal retentive sales time expenditure is 5 percent per month. If the optimal allocations are 10 hours and 2 hours respectively, the average sales expense per month per customer (in terms of the dollar cost of a salesman c) is

$$c \, \frac{10T_c + 2T_p}{T_c + T_p}$$

where T_c is the expected number of months required for conversion and T_p is the expected holding time. What is the numerical value of this quantity if $c = \$7.57$ per hour?

If P is the *total potential* monthly revenue from a customer (all that he will normally spend for products of our type), the average monthly revenue from a customer is

$$\frac{PT_p}{T_c + T_p}$$

times some fraction of his revenue which we expect to get while he is a customer, say 0.05.

Now suppose that profits are such that we can afford to spend

a fraction f of additional revenue on sales effort to produce that revenue. The smallest customer worth promoting is then determined by equating the average sales expense per month per customer to f times the average monthly sales revenue for a customer (and solving for P). What is that revenue rate if f is ½?

12. Can you indicate why the salesman assignment algorithm (discussed in a technical note) was developed in terms of minimization rather than maximization?

13. Solve the salesman assignment problem represented by the expected profit array shown below without resort to either the solution algorithm or to an exhaustive enumeration of strategies. Do this by considering in which territory each salesman performs best and which salesman is best in each territory.

EXPECTED PROFITS

Salesman	Territory		
	I	II	III
A	7	8	6
B	9	5	6
C	6	4	7

14. Use the algorithm to solve the assignment problem of exercise 13 and compare the solution with that obtained there.

15. Our company has a salesman located in each of four cities—1, 2, 3, and 4—and calls need to be made on four potential customers who are located in cities 5, 6, 7, and 8. If the table below gives the mileage from each salesman's location to each potential customer, which salesman should we ask to call on each potential customer?

MILEAGE CHART

From city	To city			
	5	6	7	8
1	41	72	39	52
2	40	40	60	51
3	39	60	51	32
4	29	40	40	60

16. What general traits might be used to characterize the individual who would be considered desirable to perform selling functions

in a playboy club, i.e., to be a bunny? Which of these traits might not be desirable in an "Avon lady"?

17. Traditionally, the prediction of sales success and the assignment of salesmen to territories or customers have been treated as two separate decision problems. Psychologists were called on to design and administer predictive tests, and other analysts developed techniques (such as the assignment algorithm) for making optimal assignments in the light of predicted success levels. What are the advantages in combining these two aspects as in the "Selection and Assignment" section?

18. Suppose that 80 percent of the candidates for sales positions turn out to be successful and that we have a predictive test which 95 percent of past sales successes and only 50 percent of past sales failures have passed. If we administer the test to 50 individuals and find that 40 pass, how many successful salesmen can we expect to have (assuming that we appoint all 40 who pass)?

19. Two predictive tests are available in the situation described in the previous exercise. The percentages of past successful and unsuccessful salesmen who have passed and failed each of the tests are given by the tables below. A sales candidate passes test 1 and fails test 2. What action should be taken?

SUCCESSFUL SALESMEN

Test 1	Test 2	
	P	F
P	85	10
F	1	4

UNSUCCESSFUL SALESMEN

Test 1	Test 2	
	P	F
P	25	25
F	25	25

20. Suppose that we have three candidates for two sales positions in the situation of exercise 19. The three candiates have respectively passed both tests, failed only the first, and failed only the second. Who should be appointed?

21. What are the steady-state market shares associated with the following brand switching arrays if the initial market shares are 50 percent and 50 percent?

Brand	Brand	
	A	B
A	0.75	0.25
B	0.4	0.6

Brand	Brand	
	A	B
A	0.1	0.9
B	0.9	0.1

What are the steady-state market shares in these two cases if brand
A initially has 90 percent of the market?

22. What product characteristics can you infer from the two brand
switching arrays in exercise 21?

23. Suppose we know that the brand switching behavior of customers
is described by the array below.

	Our brand	Other brands
Our brand	0.6	0.4
Other brands	0.3	0.7

Further, suppose that a promotion costing $500,000 is expected to
change this array to:

	Our brand	Other brands
Our brand	0.6	0.4
Other brands	0.4	0.6

If no other changes occur and we deem a 1 percent gain in market
share to be worth $10,000 in annual profits, should we embark
on the promotion?

24. Give an argument as to the advantage of a factorial design over
a one-factor-at-a-time approach (*a*) if no interaction is known to
exist and (*b*) if an interaction is believed to exist between the
factors.

25. The *Wall Street Journal* reports that one auto dealership pays salesmen in the following fashion: "a minimum . . . commission of $20
for any thing under $150 profit, 25% of the first $150 profit ($37.50),
and 35% of any profit over $150 . . . a commission of $2.50 to
$20 for financing depending on how long it was for, and 10% of
any premiums received on insurance offered by the dealers. Also,
there was a bonus of $100 for selling at least six cars in a two-week
pay period."[36] How well do you think such a compensation plan
satisfies the objectives of the auto dealer?

26. Assume that the probabilities of Table 10-17 have been determined
but we cannot make a meaningful estimate of the average profit
difference resulting from successful and unsuccessful performance
in each of the three positions. In this case, we might seek to maxi-

[36] "It's a Deal: Savvy Buyers Haggle with Shrewd Salesmen in Auto Show-
rooms," *Wall Street Journal*, June 13, 1966, p. 1.

mize the joint probability of successful performance in all three positions. How might we do this?

REFERENCES

Brown, A. A., F. T. Hulswit, and J. D. Kettelle: "A Study of Sales Operations," *Operations Research*, vol. 4, no. 3 (June, 1956).

Brunk, H. C.: *An Introduction to Mathematical Statistics*, Ginn and Company, Boston, 1960.

Canfield, B. R.: *Sales Administration*, 4th ed., Prentice-Hall, Inc., Englewood Cliffs, N.J., 1961.

Cochran, W. G., and G. M. Cox: *Experimental Designs*, 2d ed., John Wiley & Sons, Inc., New York, 1957.

Feller, W.: *An Introduction to Probability Theory and Its Applications*, 2d ed., vol. 1, John Wiley & Sons, Inc., New York, 1957.

Harary, F., and B. Lipstein: "The Dynamics of Brand Loyalty: A Markovian Approach," *Operations Research*, vol. 10, no. 1 (January–February, 1962).

Herniter, J. D., and R. A. Howard: "Stochastic Marketing Models," chap. 3 in D. B. Hertz, and R. T. Eddison (eds.), *Progress in Operations Research*, vol. 2, John Wiley & Sons, Inc., New York, 1964.

——— and J. F. Magee: "Customer Behavior as a Markov Process," *Operations Research*, vol. 9, no. 1 (January–February, 1961).

Howard, R. A.: "Stochastic Process Models of Consumer Behavior," *Journal of Advertising Research*, vol. 3, no. 3 (September, 1963).

Jucius, M. J.: *Personnel Management*, 4th ed., Richard D. Irwin, Inc., Homewood, Ill., 1959.

Kahn, G. N., and A. Shuchman: "Specialize Your Salesmen!" *Harvard Business Review*, vol. 39, no. 1 (January–February, 1961).

King, W. R.: "A Stochastic Personnel-Assignment Model," *Operations Research*, vol. 13, no. 1 (January–February, 1965).

Little, J. D. C., K. G. Murty, D. W. Sweeney, and C. Karel: "An Algorithm for the Traveling Salesman Problem," *Operations Research*, vol. 11, no. 6 (November–December, 1963).

Maffei, R. B.: "Brand Preferences and Simple Markov Processes," *Operations Research*, vol. 8, no. 2 (March–April, 1960).

Nunnally, J. C.: *Tests and Measurements*, McGraw-Hill Book Company, New York, 1959.

Sasieni, M. W., A. Yaspan, and L. Friedman: *Operations Research: Methods and Problems*, John Wiley & Sons, Inc., New York, 1959.

Still, R. R., and E. W. Cundiff: *Sales Management*, Prentice-Hall, Inc., Englewood Cliffs, N.J., 1960.

Waid, C., D. F. Clark, and R. L. Ackoff: "Allocation of Sales Effort in the Lamp Division of the General Electric Company," *Operations Research*, vol. 4, no. 6 (December, 1956).

CHAPTER 11

Distribution Decisions

A primary objective of the marketing system is the efficient and effective *distribution* of products from their place of production to their place of consumption. All other activities which may claim to be primary elements of marketing are, in the broadest context, only incidental to this function. Thus, advertising, sales promotion, pricing, and market analysis, important though they may be, are merely techniques which serve to facilitate the basic distribution objective of marketing.

Of course, the distribution objective is not solely oriented toward the addition of location utility to the product. Not only must the product be located so that the consumer's satisfactions are as great as feasible, but the time and ownership must be such that the product is available when the consumer desires it and where he can obtain it through convenient purchase transactions. Thus, the distribution system contributes location, time, and ownership utility to the product.

The *distribution channels* for a product are the sequence of institutions through which title to and/or control of the product is passed as it moves toward ultimate consumption. The emphasis in defining the channels of distribution is on *ownership and control.* rather than physical movement. The *physical distribution* of a product is also an essential element of the overall distribution system. Of course, although the distribution channels and physical distribution system are often identical, they are best treated separately because of the nature of the distribution decisions involved in each.

Distribution Channels

The manufacturer of a product views distribution channels as the means by which products get to consumers. Since the industrial revolution few manufacturers have had direct contact with consumers in the sense that they sell directly to the ultimate consumer. Moreover, most producers are forced, by the economics of large-scale production, to manufacture in anticipation of demand rather than in direct response to it. Thus, the modern manufacturer cannot simply set up shop and wait for the customers to arrive as did the local shoemaker of old. Rather, he must rely on nonmanufacturing institutions to serve as intermediaries between the producer and the consumer.

A new company entering the marketplace or an old company selling a new product is not entirely free to choose its own distribution channels. The organization of distribution channels is usually highly institutionalized so that it is impractical to organize entirely new channels for a new product. In practice, new products are generally distributed either through existing channels or through channels which represent minor modifications of existing ones.

The decisions involved in the establishment of distribution channels for a product can be of primary importance to the degree of sales success which it achieves. In any particular situation a selection of more than one channel may be deemed best. The manufacturers of tires simultaneously sell through auto manufacturers for original factory-installed equipment and through retail outlets to consumers for replacement purposes, for instance. Yet, however complex may be the range of channel alternatives and the importance of the choice, the *bounded rationality* of the decision maker is of paramount importance. Bounded rationality means that the decision maker cannot consider all of the possible channel alternatives which might be used. What he will do is to consider some restricted range of alternatives and choose one which is "good enough." In doing so, he may miss the best of all possible channels. Yet, to seek the optimal may require him to spend a lifetime searching for an answer. Operationally, the range of channel alternatives which are considered usually represents minor variations of existing arrangements differing quantitatively, but not qualitatively, from existing structures. Thus, a new product will be sold through an existing system, with the number and location of retailers and wholesalers being determined by the channel decision maker.

But just who is the channel decision maker? It is easiest to discuss channels as though it were the manufacturer who could dictate the form and organization of distribution channels. Of course, this is not always so. Sears Roebuck clearly dominates its suppliers, even to the point of design specification and profits, whereas auto manufacturers effectively dominate dealers through the franchise system. *Usually the channel decision maker is one organization, either a manufacturer or a strong middleman, who adopts the leadership role and determines, to a great extent, the policies and practices of the other elements of the channel.* Thus, while no generally valid description of the channel decision maker is possible, the identification of the organization which has assumed leadership is usually clear in any particular circumstance.

This rather peculiar situation imposes special constraints on the decision maker, for he must operate and make decisions in concert with the other elements of the channel, each of which has individual goals and some degree of freedom to pursue those goals. Thus, the capabilities and desires of the other elements must always be carefully considered by the decision maker, whatever level of the distribution system he may occupy.

DISTRIBUTION CHANNEL OBJECTIVES

The objectives which are (or should be) sought in the selection of distribution channels for a product are of basic importance; yet, because of the complexities of channels and channel decisions, they are often not clearly delineated by the decision maker. The basic objective related to distribution channels is to provide the structure for transferring the ownership and possession of a product from the producer to the consumer. Although we have conceptually distinguished between ownership and physical movement of the product, it is not so easy to do so when this objective is considered. Closely associated, and often in conflict, with this objective is the problem of *communications,* for, not only must the product be transferred, but also information must flow through the channels of distribution.

The Basic Distribution Task. The efficiency of various distribution channel alternatives, *in terms of the basic distribution task,* is summed up in the profitability which the channel produces. The effect of the distribution channel on profit depends on characteristics of the product, the potential buyers, and competition. The general profitability rationale for the existence of something other than the direct distribution of products from manufacturer to retailer can be illustrated by consideration of the utility and costs of using middlemen in distribution channels. Figure 11-1*a* illus-

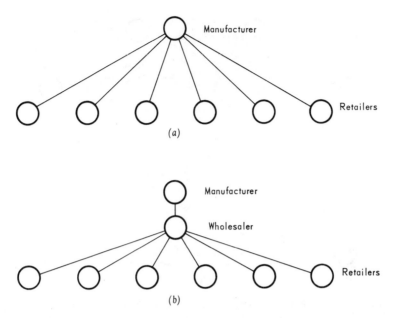

FIGURE 11-1 DISTRIBUTION CHANNELS FOR ONE MANUFACTURER

trates the situation which exists when no intermediaries operate between the manufacturer and the six retail outlets for a product. The manufacturer must have a contact line to each of the retailers. Each contact line involves direct written and verbal communications, calls by salesmen, etc. If one middleman is introduced into the same system, the situation is as depicted in Figure 11-1b. The significant difference between the two situations is the reduction in the number of contact lines which must be maintained by the manufacturer and the consequent *increase* in the total number of contact lines in the entire system.

If we assume that there is a fixed cost involved in each transaction which the contact lines depict, then for a given transaction frequency, the situation in Figure 11-1a—without the middleman—can result in greater costs to the manufacturer. When the middleman is introduced, the manufacturer must maintain only a single contact line; however, the total number of contact lines in the system is larger.

Now consider the entire complex of competitive manufacturers for a given product. With three manufacturers and six retailers, the situation without a middleman is as depicted in Figure 11-2a. There, each of the manufacturers must maintain a channel to each retailer. Figure 11-2b shows the same situation with a single middleman being used. The complexity of the overall system is obvi-

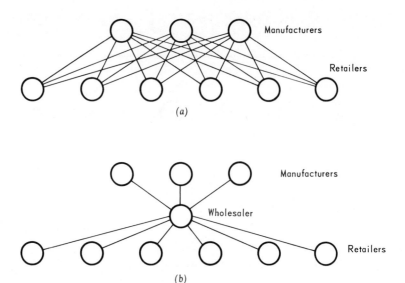

FIGURE 11-2   DISTRIBUTION CHANNELS FOR SEVERAL MANUFACTURERS

ously greatly reduced; and since each manufacturer needs only to maintain a single contact line and the middleman needs only one contact line per retailer, we can argue that the existence of the middleman imposes natural efficiencies on the total system. In essence, the selection of such a system as preferable to direct contact of each manufacturer with each retailer revolves about the absolute efficiencies to each manufacturer and a recognition by each element of the system of the goals of others. For instance, the manufacturers, whom we can assume for the moment to be the decision makers, recognize that the system of maintaining individual contact lines to each retailer is costly not only to themselves but to the retailer, for a contact line operates in two directions, and even though the manufacturer may bear much of the cost burden (such as salesmen's salaries), the retailer must spend time with salesmen, submit orders to many different manufacturers, etc.; all of this is unnecessarily costly if a middleman can be introduced. Moreover, the economic aspects of the physical distribution of products serve to amplify these efficiencies. We shall discuss these points in a later section dealing with physical distribution.

Product characteristics. The nature of the product has an important bearing on the profit which will result from various distribution channels. If the product is perishable, relatively short channels are necessary so that high spoilage costs will not result, and if the product is physically bulky, short channels will probably reduce handling costs. Indeed, virtually any of the many characteristics which any product possesses can be said to have a potential effect on the profit which various channel alternatives will generate and hence, on the degree to which a channel attains the basic distribution goal.

Market characteristics. The size and location of the potential market for a product also relates to the level of profitability which various channels will achieve. If the product is toothpaste, every household is probably a potential customer. If industrial chemicals are to be distributed, the market is numerically smaller and may well be confined to a limited geographical area; the market for chemicals used in the manufacture of steel, for instance, is confined largely to the area between Chicago and Pittsburgh.

Market characteristics, in the sense of the typical behavior of consumers, also relate to the degree of intensity of distribution. If the product is one which is purchased largely on an impulse basis, total profit will undoubtedly be enhanced if mass distribution, which provides maximum exposure, is undertaken. If the product is a kind for which customers shop, it must be available in locations where this is possible, e.g., at department stores. Of course, the expectations of consumers are also important. A product must be distributed through channels which permit it to be found where consumers expect it to be; a distribution system must put home tools in hardware stores, for example. Channels may also be selected to place tools in modern discount drug outlets, discount department stores, service stations, etc., but none are so necessary as the hardware stores to which consumers have become accustomed. In essence, this point is simply a reflection of the one made earlier dealing with the limited range of channel alternatives which are actually available. One reason for such limitation is the habits of consumers.

Competitive characteristics. The nature of competition in a particular marketing situation serves as a constraint on the selection of channels and hence, on the profitability produced by using those channels. It is often considered necessary to have a product distributed in each of the outlets which carry competitive products. This is the case in the food and drug industries, for example. In some cases the competitive factor is so great that producers are reduced to little or no selectivity in terms of outlets. The assumption in this case is that every possible outlet is profitable and therefore any possible outlet should be sought. Although this may be fallacious from the point of view of the profit generated in each outlet, from an overall point of view it may well be that the appearance of a product on the shelves of many outlets is, in itself, a form of promotion which contributes a profit response.

Channel Communications. The other basic objective related to the choice of distribution channels has to do with the information capability of the various alternatives. If the channel decision were simply to determine the channel or combination of channels which would produce the greatest profit, the problem would be sufficiently complex to warrant detailed analysis. Such an analysis would necessarily consider the cost efficiencies of various channel alternatives and their effect on sales and profits. Furthermore, the question of whose costs are relevant is an important one, for although the decision maker must necessarily consider the impact of his choices on the other organizational elements of the distribution system, he should not necessarily minimize total system cost if, in doing so, he does not minimize his own costs.

The realization that the channel of distribution also serves another primary objective—that of communications—complicates the decision problem even further. The manufacturer who is not in direct contact with the point of purchase must rely on the other elements of the system to generate and process all of the sales information which is so necessary for decision making. Previously, in discussing sales analysis, we have alluded to this basic information gathering function. There, we were primarily concerned with the impact of sales information on promotional decisions. In considering channel decisions, we are presented with the opportunity to design an information collection system within the confines of the distribution system. It is possible that it may not be desirable to do so in particular instances. Although reliance on extraorganizational store audits and surveys may be sufficient for nonrepetitive decision situations, such methods do not generally provide an adequate information base for the continuing analysis of management decisions. This is the aspect which has most frequently been neglected in the selection of distribution channels for a product. When this is the case, the communications network, which will inevitably grow on an *ad hoc* basis, is usually inefficient, and often ineffective, in performing the communications functions required of it.

In any case, the most significant communications aspect of channel selection is that *the manufacturer is deprived of basic sales information unless he takes direct action to choose channels which will provide him with information.* This is an important part of the selection of channels even if the manufacturer is not the decision maker, for it would be unwise for him simply to accept the dictates of a stronger distributor. In fact, most situations require that each element of the distribution channel have adequate information to provide a sound basis for the negotiations which invariably precede any distribution agreement.

Sometimes, in special circumstances, information on the sale of products to ultimate consumers may be made available through such devices as warranty cards. Many appliance purchasers, however, do not bother to send them in. Dealers may be asked to submit sales reports as a way of providing the basic sales data for some kinds of products, or salesmen who call on retailers may be required to submit reports on inventories, sales, shelf space allocations, etc. The use of such devices must necessarily add to the cost of maintaining channels. Hence, the channel decision maker is faced with a pervasive problem structure involving conflicting objectives. He must operate low-cost channels, which distribute the product in a profitable fashion, and at the same time he must develop information sources and handling procedures so that valid information is available for future decisions. In doing so, he is seeking to optimize long-term profits.

DISTRIBUTION CHANNEL DECISIONS

To arrive at distribution channels which will achieve all of these objectives would require a kind of global optimization which is usually not operationally practicable. Consequently, channel decisions are usually approached piecemeal. Various aspects of the system are viewed and suboptimized. Then, the set of suboptimum choices is viewed as a whole, and some evaluation of the interactive effects is made. Thus, one may independently determine the optimum number of retailers and the optimum number of wholesalers and then view the two jointly in the light of their possible conflict, the practicability of actually integrating them into an overall system, etc. If no significant practical difficulties and no joint effects which contribute negatively to the overall profit result are found, the organizational procedures under which they operate can be considered. Or, alternatively, we may determine the number and location of retailers; *then* determine the best wholesale setup to serve those retailers; *then* fix the operating procedures which are best to serve those retailers and wholesalers. In using either of these approaches, although the achievement of an overall optimal solution is not assured, we are quite likely to achieve a very good solution within the bounds of our analytic capability. The alternative to this is an expensive, and perhaps unproductive, total systems analysis or a rule of thumb which, while apparently reasonable, may be simply a device for concealing our inability to determine the best channel alternative.

Qualitative Decisions Regarding Retailers and Wholesalers. The channel decision problem which should be considered first in the piecemeal approach is the determination of the nature and characteristics of the retail and wholesale outlets to be used. At the retail level, for example, the basic alternatives for consumer goods are retail stores and nonstore procedures such as door-to-door sales, mail orders, and vending machines. Within the category of retail stores is a wide range of outlets varying from single-line and specialty stores to variety and department stores, which carry many unrelated lines. Although the precise categorization of a particular outlet as one of these types is not always possible, it is important to remember that the sales of a product can be greatly affected by its availability and that the meaning of availability is not an absolute. The important aspect of the operational definition of availability to product sales is based on the *anticipations* of potential consumers, i.e., a *perceived availability*. Thus, the product must be available where the potential consumer anticipates that he can find it. Modern supermarkets probably illustrate this point best. Many supermarkets carry a wide range of product lines which are related only tenuously to their primary food lines. Moreover the variations among supermarkets are extremely wide; patent medicines may be carried by some, children's clothing by others, etc. These markets rely on the high frequency of grocery shopping trips and impulse buying to move this merchandise. Certainly they do not expect that someone who does not shop for groceries in their establishment will seek them out in search of these nongrocery products. On the other hand, the potential customer would expect to find patent medicines in a drugstore and film in a camera shop, whether he has ever purchased anything there before. Moreover, he would expect to find both in a large department store.

Thus, the qualitative decision concerning the nature of retail outlets is affected by the general characteristics of the market, the product, and competition and the specific influence of the anticipations of potential customers. The determination of the nature and characteristics of wholesale outlets is somewhat more straightforward. The basic considerations in the wholesale aspect of the channel decision are the *ownership* and the *wholesaling functions* which each will perform. The manufacturer may choose to use wholly owned wholesale sales branches or to have independent intermediaries, or both. If he decides to use both, he will probably be utilizing two wholesale levels in the distribution system; e.g., the sales branch will sell to the independent wholesaler, who will then sell to the retailers. Of course, it may be the case that decisions involving ownership may be necessary at the retail level also. In practice, however, few manufacturers own retail outlets, although many own some portion of the wholesale outlets which they use. The reasons for this are economic ones which are rather pervasive.

For instance, the assortments of products which are most economic at the manufacturing and retail levels are so vastly different that the operation of retail outlets by a manufacturer represents a costly and resource-consuming function requiring him to perform tasks which are beyond the scope of his technology. Thus, he usually prefers to pursue his primary manufacturing capability to its utmost. At the wholesale level a specialist in distribution can greatly simplify the manufacturer's problems and allow him to concentrate his efforts on production.

The functions which the wholesale intermediary is to perform may vary from "full service"—i.e., warehousing, delivery, personal sales, credit, promotion, etc.—to "limited service," in which distributors may use a driver-salesman to deliver and sell beer or merchandisers may install and maintain magazine displays in supermarkets. Again, just as in the case of the ownership factor, the degree of service may also be a variable which is important in describing retail outlets. Yet, the retailing functions are less variable within each retail category than they are within wholesale categories. Thus, most department stores provide roughly the same level of services as most cash-and-carry stores, etc.

Other Retailing Decisions. After the types of retailers to be used have been decided, it becomes necessary to determine which particular retailers are to be selected as sales outlets. We shall assume that this is a question of selecting from an existing set of retail outlets which are located in various trade centers. For some new products it may be necessary to utilize new retailers, of course. However, the basic decision problem is of the same form; the numerical values of the costs which are involved are the only significant differences.

Retailing decisions reflect the degree of *intensity* of distribution which is to be used. To have intensive distribution is to have the product carried in as many different outlets as possible,[1] e.g., the kind of system through which cigarettes are distributed. At the other extreme is an exclusive distribution scheme such as that commonly used by franchisers.

The basic principle governing the choice of the number of retail outlets is the familiar one of declining marginal return. The manufacturer desires to have a sufficient number of outlets to perform adequately the task of making the product available to consumers, and yet not so large a number that the costs and communications difficulties become cumbersome. Of course, the number which is

[1] In the extreme, intensive distribution also implies the use of as many different *types* of outlets as possible. Here, we are considering that the types of retailers have already been selected.

adequate to make the product available is dependent on the locations chosen, and those locations are, in turn, dependent on the geography and population of the areas in which the product is to be sold. In Chapter 6 we discussed the establishment of market areas on bases which were similar to those one might use in taking a "first cut" at this retail outlet location problem.

One approach to this problem comes from an empirical study conducted some years ago. The result of the study is often referred to as Reilly's law of retail gravitation, although the term "law" is rather inappropriately used. The basic premise is that the larger the population of a trading center, the greater is its pulling power in terms of attracting shoppers from outlying areas. We may paraphrase the basic law as expressing the proposition that *two trade centers attract retail trade approximately in proportion to the population and in inverse proportion to the square of the distance from the centers.* Symbolically, this can be expressed in terms of the proportion of the retail trade from some intermediate point, say a small town, which is attracted by centers A and B—T_A and T_B respectively—

$$\frac{T_A}{T_B} = \frac{P_A}{P_B}\left(\frac{D_B}{D_A}\right)^2 \qquad\qquad (11\text{-}1)$$

In expression (11-1), P_A and P_B represent the populations of A and B, and D_A and D_B represent the respective distances to A and B from the intermediate point. Thus, if the situation were that described in Figure 11-3 for an intermediate town at the point

FIGURE 11-3   RETAIL TRADE ATTRACTION OF TWO TRADE CENTERS

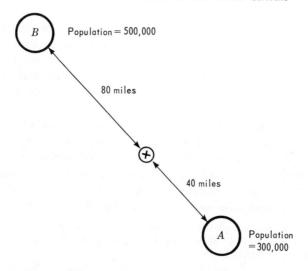

labeled with a cross, expression (11-1) would be

$$\frac{T_A}{T_B} = \frac{300,000}{500,000} \left(\frac{80}{40}\right)^2 = 3$$

and trade center A would be predicted to attract three times as much trade from the intermediate town as B.

This basic model of retail gravitation may then be manipulated to yield the breaking point between A and B. This breaking point serves to define the boundary of the area from which the two trade centers attract shoppers. At the breaking point

$$\frac{T_A}{T_B} = 1$$

This, together with a little algebra, yields

$$D_{B0} = \frac{D_A + D_B}{1 + \sqrt{P_A/P_B}} \qquad (11\text{-}2)$$

where D_{B0} is the distance from B to the breaking point and the term $(D_A + D_B)$ is obviously just the total distance between A and B. In Figure 11-3 the breaking point is

$$D_{B0} = \frac{120}{1 + \sqrt{3/5}} = 68$$

or about 68 miles from B. Thus, the town indicated by a cross in Figure 11-3 is not within the general trading area of B (since it is 80 miles from B).

The similarity, and indeed the direct relevance, of this result to the determination of market areas as described in Chapter 6 is clear. There, we considered that the final evaluation of market area boundaries would depend both on the extent of media coverage and on surveys made in the areas *near* the breaking point. The same survey technique might be used here to complement or test the result of the retail gravitation model.

To put this information to work in selecting a location for a retail outlet, one must consider two additional things: the existing competition and the population or buying power of the trade areas being considered. Through extensive product experience and market analysis, many organizations have developed simple decision rules for determining whether a trading area would support a retail outlet. Usually these rules are expressed in terms of minimum numbers of potential customers and buying power which a successful retail area must achieve.[2] For instance, one company

[2] The buying power measure may be *Sales Management*'s "Index of Buying Power," for example. See Chap. 6.

has minimum standards for population, total retail trade, and per capita income which they consider to be necessary to support a retail outlet for their product.[3]

There are some obvious flaws which might be inherent in the minimum standards to be applied. Yet, the basic predictive ideas which were used to determine market areas for sale of a product by using test market data are usually not available at this level of distribution decisions. Here, we are concerned only with the basic and more nebulous question of *which* retailers to use. Having made this determination, we might then go on to test-market the product by using the chosen retailers. In fact, if we lack confidence in the basic degree of distribution intensity which is inherent in our minimum standards, we may use test markets having various degrees of distribution intensity as experimental units whose objective is the determination of the "best" intensity. In such a case, the intensity of distribution would simply be another of those controllables whose effect was to be estimated in the test marketing program.

On the other hand, the use of empirically derived standards in this retailing decision situation again illustrates one of the most valuable aspects of quantitative analysis. Whatever may be the limitations of objective analysis, it can usually be combined with subjective rules of thumb and nonscientific analysis to provide an outcome which is no worse, and probably much better, than that which could be arrived at on the basis of experienced judgment alone. In effect, objective analysis is used to deal with those portions of the problem for which it is best suited (and which do not exceed its limitations), and the experienced judgment of the decision maker is called upon to use the results of the analysis. In this particular situation, the complementary effects are apparent. We can determine trading area boundaries analytically by using the ideas of retail gravitation, whereas it would be difficult to do this subjectively. On the other hand, the basic data which are necessary for statistical predictions may not be available, and so the judgmentally developed minimum standards are needed to complement the objective analysis in determining good retail locations.

Of course, the gravitation model is only an aid to the location of retail outlets. In the final analysis, many other factors need to be considered, for the model says nothing about the neighborhood in which to locate, the side of the street to seek, etc. Moreover, the limitation of the model to specific goods—e.g., shopping goods, convenience goods, and specialty goods—is apparent. As for any model, the results of this model should be carefully interpreted.

[3] See the Schneider reference.

Perhaps it should be even more carefully scrutinized in this instance because of its totally empirical basis. Other alternative models which may be more appropriate to particular products are easily derived. For instance, one should probably consider actual travel distance by highway rather than straight-line measures, as implied by Figure 11-3. Other descriptors of the pulling power of a trade center may be used. The Curtis Publishing Company used a variant of the basic model in defining the market areas which we referred to in Chapter 6. There, the descriptor "shopping line sales" was used in place of population in the model. Thus, the implicit assumption was not that population itself attracts shoppers, but rather that high levels of shopping activity tend to promote further activity.

Other Wholesale Decisions.[4]   Once the retail outlets to be used for a product have been determined, the question of wholesalers becomes of paramount importance. Previously we have given a discussion of the pros and cons of using a wholesaler in the distribution channel. Here, we shall provide a quantitative basis for the decision concerning the use of wholesalers.

To illustrate this, consider splitting each of the contact lines between manufacturer, wholesaler, and retailer into its cost components. We shall use only costs related to the personal selling function in the simple example to be given here. In any real situation it might be necessary to consider the level of communications which can be achieved under each channel alternative and other relevant costs such as transportation, ordering, etc.

Distribution cost analysis.   A detailed cost analysis of all direct selling costs represents a first step in evaluating various channel alternatives on a sales-cost basis. Travel costs from the salesman's location to each trade center, intracity travel costs, and cost of waiting time should be empirically determined. If this is done under some reasonable assumptions concerning the working relationships which would exist between the retailer, wholesaler, and manufacturer, it is a simple matter to sum the relevant costs for the alternative strategies,

S_1: sell directly to retailers
S_2: use one wholesaler
S_3: use two wholesalers
S_4: use three wholesalers

etc., and to select the least costly strategy.

[4] Much of this section is adapted from the Artle and Berglund paper in the references.

The decision maker should perform such a cost compilation in at least two forms, once using only his costs and once using his estimates of the total costs of each of the organizations in the distribution system. We have already pointed out the reason for this. It is possible that the cost-minimizing strategy for the decision maker is associated with unreasonably high costs for the retailer. If this is so, the system will eventually break down, for the retailer is also a profit seeker who will not long tolerate operating under a system which is favorable to another element of the distribution channel but unfavorable to him.

A comparison of alternative forms of distribution on the basis of a cost compilation, even if the costs are more exhaustive than those treated here, is not a complete solution of the wholesaling problem, however. The reason that it is incomplete is its total neglect of the *sales response* to distribution channels.

Sales response. It is a pervasive belief that the choice of a distribution channel alternative affects total sales revenue just as the selection of an advertising strategy does. It may be the case, for instance, that a retailer receives the same number of calls by salesmen under the two strategies S_1 and S_2 and that S_2 is the least-cost alternative. The further question is, however, whether the *effect* of these calls is the same. In one case (S_1) the calls are made by a manufacturer's representative, but in the other (S_2) the calls are made by a wholesaler's salesman. The potential superiority of S_1 to S_2 in terms of the sales response produced, and hence, in terms of overall profitability, is readily apparent, for the manufacturer's salesman might be thought likely to do a better job of selling his own product than the wholesaler's sales representative.

It is difficult to estimate quantitatively the sales response to distribution channels, especially in the case of selecting distribution channels for a new product which has not previously been marketed. The best way to handle this is to calculate the *difference which would be necessary to make two channel alternatives equally good.* Suppose that S_2 (one wholesaler) has been selected as the least-cost channel alternative on the basis of a cost analysis which did not incorporate sales response considerations. If it is felt that S_1 (direct sales to retailers) may be better from the standpoint of greater sales revenue generated by manufacturers' representatives, we may calculate *the increase in sales revenue per call which would be necessary to make the two channel alternatives equal.* If T represents total fixed costs, A represents the total costs which have been compiled for S_1, and B represents the total costs which have been tabulated for S_2, our assumption of the previously determined superiority of S_2 implies that

$$T + A > T + B$$

i.e.,

$$A > B$$

The unit cost of sales under S_2 is $T + B$ divided by the sales quantity d, i.e.,

$$\frac{T + B}{d} \tag{11-3}$$

If sales were to increase from d to u because manufacturers' representatives are used under S_1, the unit cost of sales (assuming the number of sales calls to be equal) would be

$$\frac{T + A + k(u - d)}{u} \tag{11-4}$$

where the last term in the numerator accounts for the increase in cost associated with selling the additional units (assumed to be linear with a unit sales cost of k).

For S_1 to be an equally good alternative to S_2, the costs given in expressions (11-3) and (11-4) should be equal.

$$\frac{T + B}{d} = \frac{T + A + k(u - d)}{u} \tag{11-5}$$

Solving for u, we find that

$$u = \frac{d(T + A - kd)}{T + B - kd} \tag{11-6}$$

i.e., the sales level which must be achieved for S_1 to be as good as S_2 is given by expression (11-6). Although this bit of information is very different from the information which would be provided by a numerical estimate of sales response, it is nonetheless helpful in the sense that we now need only judgments concerning the likelihood of achieving a sales level of u. With such judgments it may be easy for us to determine which of the alternatives is better when both costs and sales response are considered. For example, the calculated value of u may be *so high* that executives are easily able to judge that it is not achievable. Or it may be *low enough* for the best experienced judgment to assert that it can readily be achieved. In either case, the best channel alternative is clear—S_2 in the first instance and S_1 in the second.

We should point out that although we have used the phrases "so high" and "low enough" in describing these two situations, the levels which are "so high" and "low enough" are usually not at all obvious, so that the calculation in expression (11-6) and the cost estimates which are included in it are necessary parts of such conclusions. It is almost never the case that the difference in the cost of two channel alternatives, calculated by neglecting sales response, is so small that it is apparent that *any* positive sales response will be adequate to compensate for the difference. Thus, although we have mentioned only extreme cases, described by "low enough" and "so high," even such extreme cases require objective analysis to complement experienced judgment.

If u is not "so high" or "low enough" that judgmental agreement is easily reached, it is not clear which of the channel alternatives is the better. However, the value of knowing this is that we intuitively feel either that the two alternatives are actually very close in their total profitability or that poor cost estimates have made them appear so. In either case, we would wish to look more closely at the cost estimates and assumptions which were used in making the estimates. If, having done this, we feel that the two alternatives are actually very close, the selection may be made almost entirely on the basis of unquantifiable factors. In one such case, the author observed that executives selected the direct sales strategy (S_1) on the basis that the relative profitability of S_1 and S_2 were determined to be very close and several executives had qualms about the ability of a particular wholesaler to perform at the level which had been assumed in the cost estimates. Thus, in the absence of any clear preference between the two alternatives, the executives adopted a conservative position which would achieve the best results if, in fact, their suspicions about the errors in cost estimates were valid.

This approach is a particularly reasonable one here since we are dealing only with two simple alternatives rather than with a wide range of quantitative alternatives such as we must consider in determining the size of the advertising expenditure. This permits us to make a qualitative evaluation of the likelihood of achieving a sales level of u and, on the basis of that, to choose between S_1 and S_2. This is the same basic procedure which we used in the product evaluation discussion in Chapter 5. There, we considered the break-even point for a product in terms of the point at which costs and revenues were equal. Here, we are considering the break-even point in terms of an equality of the costs of two channel alternatives. In both cases this is a particularly meaningful approach because it is difficult to make a direct estimate of sales response and because we have only a small number of qualitatively different alternatives.[5] Such decision situations represent the contexts in which judgment is most easily used as an aid to quantitative analysis.

Communications Constraints. Superimposed on the cost analysis and the sales response must be a consideration of the other major objective of distribution channels—communications. In the previous analysis we have given attention only to the role of channels in providing a structure for transferring the ownership and possession of the product. We shall now consider the communications constraints which might preclude our actually making use of the least-cost or maximum-profit alternative as previously considered. In other words, our procedure for determining the channel alternative is a sequential one. We choose the retail setup, then the system of wholesalers which best fits the retail outlets, etc.—and we do all of this by giving consideration only to the costs or profits which are involved in the various alternatives. Having arrived at a tentative "best" channel alternative on this basis, we then consider the degree to which the alternative satisfies our communications objectives. If these objectives are not satisfied, some adjustment in the selected channel alternative can be instituted in the direction of greater fulfillment of the communications goals.

If, for example, the product in question is one in which there is little standardization, the importance of direct contact between manufacturer and retailer may well be so great as to void relatively small cost (or short-run profit) superiorities of channels involving middlemen to direct manufacturer-retailer contact. Custom-built machinery and fashion goods offer examples of such products. In both cases, the manufacturer's overall utility is maximized when he has direct contact with the retailer. In the case of fashion goods this contact is necessary so that the manufacturer can keep his finger on the market's pulse to determine changes in fashion preference and consumer reaction to new styles. Without this information, or if the information had to be filtered along a long channel in a time-consuming way, the manufacturer's product planning would be carried on in ignorance of the current status of the market. Clearly, whatever are the added costs of dealing directly with the retailer, the long-run profit outlook is enhanced if timely information is available to the manufacturer.

Moreover, the vagaries of operating procedures and laws may well prevent the manufacturer from using a particular channel,

---

[5] Note also the similarity between this approach and the "sensitivity analysis" described in Chap. 3. There, we were analyzing the sensitivity of the solution to errors in our probability estimates. Here, we are doing much the same thing in terms of the sales level.

or at least they may require him to adopt special information collection procedures to parallel the operating procedures. States in which a tax is assessed on retailers' inventories which are held on a particular date furnish a good example of this. In such instances, retailers schedule their orders so that their inventories will be as low as possible on the taxation date. If the manufacturer relies on data concerning "shipments to retailers" as being indicative of sales, as many actually do, the true picture of sales to consumers may be greatly distorted because of the inventory which intervenes between wholesaler and consumer. In such instances, the manufacturer usually chooses a channel which enables him to garner consumer sales data directly at the retail level, or he utilizes market research data collection organizations to obtain retail sales information. Other operating procedures used by intermediaries can also greatly distort the degree of representativeness of the sales information which is available to the manufacturer. If, for example, the wholesaler holds speculative inventories, the purchase pattern which the manufacturer sees may not be at all representative of the pattern of retail sales.[6]

Even if there are no special operating procedures or laws which distort the significance of the information which is available to the manufacturer, the mere existence of inventories between the manufacturer and consumer is normally sufficient to assure that distortions occur. If retail sales are subject to seasonal, cyclical, and random variation, ordering policies of retailers and wholesalers usually lag behind happenings in the marketplace, so when retail sales drop, retailers may not reflect this change in their orders to wholesalers for some time. Then, when wholesalers detect a change, they will not reflect it in their orders for some further time. By the time the manufacturer detects a change in the sales pattern, the entire situation may have changed at the retail level. If this is the case, both retailers and wholesalers will desire immediate delivery to meet increased demand. The manufacturer's ability to maintain stable production levels so that he will not be required to be hiring and laying off workers continuously is thereby impaired by the delays in obtaining information on happenings which have occurred in the distribution system.[7]

Such qualitative aspects of channel operations clearly relate to the communications goals of channel decisions, and they must therefore be considered in those decisions. Since it is difficult to

---

[6] See the Bucklin reference for a discussion of the role of speculation in determining the form of distribution channels.

[7] Forrester has worked extensively along the lines of simulating patterns of demand as a basis for the decisions which the manufacturer must make. See the Forrester reference.

quantify most of these elements, their effect on the attainment of communications goals, and hence, on the channel alternative which is selected, must usually be judged on nonquantitative bases.

Channels and the Marketing System. In adopting the systems viewpoint of marketing, we must consider the interactions of the various controllables which are available to the marketing manager. One cannot view channel decisions solely in terms of the cost and communications objectives; the impact of channels on promotion, pricing, and the other decision areas must also be considered. For example, price and channel decisions interact in one way in terms of the potential for speculative profits by intermediaries in the distribution system. Product decisions also relate to channel decisions, as illustrated in Chapter 5, because it may be uneconomic to distribute a narrow product line using particular channels (e.g., direct sales to retailers). Thus, the channels selected bear importantly on the product lines which it is desirable for the firm to distribute.

The relationship between channels and physical distribution is obvious, but not so apparent is the interaction between channels and promotion. A hard-sell "push" style of promotion which utilizes display, point-of-purchase advertising, coupons, and personal selling can be carried on only through outlets which can (and will) implement such a program. Moreover, some channels require that promotions be largely directed at retailers rather than at consumers. Much insurance, for instance, is sold through independent agents (who are not employees of any insurance company). Insurance companies must therefore conduct promotions aimed at persuading the agents to recommend their brand to potential customers. Only recently have large-scale "pull" promotional strategies been undertaken. Their aim is to persuade the potential insurance customer to ask for a particular brand. If he does, the agent, it is believed, is unlikely to try to switch a customer who is already partially sold on the idea of insurance to another company's insurance products.

Thus, the systems viewpoint requires that we consider each of these aspects when channel alternatives are being evaluated. It is usually not possible to do this comprehensively. Yet, some consideration should indeed be given to the interactions of each of the decision areas, for to fail to do so is to take the myopic viewpoint which is likely to lead to undesirable results. Of course, in making these evaluations, we must invariably adopt the same procedure which we used in considering communications objectives in channel decisions; i.e., we choose a channel alternative as "best" on objective grounds and then subjectively weigh the impact of the alternative on the other decision areas. If no overwhelmingly negative interactions are found, the channel alternative may be adopted. If some aspect of an interaction is negative, consideration can be given to revisions in the chosen alternative which serves to lessen the negative impact.

Physical Distribution

The basic function of marketing is to provide the consumer with the products which he desires. As the somewhat trite, but nonetheless valid, pedagogic illustration says, "business involves only cost (and no revenue) until the product is in the hands of the consumer." To accomplish this transfer, goods must be physically moved from the hands of the manufacturer to the ultimate consumer. In some cases, such as the individual who picks up his new automobile at the factory, this may be a simple process. In others the goods may pass through many hands and be stored for long periods before being delivered to the consumer. The level of complexity of our economy, which requires that distribution of most products be accomplished on some basis other than production to a customer's order, can be blamed for this great variance.

Most products are economically produced in quantities which, owing to the discrepancy in the economics of aggregation at the producer and retail levels, necessitate the use of distribution intermediaries. Thus, although large quantities are economic at the manufacturing level, the retailer requires only relatively small quantities (with consequently greater variety). This necessitates major *storage* at one or more points in the system and the *holding of inventories* in anticipation of demand. *Transportation* of the product is another important factor for consideration in physical distribution, for various carrier alternatives are available and shipments must be planned and scheduled so that the appropriate quantities arrive at their destinations at the proper time.

The relationship between physical distribution decisions and channel decisions is apparent. Although we differentiate between ownership and control (channels) and possession (physical distribution) for purposes of presentation, it is clear that the two aspects of distribution often involve the same organizations, personnel, and policies. In selecting distribution channels, one must consider the problems of physical distribution which are inherent in the channels. Indeed, it may be impossible to separate the two. Thus, our simple expository model which is necessary to categorize the decisions of marketing into neat boxes labeled "channels" and "physical distribution" may be so great a distortion of the real world as to be meaningless. In organizing the discussion in this traditional fashion, we have implicitly assumed that one *first* chooses distribution channels and *then* seeks physical distribution strategies which complement the channels. In practice, one would be foolish to choose a channel alternative before giving careful consideration to the physical distribution alternatives.[8] Happily, it is not difficult to anticipate the latter. One way to do so is to include the estimated costs of various physical distribution alternatives in our cost calculations for the selection of channels. Thus, although in the example of the previous section we spoke only of direct sales costs, we might also include transportation, warehousing, and other distribution costs. In doing this, we would be integrating channel alternatives and gross-level physical distribution alternatives into a single cost analysis. In simplest form we might do this as illustrated in Table 11-1. There, we consider

TABLE 11-1   COST OF VARIOUS DISTRIBUTION ALTERNATIVES

Channels	*Physical distribution*	
	Wholesale warehousing	*Public warehousing*
Direct sales to retailers	$\times$	c_{12}
Wholesale middleman	c_{21}	c_{22}

two channel alternatives—direct sales to retailers and the use of a single wholesaler—and two physical distribution alternatives— wholesaler warehousing and public warehousing. Only three of the four combinations in this table are feasible. Thus, when channel alternatives are being compared on a cost basis, the costs c_{12}, c_{21}, and c_{22} would be estimated and compared; i.e., *three* alternatives would be considered rather than just the two "pure" channel alternatives. It would not be meaningful to do otherwise, unless physical distribution costs were totally ignored, for to estimate the cost of each channel alternative requires that some assumption be made concerning physical distribution. If we were to assume that wholesale warehousing is inherent in the use of a wholesale middleman, we would compare only the situations involving the costs c_{12} and c_{21}. It is easy to see in Table 11-1 that to do so is to make a meaningless comparison of "apples and oranges." For instance, if

c_{21} is determined to be much less than c_{12}, it may well be that the difference has nothing whatever to do with the channel alternative, but rather is entirely due to the high cost of public warehousing. Therefore, the only meaningful comparison of channels is between the combinations involving the same physical distribution setup (i.e., c_{12} versus c_{22}) or all three channel–physical distribution combinations.

WAREHOUSING DECISIONS

One of the most important elements of a physical distribution system is the provision which is made for the storage of goods. The availability of storage is one of the simple, yet essential, features which makes a mass production market economy work, since it is not possible economically to produce most goods so that they are available when and where consumers desire them.

The term *warehousing* is usually applied to the functions involved in managing the storage of goods. The managerial problems involved in warehousing are of importance in determining the degree of efficiency with which the system will operate and hence, in determining the degree to which the physical distribution system contributes to the attainment of overall distribution goals.

Warehousing Objectives.   To provide retailers with a readily available supply of products, most manufacturers utilize warehouse stocks of finished products which may be drawn upon to fill orders. The number of warehouses required to attain such availability depends on the distribution objectives which are being sought, the costs involved, and the available transportation facilities.

The distribution objectives of warehousing are primarily related to the timeliness of delivery. One major brewer adopted this idea as the theme for an advertising campaign, i.e., "Wouldn't it be wonderful if everyone lived next door to a brewery so that he could have really fresh beer?" Of course, the theme develops along the lines of the ever-growing number of breweries which this brewer uses; hence, by implication, fresher beer from him than from anyone else. As a general rule, the greater the number of distribution points, the shorter will be the delivery lag to retailers. Indeed, if we could each have a warehouse located next door, we would need only to call across the back fence to have delivery made; but this would obviously be very costly for the distributor.

Implicit in consideration of the timeliness of delivery is the idea

[9] Of course, he fails to mention the many local brewers, who serve only a restricted area but should be able to provide beer to the consumers in that area in the shortest possible time.

of lost sales. We assume that if delivery cannot be made within a given time span, the buyer will go elsewhere, or at the very least, after some number of experiences with unavailabilities and late deliveries, he will eventually look elsewhere. In the case of certain kinds of production machinery during a recent boom period, this sort of behavior was apparent. Producers of machinery who could normally deliver within one or two months had such large order backlogs that delivery could not be promised on many items for over a year. Since virtually all heavy-machinery manufacturers were in the same situation, it might appear that the potential customer could not go elsewhere. Yet, the need to have machines to produce products to meet rising demands was so great that many buyers turned to the used-equipment market. The result was that the prevailing market price for used machinery rose until the prices paid for used equipment were higher than the list prices of comparable new items. The premium, of course, was warranted by the immediate availability of the used items.[10]

The determination of the costs involved in lost sales is the most difficult problem in measuring the degree of attainment of warehousing objectives. Of course, in most instances, one is not even certain of the existence of many of the cost elements which are believed to come into play. For instance, the long-range negative effect on revenues which is caused by minor delivery delays may not be significant enough to warrant consideration. In one case a mail order sales organization conducted a simple analysis to gain some information concerning the effect of lost sales. Its procedure involved a study of returned merchandise in which the relationship between speed of delivery and returns was of paramount importance. A (smoothed) version of the empirical relationship between the shipment delay and the percentage of returns which was found is shown as Figure 11-4. The important characteristic of that curve is the relatively small percentage of returns which occur for short shipment delays (less than t_1) and the large increase in this percentage which is displayed for each increase in delay beyond this short period. This undoubtedly reflects the nonessential nature of much of the merchandise. In regard to such goods, it is likely that the perceived "need" on the part of the customer is forgotten if a long delivery delay ensues. This particular analysis went on to other aspects, such as relating the value of an order to the return percentage, but no other factor was found which negated the basic relationship depicted in Figure 11-4. In this case, the impli-

[10] In one instance known to the author, a buyer paid a 25 percent premium over the then-current list price for a five-year-old machine on an "as is" basis. He then had to tear down (and rebuild) a wall to get the machine out and to pay for transporting it to his factory.

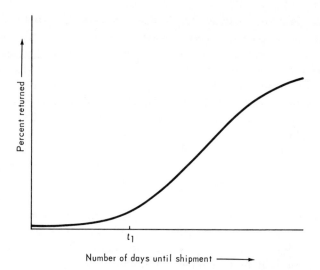

Number of days until shipment ⟶

FIGURE 11-4   MERCHANDISE RETURN–SHIPMENT DELAY RELATIONSHIP

cation of delivery delays is clear, and if one can determine such factors as the salvage value of returned merchandise—since much of it cannot be resold as new—and the average cost of an order, the imputation of a cost to delivery delays is relatively straightforward. Of course, in other types of distribution systems, the cost estimating task may not be so simple.

Some of the costs involved in maintaining stocks at various locations are readily apparent, e.g., the rent, heat, light, salaries, handling equipment, etc., which involve outlays at each location. Others are not so obvious. For instance, if the level of demand is uncertain, *a greater number of distribution points* will require a larger total inventory to fulfill a given demand than a lesser number will. The reason for this concerns the uncertainties in the quantity to be demanded at any single warehouse. If all of the uncertainty is aggregated into a small number of distribution points, the total uncertainty is less than the sum of the uncertainties at each of a large number of points. Thus, a smaller safety stock is required to fulfill uncertain demands.[11]

[11] Anyone who is familiar with the statistical measure of variability called *variance* can easily demonstrate why this is true. If x and y—the uncertain demand levels at each of two warehouses—are aggregated so that only one warehouse is used, the variance of the total demand is simply the sum of the individual variances (assuming independence). The standard deviation (square root of variance) of the total demand is therefore less than the sum of the standard deviations of the individual demand levels. For instance, if the variance of x is 9 and the variance of y is 16, the standard deviation of $x + y$ is $\sqrt{25}$, or 5. The sum of the standard deviations of x and y

Number and Location of Warehouses. Having determined some valid measures of the degree of attainment of the warehousing objectives related to timeliness of delivery and costs, the organization must focus attention on the number and location of the distribution points—the strategies by which the goals will be sought. These choices are interrelated with those of transportation, but we shall defer a discussion of this aspect until later. Here, we shall assume that the transportation modes and costs are predetermined.

Number of warehouses. Most organizations can quite easily determine the average delivery delay which is inherent in setups involving various numbers of distribution points. For example, if California is served by a centrally located warehouse, the average delivery delay may be three days, whereas if two warehouses are used—one in the north and one in the south—the average delay may be reduced to two days. Such data are rather easily determined from a knowledge of the transportation carriers, schedules, etc. Since the added cost of additional warehouses is also relatively easy to estimate, if the "cost of lost sales" (as discussed in the previous section) can be determined, the question of determining the number of warehouses can be approached by using a simple cost trade-off.

Most companies have determined a minimal level of service which they are willing to accept. This level serves to determine the smallest number of warehouses which will be feasible. For instance, delivery to 90 percent of one's customers within three days might be a valid minimal service level which could be accomplished only with ten warehouses. In most cases, it is assumed that this minimal level is the desired level, so that no further increases in the number of distribution points, and consequent decreases in delays, are considered. In effect, the incremental cost of delays is being treated as infinite if the service falls below the intended level, and zero at levels of service which are above the "target."

Most firms tend to err on the side of too much distribution rather than too little; i.e., they tend to have more distribution points than are economically justified. This is a natural tendency because of the large degree of uncertainty in estimating lost sales costs. Thus, although the cost of establishing a new warehouse may be large, it is known with a good deal of certainty, and the

---

is $3 + 4$, or 7. Since the likelihood that the demand exhibited at any warehouse will be less than any fixed quantity is directly related to the standard deviation, a greater total safety stock would be required in two warehouses than in a single warehouse. The direct extension of this argument to more than two distribution points is also valid.

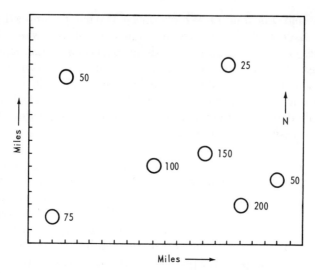

FIGURE 11-5   MAP SHOWING LOCATION OF RETAILERS TO BE SERVED BY WAREHOUSE

cost of not doing so is at best nebulous. The safe thing to do has therefore been to follow the rule: "when in doubt, build a warehouse." This trend has decreased in recent years because of a growing recognition of the tremendous costs involved in operating warehouses. The recent tendency has been toward consolidation and reliance on other transportation methods to achieve delivery goals at lower cost.

Warehouse location. The general location for a warehouse which is to serve a predefined area may be selected on an objective basis. Suppose, for example, that the area to be served is represented by the map shown in Figure 11-5 and that the average demand which is forecast over some planning period (say five years)[12] has been determined for each customer to be served. This quantity is indicated by the number at the indicated location of each customer. The thing to be determined is the location of the warehouse which is to service these customers and fill these expected demands.

The location which would best serve the customers in such a case depends on the *demand* at each point in the *direction* of each point from the warehouse, but *not on distance,* as such.[13]

[12] This period might be determined by the useful life of the warehouses, if owned, or the lease length, if rented, or it may simply be the organization's planning horizon for distribution.

[13] The most commonly used algorithm for determining the best location is one which defines the location of the best distribution point as the weighted arithmetic mean of the customer locations, the weights being the demands. Although this approach is simple to apply, it is incorrect in terms of the criterion applied here.

This is demonstrated in the technical note to follow. Thus, the idea of defining the limits of the area to be served on other bases, such as organizational structure and geographical factors like lakes and rivers, is of great importance.

The analytic solution which is given in the subsequent technical note is difficult to apply, but a simple mechanical model can be used to approximate the best location. The basic problem is equivalent to determining the equilibrium point for a system of weights (the demands) which are located at each customer's location. A flat board on which the appropriate map has been superimposed can be used as an analog model of this situation. A weight which is proportional to the demand at each point is tied on the end of a light string, and each string is inserted through a hole at the appropriate point representing a customer's location. The resulting device has a number of strings connected at a single point by a knot appearing on top of the board, each of which goes through a hole and has a weight suspended at its end under the board. The best location may be approximated by allowing the point at which all of the strings are connected to reach equilibrium. Because of the friction involved in the movement of the strings through the holes, it is best to do this several times, beginning each time with a different starting point for the knot at which the strings are connected. The minor differences which will result on each trial are largely irrelevant, since the technique does not consider such factors as roads, physical geography, etc. Thus the "best" location arrived at by using this model indicates only the general area of the best location. The particular area would then need to be analyzed carefully to take cognizance of the factors which are omitted from the simple analog model.

Technical Note on Warehouse Location.[14] The problem of finding the coordinates at which the location of a warehouse will minimize the cost of delivering a known quantity of goods to each of a finite number of points, if the cost is a linear function of distance, may be formalized in terms of the minimization of total cost, i.e.,

Min (TC)

The total cost expression may be represented as

$$TC = k \sum_{i=1}^{n} d_i \sqrt{(x_i - x)^2 + (y_i - y)^2} \qquad (11\text{-}7)$$

where k is the unit cost of delivery, d_i is the demand at customer location i, (x_i, y_i) designates the coordinates of the ith customer

[14] This technical note requires only a knowledge of optimization using partial derivatives.

location, and (x,y) designates the coordinates of the warehouse. Hence, the coordinates (x,y) are the decision variables.

Differentiating with respect to x and y to determine the best location, we get

$$\frac{\partial TC}{\partial x} = k \sum_{i=1}^{n} d_i \frac{x_i - x}{\sqrt{(x_i - x)^2 + (y_i - y)^2}} = 0 \qquad (11\text{-}8)$$

and

$$\frac{\partial TC}{\partial y} = k \sum_{i=1}^{n} d_i \frac{y_i - y}{\sqrt{(x_i - x)^2 + (y_i - y)^2}} = 0 \qquad (11\text{-}9)$$

These equations can be written in terms of the *directions* of each customer location from the warehouse as

$$\sum_{i=1}^{n} d_i \cos A_i = 0 \qquad (11\text{-}10)$$

and

$$\sum_{i=1}^{n} d_i \sin A_i = 0 \qquad (11\text{-}11)$$

where the tangent of the angle A_i is the slope of the line joining the warehouse and ith customer location. The reader should note that this formal solution is indeed *independent of distance;* it depends only on the demand d_i and the *direction* which is prescribed by the angle A_i.

Warehouse Operations.   In each of the previous sections dealing with warehouse decisions, we have dealt with *design* aspects of warehousing, i.e., with problems involved in establishing warehouse systems. The managerial problems involved in *operating* the warehouse are also significant, from the viewpoint of both the independent operator and the manufacturer. As we have illustrated, even if the warehouse is not controlled by the manufacturer, he is greatly concerned with its method of operation. His concern is motivated by his distinct self-interest in the maximum efficiency of each of the elements of the distribution system which operate between him and his customers, who affect *his* profit result. If their operation is efficient, the tendency will be toward stability of the distribution system which, although not an end in itself, usually results in better product distribution than does a "dynamic" system in which changes are caused by bankruptcies and business failures. If the manufacturer operates his own warehouses, his interest is a direct one, of course.

Inventory management. One of the basic functions of warehousing is the holding of inventories. In the purchasing decision models of Chapter 7 we have already treated many of the aspects of inventory management decisions which are applicable to the warehouse environment. Early in that chapter we discussed the inherent interrelationship between purchasing and inventory decisions. The reader is advised to refresh his memory of this argument at this point by rereading at least the introductory remarks to that chapter.

The ideas of Chapter 7 are directly applicable to warehouse inventory management, but some special aspects of the problems which are faced in this context warrant special attention. Among the most significant of these is the *decision criterion* which is used. In the basic static model described in Chapter 7, for instance, the criterion used was the minimization of cost. Later in this chapter a purchasing decision problem under risk was illustrated. The natural extension of these two ideas is a criterion involving the minimization of expected cost. In many warehousing situations, however, it may be very difficult to estimate the relevant cost parameters, e.g., inventory carrying costs and order costs, or in particular instances the best strategy may be so sensitive to errors in these cost estimates that we are unwilling to rely on them. If a shortage cost is included in a particular analysis, the estimation problem is particularly vivid, since the cost of a shortage involves many intangibles such as goodwill, administrative difficulties, etc.

In such instances, the warehouse manager may be well advised to use a criterion which utilizes the information available to him concerning the likelihood of various occurrences but which does not require precise cost estimates. A simple criterion possessing these properties is one which *directs the holding of an inventory which is sufficient to ensure that the probability of a shortage is no greater than some fixed number,* say 0.05. This idea is similar to that which we alluded to in the previous discussion of relative quantities of stock which must be held at one distribution point and at two points in order to satisfy an uncertain demand. Thus, if the manager is restricted to receiving monthly orders, he can determine the order quantity from an estimated probability distribution of monthly demand. If the probability distribution is that shown in Figure 11-6, for example, and if he feels that a probability of shortage of 0.05 is sufficiently small to be tolerated, he will select the stock level to be q_0 (as shown in Figure 11-6). This level is that above which monthly demand will be observed only 5 percent of the time. He is assured that this is so because the *area under*

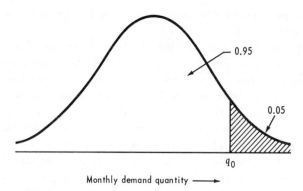

Monthly demand quantity ⟶

FIGURE 11-6   PROBABILITY DISTRIBUTION OF MONTHLY DEMAND QUANTITY

the curve to the right of q_0 is 5 percent of the total area under the curve. (Since this is a probability distribution, the total area under the curve is unity.)

Multiproduct inventory management. Generally, warehouses serve as storage locations for many different products at the same time. If there is no interdependence of the various products, all inventory management analyses need only to be aggregates of single-product analyses. In such a case the only difficulty encountered is size. Often it is simply not feasible to perform an analysis of each and every product held in a warehouse. If this is the case, the products may be categorized into value-use classes, and analyses may be performed for each *class* rather than for each product. For example, one might use four categories:

High value—high usage rate
High value—low usage rate
Low value—high usage rate
Low value—low usage rate

Having made this classification, one would estimate cost parameters which adequately depict each category and apply a model, such as the "basic model under certainty" of Chapter 7, to determine optimal purchase quantities. In doing this, one would need to solve only four problems rather than a different problem for each of the many products. Of course, one pays the price in accuracy for these savings, but some categorization of this sort is usually a practical necessity when many products are involved.

If, as is more commonly the case, some interaction exists between the various products with respect to inventory management decisions, it is not possible to treat each product, or class, as a separate entity. To illustrate this, let us consider the multiproduct version of the seasonal purchasing (inventory) problem already treated in Chapter 7.[15] There, we discussed a single-shot decision concerning the quantity of a product to be stocked in the face of a probabilistic demand. Here, we shall consider three such products. The problem is to determine the order quantity for each—q_1, q_2, and q_3—given the knowledge that any items left over at the end of the season must be disposed of at a price which is less than the in-season price. Table 11-2 gives the unit total cost,[16] in-season

TABLE 11-2 PRODUCT DATA

Product	Unit total cost	In season selling price	Post season selling price
1	$ 20	$ 30	$ 10
2	20	40	20
3	20	40	0

selling price, and postseason selling price which are applicable to each of the three products.

If the probability distribution of in-season demand for each of the products is symbolized as $f_1(Z_1)$, $f_2(Z_2)$, and $f_3(Z_3)$ respectively for in-season demand quantities Z_1, Z_2, and Z_3, and if any quantity can be disposed of at the postseason price, we may follow the procedure used in Chapter 7 to calculate the expected profit contribution for each product as[17]

$$E_1(P) = \sum_{z_1=0}^{q_1} (20Z_1 - 10q_1)f_1(Z_1) + \sum_{z_1=q_1+1}^{\infty} 10q_1 f_1(Z_1) \qquad (11\text{-}12)$$

$$E_2(P) = \sum_{z_2=0}^{q_2} 20Z_2 f_2(Z_2) \qquad (11\text{-}13)$$

$$E_3(P) = \sum_{z_3=0}^{q_3} (40Z_3 - 20q_3)f_3(Z_3) + \sum_{z_3=q_3+1}^{\infty} 20q_3 f_3(Z_3) \qquad (11\text{-}14)$$

[15] The remainder of this section is an extension of the model under risk which was discussed as a part of the "Purchasing with Fixed Prices" section of Chap. 7. Although the basic necessary information is summarized here, the reader would be well advised to reread that section.

[16] Including the inventory carrying cost per unit.

[17] Note that these expressions are developed by using the same arguments which are relevant to expression (7-23).

The total expected profit resulting from stocking quantities q_1, q_2, and q_3 of the three products is

$$TEP = E_1(P) + E_2(P) + E_3(P) \qquad (11\text{-}15)$$

Now, suppose that only a fixed investment amount is available, i.e., that we are restricted from stocking as much of each of the products as we might desire because of budgetary constraints. If this is the case, we cannot treat the three products independently, but we must consider the determination of all three order quantities simultaneously. If we assume that we know it is best to stock at least as much as is permitted by our budget B, we can formalize this constraint as

$$50q_1 + 40q_2 + 10q_3 = B \qquad (11\text{-}16)$$

Of course, since we are operating with discrete units, we will interpret this constraint as meaning "come as close to B as is possible."

To determine the best order quantities, we shall use \$20 expenditure increments and determine the *expected profit contribution* and marginal expected profit contribution for each of the three products. We shall tabulate these as shown in Table 11-3. There,

TABLE 11-3   EXPECTED PROFITS AND MARGINAL EXPECTED PROFITS
AS A FUNCTION OF EXPENDITURE LEVEL

Level of expenditure	$E_1(P)$	$\Delta E_1(P)$	$E_2(P)$	$\Delta E_2(P)$	$E_3(P)$	$\Delta E_3(P)$
\$ 20	.	.	.	.	.	.
\$ 40	.	.	.	.	.	.
\$ 60	.	.	.	.	.	.
\$ 80	.	.	.	.	.	.
\$100	.	.	.	.	.	.
\$120						
\$140						

the first column is the expected profit contribution of the first product at the expenditure levels \$20, \$40, \$60, etc., i.e., one, two, three, etc., units. The second column is the marginal expected profit calculated as the difference between levels of profit contribution at the various expenditure levels for the product. The remaining columns represent the same information for the other two products.

With this information we can begin with an arbitrary allocation of the total sum B to the three products. One of the expenditure levels for a product will show the smallest marginal loss in expected profit if the investment level is reduced by one $20 increment. If the expenditure in that product is reduced and a corresponding $20 increase is made for that product having the highest marginal gain in expected profit, the total expected profit resulting from the change will be increased. Repeated application of this idea will lead to an allocation of the total amount B which is optimal. The end point is indicated *when the smallest marginal loss is greater than or equal to the greatest marginal loss.* The accuracy of the solution for general problems of this type depends on the incremental level of expenditure which is used (here it is $20). In most cases in which the amount B is not adequate to achieve the maximum expected profit for each product, this method will lead to a very good solution.

TRANSPORTATION DECISIONS

The movement of goods between the manufacturing plant, wholesaler, retailer, and consumer involves the use of transportation facilities. Although many retailers and wholesalers operate their own delivery fleets, it is simplest to think of the problems of transportation in terms of using nonowned facilities. The ideas and methods of analysis may be equally well applied, however, to wholly owned facilities.

Of course, the transportation decision maker cannot deal with transportation problems in a vacuum. As we have pointed out, the transportation mode which is selected, the manner of routing goods, and the scheduling of deliveries all have a great impact on the other decision problems of distribution. In the previous treatments we have assumed that these factors are predetermined. Here, in the light of the choices previously made, we shall analyze the transportation system to be used. If we find that this analysis bears importantly on the choices previously made, we should go back, review the effect, and make any necessary changes in the tentative distribution alternatives which have already been selected.

Selecting the Mode of Transportation.   Products are usually borne from the point of manufacture to the places of storage by organizations that specialize in transportation. The basic transportation alternatives which are available are common carriers, which operate on fixed schedules at fixed rates, and contractual carriers, with whom negotiated agreements involving specialized transportation may be reached.

Railroads and trucks are the two best-known modes of transportation for most products, although pipelines, waterways, and other facilities have obvious uses in particular circumstances. In recent

years air transport has become economically feasible to the extent
that its use has increased for more and more different products.
Phonograph records, electronic parts, replacement parts for equipment, and cut flowers are only a few of the items now being trans-
ported regularly by air. This illustration is particularly appropriate
because one suspects that it represents, for many products, an alternative which is not considered by many transportation managers
(or traffic managers, as they are often called); i.e., it is a neglected
strategy.[18]

The choice of a mode of transportation should be based on the
degree of attainment of the basic distribution goals of timeliness
and cost. At one extreme is a wholly owned system which would
presumably be most convenient and involve the highest cost. At
the other is total reliance on slow common carriers, e.g., railroads,
with low cost and consequent reduced flexibility. For different
products different modes are best. Obviously, bulk raw materials
are more suitable for shipment via train or barge than by air freight,
except, of course, in emergencies when speed becomes overwhelmingly significant, e.g., in the Berlin airlift or a flood.

The important analytic point for the marketing manager to consider is that *the evaluation and comparison of the various strategies
involving modes of transport should be made on the basis of total
cost, rather than just "pure" transportation cost.* This point illustrates the interaction of transportation and warehousing decisions,
for the use of total systems costs may well prove that a costly
transportation mode is a necessary part of a least-cost distribution
system. One manufacturer of vacuum tubes, for example, has found
that relatively high-cost air transport has resulted in a consolidation
of warehouse facilities at substantially reduced total costs for physical distribution.

The total distribution costs which need to be considered are
transportation, inventory, warehousing, packing, insurance, docu-
mentation, etc. In one sampling study of such costs (see the Sealy
and Herdson reference) it was shown that only 2 percent of the
consignments of goods for various shipment distances, product
classes, and package weights justified air shipment on the basis
of transportation costs alone, whereas 18 percent qualified when
judged in terms of a total cost criterion. The implication of this
finding is clear: the systems viewpoint, which we often neglect
on the basis of an argument dealing with our analytic inadequacies,

---

[18] The degree to which we depend on airline transport of some products
was made clear during a recent major airline labor strike. Such items as
quality cut flowers were practically unobtainable in the midwestern United
States. No very good alternative transportation mode was found for this
product—not even special refrigerated trucks.

is the only valid one, and we may well be led astray by attempting to consider transportation and warehousing as separate entities.

Xerox Corporation offers a prime illustration of a distribution system which has evolved from a *total systems cost* analysis. Xerox needs to supply vast quantities of materials for its office copying machines; it formerly used 40 sales branches, each with its own supply of paper, chemicals, and machine parts, to do so. Now, only seven distribution centers are used, since 80 percent of the items in the inventory were found to be relatively slow movers which could be furnished by air freight when needed. The resulting increase in transportation cost is more than offset by the savings garnered from a halving of inventories in the distribution pipeline, the overall result being an estimated additional $9 million net profit over the first three years.[19]

Transportation Planning. If the carriers to be used for transporting goods have been determined or are imposed, the efficient planning of the use of the transportation system is the chief area for achieving economies of operation. To illustrate this, let us consider a physical distribution system involving a number of product plants and warehouses, each located in a different city.[20]

Suppose that the production quantities which will be available at each of the plants during some planning period are known and that we can estimate the quantities which will be required at each warehouse during the same period. The transportation planning problem involves the allocation of finished products to warehouses in a way which minimizes total transportation costs, i.e., the linking up of plants and warehouses so that costs are minimized. The reader will note that this problem is similar to the salesman assignment problem which we dealt with in the previous chapter. Indeed, as we shall see, if it were not for the variations in the quantities which are available from the plants and those which are required at the warehouses, the structure of the two problems would be identical.

If we are dealing with three plants and five warehouses, we might find that the unit transportation (and handling, etc.) cost for each plant-warehouse link is represented by the dollar amounts in Table 11-4. The entries in the last column represent the quantities of the product which will be available at each of the plants, and the entries in the last row indicate the quantities which are expected to be required at each of the warehouses. For simplicity

[19] "New Strategies to Move Goods," *Business Week,* Sept. 24, 1966, pp. 112–136.
[20] We might also consider the same problem involving wholesalers and retail outlets, using exactly the same approach to be demonstrated here.

we have made the total availability and requirements equal, i.e., 900 units.

Determining a good allocation.   One way to find a good (though not necessarily optimal) allocation of items from plants to warehouses is to consider the *difference between the lowest unit cost and next-lowest cost for each plant and each warehouse.* For instance, in Table 11-4 the lowest unit cost for plant 1 is the link to warehouse B. The next-lowest unit cost for that warehouse is

TABLE 11-4   UNIT TRANSPORTATION COSTS FOR VARIOUS
PLANT-WAREHOUSE LINKS (DOLLARS)

	Warehouse					*Quantity available*
	A	*B*	*C*	*D*	*E*	
Plant 1	5	2	6	4	7	200
Plant 2	11	9	8	5	7	300
Plant 3	7	2	3	4	6	400
Quantity demanded	100	150	200	250	200	

TABLE 11-5   COST ARRAY AND COST PENALTIES

	Warehouse					*Quantity available*	*Cost penalty*
	A	*B*	*C*	*D*	*E*		
Plant 1	5	2	6	4	7	200	(2)
Plant 2	11	9	8	5	7	300	(2)
Plant 3	7	2	3	4	6	400	(1)
Quantity demanded	100	150	200	250	200		
Cost penalty	(2)	(0)	(3)	(0)	(1)		

the link to D. The difference between the lowest and next-lowest unit cost is $2. This represents *the extra penalty which must be paid for not making use of the lowest cost linkage for this warehouse.* Similarly, the numbers in parentheses at the right and bottom of Table 11-5 represent these differences for each plant and each warehouse.

The highest penalty is associated with warehouse C. If we do not allocate as much as is possible to the lowest cost link for this warehouse (plant 3–warehouse C), we will pay a $3 per unit penalty for using the next-lowest cost link. Therefore, it seems reasonable to avoid this severe penalty by including an allocation of as much as possible to the 3-C link. The greatest such allocation is 200 units, since this is what is demanded by warehouse C; i.e., it is the lesser of the quantity demanded by the warehouse and the quantity available at the plant corresponding to that link. Note in Table 11-5 that two of the warehouses have zero penalties for not choosing the lowest cost link. This indicates that at least two plants have linkages to those warehouses which have the same unit transportation cost. Consequently, we have greater flexibility in making allocations to those warehouses since, even if we do not use one of the lowest cost linkages, we will have available the other linkage at exactly the same lowest unit cost. Because greater flexibility is associated with these warehouses, it seems logical to defer allocations to them until those warehouses with less cost flexibility have been accounted for.

Table 11-6 indicates that the allocation of 200 units to the 3-C link has been made. This partial allocation reduces the problem which we must face to one in which plant 3 has only 200 units available and only four warehouses need to be considered, since warehouse C's requirement is completely filled. This is indicated by the change in the availability column for plant 3 and the crossing off of the entire column corresponding to warehouse C. The cost penalties have been *recalculated* on the basis of this partial allocation and inserted in parentheses in the last row and last column of the table. The reader should note that the cost penalty associated with plant 3 is the only one which has changed. This is the case because the deletion of warehouse C used up the link involving the second-lowest unit cost for plant 3, thus changing the second-lowest cost and consequently the cost penalty. It is important to note that any of the other plants' penalties might also have been changed in going from Table 11-5 to 11-6. If the deleted warehouse had involved links which had either the lowest or second-lowest unit cost for any plant, that plant's cost penalty would probably have changed also.

In Table 11-6 four of the cost penalties are equally the highest for their respective plant or warehouse ($2). Hence, we can choose a second partial allocation from the lowest cost link for either plant 1, 2, or 3 or warehouse A. We shall arbitrarily select the 1-B link and allocate as much as possible there. The maximum amount is 150 units, since that is as much as is demanded by B. This allocation is indicated in the revised cost array of Table 11-7.

Table 11-7 gives also the revised cost penalties which are recalculated on the basis of the two allocations already made. Table 11-8*a* through d indicates the remaining steps, all done in exactly the same fashion, so that each successive table results from the repeated application of the procedure described. The arrow in each array indicates the plant or warehouse which was chosen as the

TABLE 11-6 PARTIAL ALLOCATION, COSTS, AND COST PENALTIES

	A	B	C	D	E	Quantity available	Cost penalty
Plant 1	5	2	6	4	7	200	(2)
Plant 2	11	9	8	5	7	300	(2)
Plant 3	7	2	3 (200)	4	6	200	(2)
Quantity demanded Cost penalty	100 (2)	150 (0)		250 (0)	200 (1)		

TABLE 11-7 REVISED PARTIAL ALLOCATION

	A	B	C	D	E	Quantity available	Cost penalty
Plant 1	5	2 (150)	6	4	7	50	(1)
Plant 2	11	9	8	5	7	300	(2)
Plant 3	7	2	3 (200)	4	6	200	(2)
Quantity demanded Cost penalty	100 (2) ↑			250 (0)	200 (1)		

one with the greatest penalty for not choosing the lowest cost link available.

The resulting feasible allocation shown in Table 11-8d can be seen to fulfill all the necessities concerning availabilities and demands. The allocation is interpreted as indicating that the demand

TABLE 11-8 FURTHER REVISED PARTIAL ALLOCATIONS

5 (50)	2 (150)	6	4	7	
11	9	8	5	7	300(2)
7	2	3 (200)	4	6	200 (2)
50 (4) ↑			250 (1)	200 (1)	

(a)

5 (50)	2 (150)	6	4	7	
11	9	8	5	7	300 (2)
7 (50)	2	3 (200)	4	6	150 (2) ←
			250 (1)	200 (1)	

(b)

5 (50)	2 (150)	6	4	7	
11	9	8	5	7	300 (2) ←
7 (50)	2	3 (200)	4 (150)	6	
			100 (0)	200 (0)	

(c)

5 (50)	2 (150)	6	4	7
11	9	8	5 (100)	7 (200)
7 (50)	2	3 (200)	4 (150)	6

(d)

of 100 units at *A* should be satisfied by shipping 50 units from plant 1 and 50 units from plant 3, the 150 unit demand at *B* should be entirely satisfied by plant 1, etc. Although this is not necessarily the best allocation, it is undoubtedly a very good one. The total transportation and handling costs associated with this allocation may be calculated to be

$$50(\$5) + 50(\$7) + 150(\$2) + 200(\$3) + 100(\$5) + 150(\$4) + 200(\$7)$$

or \$4,000, i.e., the unit cost of transportation along each link times the number of units routed over that link (summed over all links which appear in the allocation).

Technical Note on Determining the Best Allocation.[21] A determination of the best allocation may be made by starting with the good one shown in Table 11-8*d* and revising it in specified ways to reduce the total cost. This two-step procedure—i.e., first to find a good allocation and then to revise it to get the best allocation— may appear at first to be a tedious one. Yet, as we shall see, the technique for determining the best allocation requires that we begin with an allocation which satisfies all of the constraints concerning the quantities available and the quantities demanded. If this feasible allocation with which we begin is a good one, rather than just any which satisfies these constraints, we shall find that the determination of the best allocation is usually far less difficult.[22]

To demonstrate the ideas involved in making a revision of the good allocation which we have obtained, consider the change caused by allocating 50 units to the 3-*B* link. If we do this, we must also change several other allocations, because we must remain within the bounds imposed by the availabilities and requirements. To satisfy the demand constraint for warehouse *B* if 50 units are allocated to the 3-*B* link, we must allocate 50 less units to some other link involving warehouse *B*. In the reproduction of the previously determined good allocation shown in Table 11-9*a*, we can see that only the 1-*B* link has such an allocation. Therefore we must reduce its allocation from 150 to 100 to make the *total* allocation to warehouse *B* remain constant at 150 units. But this means that the total amount available at plant 1 will not be allocated, and so we must make a change there too. We can do this by

[21] This section is set off as a technical note only because it deals with a solution algorithm which need not be covered on first reading and may not be of interest to some readers.

[22] We should emphasize that it is *not necessary* to have a good allocation to begin the search for the best one. We might well simply use any allocation which satisfies constraints to begin the procedure to be described in this section.

increasing the only other allocation involving plant 1—the 1-*A* link—to 100 units. However, this changes the situation with regard to warehouse *A*, so that we must also make a change in some allocation in the first column. We can do this by allocating nothing to the 3-*A* link. The resulting allocation is shown in Table 11-9*b*. The positive and negative signs in Table 11-9*b* indicate the links which require allocation changes as a direct result of the arbitrary decision to reallocate 50 units to the 3-*B* link. The elements indicated by minus signs lost 50 units because of this induced change, and those indicated by plus signs gained 50 units. The important

TABLE 11-9   GOOD STARTING ALLOCATION AND REVISED ALLOCATION

	A	B	C	D	E
1	50	150			
2				100	200
3	50		200	150	

(a)

	A	B	C	D	E
1	+ 100	− 100			
2				100	200
3	−	+ 50	200	150	

(b)

point demonstrated here is the impact of any single change in a feasible allocation, i.e., a change in an allocation which satisfies all of the requirements and availabilities. *To change a feasible allocation and have another feasible allocation result requires that changes in the allocation to all affected plants and warehouses also be made.*

The evaluation of such a change also involves all of the concerned plants and warehouses. In the change from the allocation in Table 11-9*a* to the one in Table 11-9*b*, the additional costs being incurred are the ones involving the shipment of 50 units along link 3-*B*, at a unit cost of $2, and the shipment of 50 additional units along the 1-*A* link, at a unit cost of $5, for a total additional cost of $350. The cost savings which are incurred by this change involve the shipment of 50 units less along 3-*A* and

1-*B*, for a total *saving* of

$$50(\$7) + 50(\$2) = \$450$$

Thus, the *net saving* introduced by the change in allocation is $100, the difference between the $450 saving and the $350 additional cost. The allocation indicated in Table 11-9*b* therefore is superior to the good one in Table 11-9*a*. It is superior by the margin of $100; this can be verified by calculating the total transportation cost to be

$$100(\$5) + 100(\$2) + 50(\$2) + 200(\$3) + 100(\$5)$$
$$+ 150(\$4) + 200(\$7) = \$3,900$$

or $100 less than the $4,000 total cost associated with the good allocation.

Of course, it is now obvious that the particular change which we chose to discuss was not an arbitrary one. There are, in fact, no other changes in the allocation of Table 11-9*a* which would reduce the total transportation cost, although there are several other changes which would not increase it. By making this change and analyzing its effect on total transportation cost, we have demonstrated the method by which one can go from a good allocation to the best one—through a sequence of changes, each of which reduces total transportation cost. Now that we know the effects of a change and the way to evaluate it, we need only to consider the key point of how to identify those changes which will indeed result in lower total costs.

The way we choose to do this is simply by *evaluating* the additional cost associated with reallocating a single unit to *each* of the "empty" links in our starting allocation. If we find that such a change will reduce total costs, we will revise the starting allocation. Then, working with the revised allocation, we do the same thing again. Eventually we will come to a point at which no cost improvements are possible, i.e., the best allocation.

Although we can develop a relatively efficient method of performing these evaluations, the method is essentially an enumerative one. Therefore, the value of beginning with a good allocation is apparent, for the farther away from the best allocation we begin, the longer we will take to get to the best allocation through a sequence of one-step changes, and the closer we begin, the quicker we will get there.

We shall first present the algorithm which we can use to evaluate the worth of making a reallocation to an empty link and then show why it works. The algorithm requires only three steps.[23] Using the starting allocation, we:

[23] The application of this algorithm requires that the good allocation with

1. Determine sets of numbers u_i, each corresponding to one of the rows, and v_j, each corresponding to one of the columns of the transportation cost array, so that for each link to which there is an allocation, the sum of the corresponding u_i and v_j is equal to the unit transportation cost.
2. For each empty link calculate the difference between the unit transportation cost and the sum of its associated u_i and v_j values.
3. Allocate as much as possible to the link having the *most negative* such difference. If no negatives exist, the best allocation has been determined.
4. Repeat these steps until the best allocation is determined.

To demonstrate the algorithm, let us apply it, using as the beginning allocation that determined to be a good one in the previous section, i.e., the one shown in Table 11-9a. In Table 11-10, since the u_i and v_j numbers are determined from the unit costs in the cells to which allocations have been made in the good allocation with which we begin, only the unit costs for those cells are inserted. We assign an arbitrary u value to some row, say 0 to the third row, and using step 1, we determine the other values of u_i and v_j as shown; e.g., if the u value for the third row is 0, the v value for the first, third, and fourth columns must be 7, 3, and 4 respectively, since the sum of 0 and the v value must equal the 7, 3, and 4 unit costs. The rest of the u_i's and v_j's follow from this same equality for the allocated links. Note that the u_i's and v_j's must only meet this restriction on their sum so that it will always be possible to assign one arbitrarily.

To calculate the differences required in step 2 of the algorithm, we view the empty links and subtract the $u_i + v_j$ sum from the relevant unit transportation cost. To illustrate this, it is simplest to write down the original cost array (omitting the cost for the allo-

---

which we start have two properties. First, the *number* of links involved in the allocation must be exactly one less than the sum of the number of plants and warehouses. Secondly, these allocations must all be in locations such that one cannot proceed along a path from one allocation to another in a series of alternating vertical and horizontal "jumps" and return to the starting point without a direct reversal of route. Such a condition must hold for every allocated link in the beginning allocation. Moreover, these conditions must hold for every revised allocation which is used in determining the best one.

These conditions are of great significance only if the reader's interest is in the mechanics of the algorithm rather than in the general methodology of solution. The "good" allocation which we use as a starting point here has these two characteristics, as do all of the revised allocations with which we will deal. Those interested in further pursuit of these points are referred to the allocation chapter of the Sasieni et al. reference.

cated links) and the array of $u_i + v_j$ sums and to subtract the two. Tables 11-11a through c represent these arrays.

Only one link has a negative dfference according to step 2 of the algorithm. Step 3 tells us to change the good allocation by

TABLE 11-10 APPLICATION OF FIRST STEP OF ALGORITHM

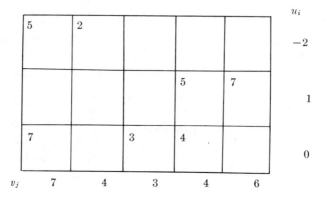

					u_i
5	2				−2
			5	7	1
7		3	4		0
v_j 7	4	3	4	6	

TABLE 11-11 APPLICATION OF ALGORITHM

		6	4	7
11	9	8		
	2			6

(a)

					u_i
		1	2	4	−2
8	5	4			1
	4			6	0
v_j 7	4	3	4	6	

(b)

		5	2	3
3	4	4		
	−2			0

(c)

allocating as much as possible to the link with the greatest negative difference. We have previously demonstrated that this can be done by reallocating 50 units to get the allocation in Table 11-9b.

If we repeat the process, using this revised allocation, we can determine the u_i and v_j values shown in Table 11-12 (beginning with an arbitrary zero for the third row). Table 11-12 corresponds to Table 11-10 in every way except that we have begun with the revised allocation and therefore have included different links. The unit cost and $(u_i + v_j)$ arrays corresponding to Table 11-11 are shown together with the array consisting of their respective dif-

TABLE 11-12 REAPPLICATION OF FIRST STEP OF ALGORITHM

					u_i
5	2				0
			5	7	1
	2	3	4		0
v_j 5	2	3	4	6	

TABLE 11-13 REAPPLICATION OF ALGORITHM

		6	4	7
11	9	8		
7				6

(a)

		3	4	6	u_i
		3	4	6	0
6	3	4			1
5				6	0
v_j 5	2	3	4	6	

(b)

		3	0	1
5	6	4		
2			0	

(c)

ferences as Table 11-13a through c. Table 11-13c indicates that there are no negative differences; hence, step 2 tells us that we have determined the optimal allocation.

Now, let us consider what we have done. First, the reader should recognize that any change in an existing allocation to any linkage will require a change in several other linkage allocations. These changes will occur along a path of links to which allocations have previously been made. To determine this path, one needs only to proceed in a series of alternating vertical and horizontal jumps from one *allocated* link to another in a path which eventually leads back to the starting point. Two such paths are illustrated in Table 11-14 for the allocation indicated. These paths represent the allocated links which would require change if an allocation were to be made to the (empty) links indicated by crosses.

TABLE 11-14   TWO REALLOCATION PATHS

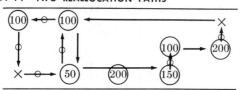

Every empty link will have a path, such as those shown in Table 11-14, which designates the links requiring reallocations if a change in the existing allocation is made to include the link. The overall cost reduction resulting from a unit change "around the loop" determines whether or not such a change would be beneficial.

To show that the algorithm we have used does indeed determine the worth of a unit change around a loop, consider that the change occurs in the simple loop shown below:

$$(m,j) \quad \cdots \quad (m,k)$$
$$\vdots \qquad\qquad \vdots$$
$$(i,j) \quad \cdots \quad (i,k)$$

If we add a unit to the link indicated as (i,j), the cost increase will be c_{ij}, i.e., the unit cost associated with link (i,j). This will require a one-unit deletion from links (i,k) and (m,j) and a one-unit addition to link (m,k). The net change in cost will be

$$c_{ij} - c_{ik} + c_{mk} - c_{mj}$$

which, in terms of the u's and v's of the algorithm, is

$$c_{ij} - (u_i + v_k) + (u_m + v_k) - (u_m + v_j)$$

or,

$$c_{ij} - (u_i + v_j)$$

This is exactly the difference which the algorithm tells us to calculate to use as an evaluation of the worth of reallocating one unit to the unallocated link (i,j). Thus, the algorithm in terms of the u's and v's is just a simple way of using the *net change in cost for allocating one unit to an unallocated link* as a measure for determining the worth of such a change.[24]

Systems and the Distribution System

Consideration of distribution channels and the physical distribution system offers one of the best opportunities in marketing analysis for the application of the *systems approach*. Distribution channels and systems are made up of networks of different organizations, each with different objectives and each performing different functions. Thus, the dangers of suboptimization are great. It would be all too easy for each organization to go off on its own, making decisions which optimize the attainment of its own objectives, without regard to the other elements and the effect of decisions on them. The result would be a costly and inefficient system.

Of course, most marketing organizations are too clever to fall prey to this extreme sort of suboptimization. Invariably, each realizes to some degree the joint interests of the group and acts only after consideration has been given to them. Indeed, as we have pointed out, it is to the distinct self-interest of each channel element to have the other organizations in the channel operate efficiently.

There is a good economic basis for this apparently intuitive conclusion. The argument is based on the idea of *substitutability*—the capacity of various business functions for being used as substitutes for one another, within certain limitations. A simple illustration of substitutability is the use of the inventory holding function to reduce the costs of production to meet a seasonal demand. Thus, if no inventory were held, production would take place only during the season of consumption. An organization which decides to spread production over a longer (or different) period of time to reduce costs essentially creates a new distribution channel for the product. If competitive conditions exist, the organizations that have not made use of this substitutability will find themselves operating

---

[24] Of course, all paths are not of the simple rectangular kind for which we have demonstrated this. In general the path connecting the empty link (i,j) to allocated links will be of the form $(i,j) \to (i,j_1) \to (i_1,j_1) \to (i_1,j_2) \to (i_2,j_2) \to (i_2,j_3) \to \cdots \to (i_n,j_n) \to (i_n,j) \to (i,j)$. For such a loop the net change in cost can also be seen to reduce to $c_{ij} - (u_i + v_j)$.

relatively high-cost channel alternatives. They will then either be forced to convert to the new procedure or be forced out of business. Thus, the competitive pressures serve to eliminate the excess profits of the innovator. Under such competitive conditions the idea of substitutability implies that organizations will allocate the work load to various functions to *minimize total channel costs* rather than to minimize the cost of a particular function.

The rationale for a systems approach and the consideration of total distribution costs is clear. As we have repeatedly illustrated, however, the distribution management decisions which must be made cannot often be approached on a total systems basis; rather, they must usually be analyzed piecemeal with consideration being given to the other aspects of the system, both before the problem is formulated and after the solution is determined. In this fashion "pure" suboptimization is avoided and (hopefully) global optimization is approximated.

One of the fertile areas for management action in furthering the systems approach to distribution *analysis* and to the *operation* of the distribution system lies at the organizational interfaces, the points of contact between organizations. To perform analyses which focus on the interfaces, however, requires that the organizations involved be less secretive about their operations than is commonly the case. Often, the secret nature of operational procedures and data emanates from a desire to preserve an effective bargaining position with the other elements of the system. The analyses of the interfaces can, in such cases, proceed only after the potential benefits have been shown to outweigh the bargaining advantage.

Another difficulty in making analyses at this level is the reaction of the people who are involved. Case Institute of Technology's Operations Research Group performed a study of railroad accounting practices dealing with the invoicing, billing, and payment procedures involved *between* railroads to account for the movement of goods on more than one, the leasing of one another's freight cars, etc. (See the Van Voorhis reference.) They demonstrated that a system of scientifically *sampling* waybills had considerable economic advantage over a 100 percent tabulation system. Further, the error in the sampling plan was of a known degree, whereas the errors in the usual laborious accounting routine were of unknown magnitude and significance. Yet, five years later, when the progress of implementing the sampling plan was studied, it was found that little application had been made. Churchman (see references) feels that the reason for this is that sampling simply did not fit the thought patterns of the people involved. To them, sampling was a way of shirking their clear duty of collecting every penny owed to the railroad, regardless, one must presume, of the cost of doing so.

So, although there are valid factors which serve to argue for the use of a systems approach to distribution analysis and the operation of a distribution system, organizational secrecy and the psychology of the people involved often serve as deterrents. Both of these aspects fall under the heading of *established patterns of thought*, so that with distribution, as with the other areas of marketing decisions, it is valid to describe the primary characteristic of the modern marketing manager as an ability to break away from established patterns.

Only when the marketing manager can lay aside established patterns of thought and ways of doing things to consider new and radical alternatives and methods of analysis will the approaches of this chapter and text achieve their potential utility.

Summary

The distribution of products from their place of production to their place of consumption is the primary aspect of traditional marketing. Were it not for the emphasis placed by modern marketing managers on the *strategic planning* aspect of their jobs, the *execution* aspect of marketing, which involves the accomplishment of product distribution, would remain of paramount importance.

However significant and "modern" is the current emphasis on planning, the marketing manager would be well advised to recall that (1) there are also strategic planning aspects to distribution and (2) the distribution system can weigh as heavily in the determination of sales and profit results as can any of the more "fashionable" areas, such as advertising, promotion, and product decisions. There is a growing realization of this by marketing managers, many of whom are returning to distribution considerations as the best area for obtaining improvements in product sales and for instituting efficiencies which contribute to profit.[25]

REFERENCES

Artle, R., and S. Berglund: "A Note on Manufacturers' Choice of Distribution Channels," *Management Science*, vol. 5, no. 4 (July, 1959), p. 460.

Beckman, T. N., and W. R. Davis: *Marketing*, 7th ed., The Ronald Press Company, New York, 1962.

[25] An excellent summary of this emphasis is provided in "New Strategies to Move Goods," *Business Week*, Sept. 24, 1966, pp. 112–136.

Bucklin, L. P.: "Postponement, Speculation and the Structure of Distribution Channels," *Journal of Marketing Research*, February, 1965, pp. 26–31.

Burstall, R. M., R. A. Leaver, and J. E. Sussams: "Evaluation of Transport Costs for Alternative Factory Sites," *Operations Research Quarterly*, vol. 13, no. 4 (December, 1962), p. 345.

Churchman, C. W.: "Sampling and Persuasion," *Operations Research*, vol. 8, no. 2 (March–April, 1960), p. 254.

Converse, P. D.: "New Laws of Retail Gravitation," *Journal of Marketing*, vol. 14 (October, 1949), p. 382.

Eddison, R. T.: "Warehousing, Distribution, and Finished Goods Management," chap. 4 in D. B. Hertz and R. T. Eddison (eds.), *Progress in Operations Research* (see below).

Forrester, J. W.: *Industrial Dynamics*, The M.I.T. Press, Cambridge, Mass., 1961.

Hertz, D. B., and R. T. Eddison (eds.): *Progress in Operations Research*, vol. 2, John Wiley & Sons, Inc., New York, 1964.

Mossman, F. H., and N. Morton: *Logistics of Distribution Systems*, Allyn and Bacon, Inc., Boston, 1965.

Reilly, W. J.: *The Law of Retail Gravitation*, William J. Reilly Co., New York, 1931.

Sasieni, M. W., A. Yaspan, and L. Friedman: *Operations Research: Methods and Problems*, John Wiley & Sons, Inc., New York 1959.

Schneider, G. W., III: "How to Reduce the Risk When Entering New Markets," *Sales Management*, Aug. 15, 1958, pp. 38–40.

Sealy, K. R., and P. C. L. Herdson: *Air Freight and Anglo-European Trade*, London School of Economics, London, January, 1967.

Van Voorhis, W. R.: "Sampling Methods in Railroad Accounting," *Operations Research*, vol. 1 (1953), p. 259.

EXERCISES

1. Why might a manufacturer choose not to include a middleman in the channels of distribution for his product?

2. What are the implications of market segmentation to the manufacturer's choice of distribution channels?

3. Describe the distribution channel decision maker's need to consider the other elements of the system in any channel decision which he may make. Why is this so? Why is this need more pronounced in the channel context than in other areas such as advertising decisions, product decisions, etc.?

4. What is the total number of contact lines associated with n manufacturers and m retailers if no middlemen are used? If one middleman is introduced, what is the number of such lines? If w wholesalers are used, what is the number of lines?

5. If you were the manufacturer of a new brand of cigarettes, what would you believe concerning the consumer's anticipations with regard to the points at which your product will be available? Would

you wish, for instance, to have them extensively carried by cigarette machines?

6. There are a large number of cigarette brands, and most distribution points cannot support more than a few cigarette machines. Suppose that you, as the manufacturer in exercise 5, have difficulty in persuading dispensing-machine operators to stock your brand. Can you suggest a satisfactory solution?

7. What is the intuitive explanation for the *direct* proportional relationship of the population factor and the *inverse* relationship of the distance factor in Reilly's law?

8. Apply Reilly's law to an intermediate point midway between two centers, one of which has twice the population of the other.

9. Apply Reilly's law to a point which is twice as far from one center as from another, assuming that the centers have equal populations.

10. Suppose that fixed costs are $10,000 and that a strategy of direct calls by manufacturers' representatives has been judged to cost $100,000 per year as contrasted with a cost of $70,000 per year for a strategy of using wholesalers' personnel to call on retailers. If the current sales quantity (using wholesaler personnel) is 100,000 units per year, what increase in sales would be necessary to justify the use of manufacturers' representatives? (Assume that an extra sales cost of $0.50 per unit is associated with the added sales.)

11. What would the likely choice be in the situation of exercise 10, assuming that the market has already been penetrated to a rather good degree?

12. Why is it usually necessary to consider *both* the channel and physical distribution aspects of distribution decisions simultaneously?

13. At a technical meeting you hear someone say that the location of a warehouse should obviously depend on the distance between the warehouse and the customers to be served. How would you respond?

14. Suppose that a portion of the empirical probability distribution of demand for a product which has been observed at a warehouse is given in the table below.

Number demanded	5	6	7	8	9	10
Relative frequency	0.20	0.20	0.20	0.15	0.10	0.10

If the manager desires to ensure that the probability of a shortage is less than 0.05, what should he do?

15. Multiproduct inventory decisions may be either a simple collection of single product decisions or a single complex decision involving all products. What distinguishes these two situations from one another?

16. Solve the multiproduct inventory management problem in the fashion described in Table 11-3 if B is $160.

17. How does the comparison of two modes of transportation illustrate the need for a systems viewpoint?

18. Give an argument concerning the validity of the approach demonstrated for finding a "good" allocation in the transportation planning decision illustrated in Table 11-4.

19. Determine a good allocation in the allocation problem described by the unit transportation cost array shown below if demands of 200, 100, 300, and 400 must be filled by availabilities of 500, 400, and 100 units.

From	*To*			
	3	4	1	6
	3	2	5	7
	6	1	1	4

Unit Transportation Costs

Least-squares Analysis

The equation

$$y = a + bx$$

represents a straight line with y-intercept a and a slope which has the numerical value b. If we wish to fit such a straight line to observed data, i.e., (x,y) pairs, a number of assumptions are commonly imposed:

1. The values of the variable x are controlled or observed without measurement error.
2. The expected value of y given x is linear in x.
3. The deviations between the observed and predicted values of y are independent of one another.
4. The variance of these deviations is not dependent on the value of x.
5. These deviations are normally distributed.

It is important that the user of least-squares analysis be cognizant of these assumptions so that cases which violate them can be more easily recognized and dealt with. However, minor inconsistencies between the assumptions and realities of the data are usually not overwhelmingly important. Since the prediction of y for a given x_i is $a + bx_i$, the sum of squares of the deviations between the predicted and observed values of y may be written as

$$D = \sum_{i=1}^{n} [y_i - (a + bx_i)]^2$$

where i is the subscript running over the n observations. To minimize D, we equate the partial derivatives of D with respect to a and b to zero:

$$\frac{\partial D}{\partial a} = 0 \qquad \frac{\partial D}{\partial b} = 0$$

The resulting equations are

$$2 \sum_{i=1}^{n} [(y_i - a - bx_i)(-1)] = 0$$

$$2 \sum_{i=1}^{n} [(y_i - a - bx_i)(-x_i)] = 0$$

These may be simplified to obtain the "normal equations"

$$na + b \sum_{i=1}^{n} x_i = \sum_{i=1}^{n} y_i$$

$$a \sum_{i=1}^{n} x_i + b \sum_{i=1}^{n} x_i^2 = \sum_{i=1}^{n} x_i y_i$$

If these are solved for a and B, we obtain[1]

$$b = \frac{\Sigma x_i y_i - (\Sigma x_i \Sigma y_i / n)}{\Sigma x_i^2 - (\Sigma x_i)^2 / n}$$

and

$$a = \frac{\Sigma y_i}{n} - b \frac{\Sigma x_i}{n}$$

In the case of multiple predictor variables $x_1, x_2, \ldots, x_p$, the sum of squared deviations may be written in the same form, i.e.,

$$D = \sum_{i=1}^{n} (y_i - a - b_1 x_{1i} - b_2 x_{2i} - \ldots - b_p x_{pi})^2$$

and the partial derivatives with respect to $a, b_1, b_2, \ldots, b_p$ may be equated to zero. The result is a set of $p + 1$ normal equations which may be written in similar form, as was done for the case of $p = 2$ in expressions (6-10) and (6-11).

For details of the computational efficiencies which may be realized in large problems of this kind, the reader is referred to M. Ezekiel, and K. A. Fox, *Methods of Correlation and Regression Analysis*, 3d ed. (John Wiley & Sons, Inc., New York, 1963).

[1] The reader should note that expression (6-6) is identical to the first of these two expressions. The significance of the expression for a should be recognized in the context of expression (6-3); i.e., note that it is equivalent to $a = \bar{y} - b\bar{x}$.

Index

Index